METHODS IN MOLECULAR BIOLOGY

Series Editor
John M. Walker
School of Life and Medical Sciences,
University of Hertfordshire, Hatfield,
Hertfordshire, UK

For further volumes:
http://www.springer.com/series/7651

For over 35 years, biological scientists have come to rely on the research protocols and methodologies in the critically acclaimed *Methods in Molecular Biology* series. The series was the first to introduce the step-by-step protocols approach that has become the standard in all biomedical protocol publishing. Each protocol is provided in readily-reproducible step-by-step fashion, opening with an introductory overview, a list of the materials and reagents needed to complete the experiment, and followed by a detailed procedure that is supported with a helpful notes section offering tips and tricks of the trade as well as troubleshooting advice. These hallmark features were introduced by series editor Dr. John Walker and constitute the key ingredient in each and every volume of the *Methods in Molecular Biology* series. Tested and trusted, comprehensive and reliable, all protocols from the series are indexed in PubMed.

Flavins and Flavoproteins

Methods and Protocols

Edited by

Maria Barile

Department of Biosciences, Biotechnology and Biopharmaceutics, University of Bari "A. Moro", Bari, Italy

Humana Press

Editor
Maria Barile
Department of Biosciences,
Biotechnology and Biopharmaceutics
University of Bari "A. Moro"
Bari, Italy

ISSN 1064-3745 ISSN 1940-6029 (electronic)
Methods in Molecular Biology
ISBN 978-1-0716-1288-0 ISBN 978-1-0716-1286-6 (eBook)
https://doi.org/10.1007/978-1-0716-1286-6

This Springer imprint is published by the registered company Springer Science+Business Media, LLC, part of Springer Nature.
The registered company address is: 1 New York Plaza, New York, NY 10004, U.S.A.

Preface

This book of protocols, devoted to the yellow coenzymes derived from Riboflavin (Rf) or Vitamin B_2 and to the hundreds of enzymes whose functionality depends on them, represents a sort of compendium of questions which can rise in a researcher's mind when working with flavoproteins or with the wide spectrum of functions that flavoproteins can drive in the cells.

Therefore, starting with Rf production in microorganisms and the chemical, optical, and redox properties of these fascinating molecules and moving along to the variety and the peculiarity of some single flavoenzymes, the reader is shown the complexity of functions and distribution of these molecules in the cell. Some still unsolved biological questions are addressed through descriptions of adequate methods to approach the problem.

Special attention in the last chapter of this book is devoted to some recently discovered human neuromuscular pathologies connected to flavoproteome derangements.

For these reasons, this book could interest a wide audience: first of all, protein chemists interested in purifying and characterizing flavoproteins, as well as microbiologists, physiologists, and clinicians, who have had the chance to study problems connected with flavoproteins.

I wish to acknowledge my master and PhD students, who contributed to different aspects of the realization of this book, and Miss. Costanza Indiveri, who helped me as an editorial secretary.

Bari, Italy ***Maria Barile***

Contents

Contributors

ALESSANDRO ALIVERTI • *Department of Biosciences, University of Milan, Milan, Italy*

MARIA BARILE • *Department of Biosciences, Biotechnology and Biopharmaceutics, University of Bari "A. Moro", Bari, Italy*

FRANCESCO BONOMI • *Section of Chemical and Biomolecular Sciences, DeFENS, University of Milan, Milan, Italy*

GIULIA CALLONI • *AB SCIEX Germany GmbH, Darmstadt, Germany*

VITTORIO CAPOZZI • *Institute of Sciences of Food Production, National Research Council of Italy (CNR), Foggia, Italy*

PAOLO COCOMAZZI • *Department of Biosciences, University of Milan, Milan, Italy*

LARA CONSOLE • *Department DiBEST (Biologia, Ecologia, Scienze della Terra) Unit of Biochemistry and Molecular Biotechnology, University of Calabria, Arcavacata di Rende, Italy*

FEDERICA COSSU • *CNR-IBF, Consiglio Nazionale delle Ricerche – Istituto di Biofisica, Milan, Italy*

NICOLA DE SIMONE • *Department of Agriculture Food Natural Science Engineering, University of Foggia, Foggia, Italy*

GENNY DEGANI • *Department of Biosciences, University of Milan, Milan, Italy*

GLORIA DEL SOLAR • *Department of Microorganisms and Plant Biotechnology, Biological Research Center – Margarita Salas (CIB-Margarita Salas, CSIC), Madrid, Spain*

STEFANIA DIGIOVANNI • *Department of Biosciences, University of Milan, Milan, Italy*

KOSTYANTYN V. DMYTRUK • *Department of Molecular Biology and Biotechnology, Institute of Cell Biology, NAS of Ukraine, Lviv, Ukraine*

DARIYA V. FEDOROVYCH • *Department of Molecular Biology and Biotechnology, Institute of Cell Biology, NAS of Ukraine, Lviv, Ukraine*

PATRICIA FERREIRA • *Departamento de Bioquímica y Biología Molecular y Celular, Facultad de Ciencias, Universidad de Zaragoza, Zaragoza, Spain; Institute of Biocomputation and Physics of Complex Systems (BIFI), Universidad de Zaragoza, Zaragoza, Spain*

KARL FISHER • *Manchester Institute of Biotechnology, University of Manchester, Manchester, UK*

MICHELE GALLUCCIO • *Department DiBEST (Biologia, Ecologia, Scienze della Terra) Unit of Biochemistry and Molecular Biotechnology, University of Calabria, Arcavacata di Rende, Italy*

GALINA Z. GAYDA • *Institute of Cell Biology, National Academy of Sciences of Ukraine, Lviv, Ukraine*

MYKHAILO V. GONCHAR • *Institute of Cell Biology, National Academy of Sciences of Ukraine, Lviv, Ukraine*

STEFANIA IAMETTI • *Section of Chemical and Biomolecular Sciences, DeFENS, University of Milan, Milan, Italy*

CESARE INDIVERI • *Department DiBEST (Biologia, Ecologia, Scienze della Terra) Unit of Biochemistry and Molecular Biotechnology, University of Calabria, Arcavacata di Rende, Italy*

TARAS S. KAVETSKYY • *Drohobych Ivan Franko State Pedagogical University, Drohobych, Ukraine; The John Paul II Catholic University of Lublin, Lublin, Poland*

HALYNA M. KLEPACH • *Institute of Cell Biology, National Academy of Sciences of Ukraine, Lviv, Ukraine; Drohobych Ivan Franko State Pedagogical University, Drohobych, Ukraine*

PIERO LEONE • *Department of Biosciences, Biotechnology and Biopharmaceutics, University of Bari "A. Moro", Bari, Italy*

DAVID LEYS • *Manchester Institute of Biotechnology, University of Manchester, Manchester, UK*

PALOMA LÓPEZ • *Department of Microorganisms and Plant Biotechnology, Biological Research Center – Margarita Salas (CIB-Margarita Salas, CSIC), Madrid, Spain*

ANABEL LOSTAO • *Instituto de Nanociencia y Materiales de Aragón (INMA), CSIC-Universidad de Zaragoza, Zaragoza, Spain; Laboratorio de Microscopias Avanzadas, Universidad de Zaragoza, Zaragoza, Spain; Fundación ARAID, Zaragoza, Spain*

STEPHEN A. MARSHALL • *Manchester Institute of Biotechnology, University of Manchester, Manchester, UK*

ELOISE MASTRANGELO • *CNR-IBF, Consiglio Nazionale delle Ricerche – Istituto di Biofisica, Milan, Italy*

MILAGROS MEDINA • *Departamento de Bioquímica y Biología Molecular y Celular, Facultad de Ciencias, Universidad de Zaragoza, Zaragoza, Spain; Institute of Biocomputation and Physics of Complex Systems (BIFI), Universidad de Zaragoza, Zaragoza, Spain*

MARI LUZ MOHEDANO • *Department of Microorganisms and Plant Biotechnology, Biological Research Center – Margarita Salas (CIB-Margarita Salas, CSIC), Madrid, Spain*

ALESSIA NISCO • *Department of Biosciences, Biotechnology and Biopharmaceutics, University of Bari "A. Moro", Bari, Italy*

LAURA POPOLO • *Department of Biosciences, University of Milan, Milan, Italy*

STEFANO QUARTA • *Department of Biosciences, Biotechnology and Biopharmaceutics, University of Bari "A. Moro", Bari, Italy*

JOSÉ ÁNGEL RUIZ-MASÓ • *Department of Microorganisms and Plant Biotechnology, Biological Research Center – Margarita Salas (CIB-Margarita Salas, CSIC), Madrid, Spain*

PASQUALE RUSSO • *Department of Agriculture Food Natural Science Engineering, University of Foggia, Foggia, Italy*

ANDRIY A. SIBIRNY • *Department of Molecular Biology and Biotechnology, Institute of Cell Biology, NAS of Ukraine, Lviv, Ukraine; Department of Microbiology and Biotechnology, University of Rzeszow, Rzeszow, Poland*

OLEH V. SMUTOK • *Institute of Cell Biology, National Academy of Sciences of Ukraine, Lviv, Ukraine; Drohobych Ivan Franko State Pedagogical University, Drohobych, Ukraine; Clarkson University, Potsdam, NY, USA*

LUCA SORRENTINO • *Department of Biosciences, University of Milan, Milan, Italy; Department of Chemistry, University of Milan, Milan, Italy*

GIUSEPPE SPANO • *Department of Agriculture Food Natural Science Engineering, University of Foggia, Foggia, Italy*

DELIA TARANTINO • *Department of Biosciences, University of Milan, Milan, Italy*

MARIA TOLOMEO • *Department of Biosciences, Biotechnology and Biopharmaceutics, University of Bari "A. Moro", Bari, Italy*

ANDRIY O. TSYRULNYK • *Department of Molecular Biology and Biotechnology, Institute of Cell Biology, NAS of Ukraine, Lviv, Ukraine*

R. Martin Vabulas • *Charité – Universitätsmedizin Berlin, Institute of Biochemistry, Berlin, Germany*
Maria Antonietta Vanoni • *Department of Biosciences, University of Milan, Milan, Italy*
Oksana M. Zakalska • *Institute of Cell Biology, National Academy of Sciences of Ukraine, Lviv, Ukraine*
Andriy E. Zakalskiy • *Institute of Cell Biology, National Academy of Sciences of Ukraine, Lviv, Ukraine*

Part I

Riboflavin Production and Export by Microorganisms

Chapter 1

Selection of Riboflavin Overproducing Strains of Lactic Acid Bacteria and Riboflavin Direct Quantification by Fluorescence

Pasquale Russo, Nicola De Simone, Vittorio Capozzi, Mari Luz Mohedano, José Ángel Ruiz-Masó, Gloria del Solar, Paloma López, and Giuseppe Spano

Abstract

Riboflavin (vitamin B_2) is a vitamin of the B group involved in essential biological pathways, including redox reactions and the electron transport chain. Some lactic acid bacteria (LAB) can synthesize riboflavin and this capability is strain-dependent. In the last years, a growing interest has focused on the selection of riboflavin-overproducing food-grade LAB for the vitamin biofortification of fermented foods, as well as for the formulation of innovative functional products.

In this chapter we report fast and inexpensive techniques in order to (1) screen LAB isolates able to produce riboflavin from different matrices, (2) select spontaneous roseoflavin-resistant riboflavin overproducing strains, and (3) quantify vitamin B_2 in culture media by fluorescence detection.

These protocols could be useful to select new overproducing strains and/or species from different ecological niches, as well as to optimize the conditions for vitamin bioproduction.

Key words Lactic acid bacteria, Riboflavin, Roseoflavin, Biofortification, Vitamins, Fluorescence detection

1 Introduction

Riboflavin is a vitamin of the B group (vitamin B_2) which is synthesized by plants and some microorganisms. Important dietary sources of vitamin B_2 are leafy green vegetables, seeds (mainly cereals and nuts), and foods of animal origin, like eggs, red meat, milk, and dairy products [1]. In addition, commensal microorganisms can also contribute to the production of vitamin B_2 in the gut environment [2]. Riboflavin is the precursor of flavin adenine dinucleotide (FAD) and flavin mononucleotide (FMN), coenzymes involved in biological redox reactions and the electron transport chain. Therefore, riboflavin deficiencies are related to several

Maria Barile (ed.), *Flavins and Flavoproteins: Methods and Protocols*, Methods in Molecular Biology, vol. 2280,
https://doi.org/10.1007/978-1-0716-1286-6_1,

metabolic diseases [3]. However, ariboflavinosis can occur in developed and developing countries due to particular physiological conditions, malnutrition and/or unbalanced diets [4].

In recent years, there is an increasing interest in the food industry of fortified foods. In this field, the employment of food-grade riboflavin-producing lactic acid bacteria (LAB) as a starter is an attractive strategy to obtain biofortified fermented foods. Nevertheless, the ability to synthesize riboflavin is a strain-dependent metabolic trait shared by some strains among several LAB species, though in general their levels of vitamin B_2 production are low and not compatible with the fortification concept.

In Gram-positive bacteria, including LAB, expression of the riboflavin biosynthesis (*rib*) operon is regulated by FMN riboswitch-mediated transcriptional attenuation [5]. These FMN riboswitches are RNA elements located at the 5′ untranslated region of the *rib* operon mRNA that consist of a conserved FMN-sensing aptamer domain, the so-called RFN element, and an expression platform exhibiting alternative terminator and anti-terminator structures [6]. Binding of FMN to the RFN element in the nascent mRNA results in a shift in the conformation of the expression platform toward the terminator structure of the riboswitch. The occurrence of spontaneous mutations in the RFN element can impair the proper regulatory activity of the FMN riboswitch resulting in riboflavin overproduction [6]. Exposure of riboflavin-producing microorganisms to the selective pressure of roseoflavin, a natural, toxic analog of riboflavin, has been reported as a strategy to select LAB with a riboflavin-overproducing phenotype [7]. In particular, by using this approach several derivative LAB strains, able to overproduce riboflavin to different extents, have been selected, and they belong to the species of *Lactococcus lactis*, *Lactiplantibacillus plantarum* (formerly *Lactobacillus plantarum*), *Limosilactobacillus fermentum* (formerly *Lactobacillus fermentum*), and *Leuconostoc mesenteroides* [7–10]. Over the years, these strains have been proposed to be useful for the manufacture of riboflavin-enriched fermented foods (i.e., pasta and bread) [9, 11], and probiotic fermented foods (i.e., yogurt, and nondairy products such as soya milk, oat-based, and kefir-like formulations) [8, 12–14]. Moreover, several riboflavin-overproducing LAB strains have been characterized for their probiotic potential by using in vitro and in vivo [15, 16], and tested for their ability to attenuate ariboflavinosis and intestinal colitis and mucositis in mice [17–19]. Therefore, also non–riboflavin-fortified food matrices (e.g., fresh-cut fruits) have now increasing interest as carriers to vehicle beneficial probiotic vitamin B_2–overproducing strains [20, 21].

Thus, future objectives in the food industry to improve the vitamin content of some food products should include the selection of new overproducing strains and/or species isolated from different ecological niches.

However, the selection of valuable riboflavin-producing strains is dependent on the capability to detect and quantify the amount of the vitamin produced. In general, riboflavin needs a previous extraction from the matrix, mainly based on conventional procedures including acid hydrolysis and enzyme treatment in combination with solid-phase extraction clean-up approaches, followed by quantification by analytical techniques based on liquid chromatography and electrophoresis [22, 23]. Nonetheless, these techniques are time-consuming and expensive, and not suitable for fast and routine analysis. In the past, a microbiological assay has been proposed as appreciably reproducible, sensitive, rapid, and inexpensive for screening riboflavin-producing strains [24]. Moreover, riboflavin can be detected quantitatively by measurement of its fluorescent emission at 530 nm after excitation at 450 nm, and this method has been recently validated for the real-time detection of riboflavin in *L. plantarum* cultures during microbial growth [10].

Here, we report how to perform a fast screening of LAB isolates from different matrices in order to select new riboflavin-producing strains. Moreover, we describe a methodological approach to obtain vitamin B_2-overproducing strains by exposure to the selective pressure of roseoflavin. Finally, we describe a method based on fluorescence detection for a fast quantification of riboflavin in culture media that could be useful to select the best producer strains as well as to determine the real-time production during the bacterial growth in order to optimize the condition for vitamin production.

2 Materials

All solutions should be prepared using ultrapure water and analytical grade reagents. Stock solutions and the chemically defined riboflavin-free medium (CDRFM) are stored at 4 °C in the dark. Under these conditions, the stability of the stock solutions and media is approximately 1 year.

2.1 Chemically Defined Riboflavin-Free Medium (CDRFM)

1. The composition of chemically defined riboflavin-free medium (CDRFM) is reported in Table 1 (*see* **Note 1**).
2. Stock solution of carbohydrates, amino acids, vitamins, and mineral salts are prepared as reported in Table 1. Store stock solutions at 4 °C (*see* **Note 2**).
3. To prepare CDRFM, the corresponding volume of each compound (as reported in Table 1) is dispensed into a glass beaker. Add water until about 90% of the final volume. Mix by using a magnetic stirrer and adjust the pH to 6.2 by using 1 M KOH. Make up to the final volume with water by using a graduate

Table 1
Analytical composition of the chemically defined riboflavin-free medium (CDRFM), and concentration of the stock solution as suggested for each compound

Compound	Concentration (g/L)	Concentration in stock solution (g/L)
Carbohydrates		
D-glucose	10	100
Amino acids		
L-alanine	0.2	10
L-arginine	0.75	10
L-asparagine	0.15	10
L-aspartic acid	0.35	4 (2.5% in 1 M HCl)
L-cysteine	0.2	5 (10% in 1 M HCl)
L-glutamic acid	0.5	10 (10% in 1 M HCl)
L-glutamine	0.2	10
L-glycine	0.5	50
L-histidine	0.5	10
L-isoleucine	0.2	10
L-leucine	0.2	10
L-lysine	0.25	10
L-methionine	0.15	5 (10% in 1 M HCl)
L-phenylalanine	0.2	10
L-proline	0.5	10
L-serine	0.4	10
L-threonine	0.35	10
L-tryptophan	0.2	10
L-tyrosine	0.2	5 (10% in 1 M NaOH)
L-valine	0.2	10
Vitamins		
4-Aminobenzoic acid	1×10^{-4}	0.1
Biotin	2×10^{-3}	1 (50% in 1 M NaOH)
Choline chloride	2×10^{-3}	1
Cyanocobalamin	1×10^{-4}	0.1
Folic acid	2×10^{-3}	1 (10% 1 M HCl)
Nicotinic acid	2×10^{-3}	1
Ca-D-pantothenate	2×10^{-3}	1

(continued)

Table 1
(continued)

Compound	Concentration (g/L)	Concentration in stock solution (g/L)
Pyridoxine hydrochloride	2×10^{-3}	1
Thiamine hydrochloride	1×10^{-4}	1
Mineral salts		
$MnSO_4 \times 4\ H_2O$	0.1	10
$MgSO_4 \times 7\ H_2O$	0.1	10
K_2HPO_4	1	50
$CaCl_2$	0.44	100
$CuSO_4 \times 5\ H_2O$	1.5×10^{-5}	0.01
$FeSO_4 \times 7\ H_2O$	2×10^{-2}	5
$ZnSO_4 \times 7\ H_2O$	1.35×10^{-4}	0.25

cylinder. Sterilize CDRFM by filtration through 0.45 μm filters (*see* **Note 3**). Store at 4 °C in the dark (*see* **Note 4**).

2.2 Screening for Riboflavin LAB Producers

1. Preculture media: de Man–Rogosa–Sharpe (MRS) broth.
2. Culture media: CDRFM broth and agar (15 g/L).
3. Saline solution (8.6 g/L NaCl).
4. Roseoflavin: a stock solution is prepared by dissolving the compound at 25 mg/mL in DMSO. Store at −20 °C.

2.3 Quantification of Riboflavin Production by Fluorescence

1. Preculture media: MRS broth.
2. Culture media: CDRFM.
3. Phosphate Buffered Saline (PBS) pH 7.4: 137 mM NaCl, 2.7 mM KCl, 10 mM Na_2HPO_4, and 1.8 mM KH_2PO_4.
4. Microplates for fluorescence-based assays, 96-well. We use Thermo Scientific™ Nunc™ MicroWell™ 96-Well Optical-Bottom Plates with Polymer Base (Thermo Fisher Scientific, Waltham, US).
5. Fluorescence microplate readers. We use a Varioskan Flash System® reader (Thermo Fisher Scientific).

3 Methods

3.1 Screening for Riboflavin-Producing LAB

1. Inoculate the LAB isolates from a cryopreserved stock in MRS broth (1:1000 v/v) and incubate at the optimal growth conditions until reaching the early stationary phase (*see* **Note 5**).

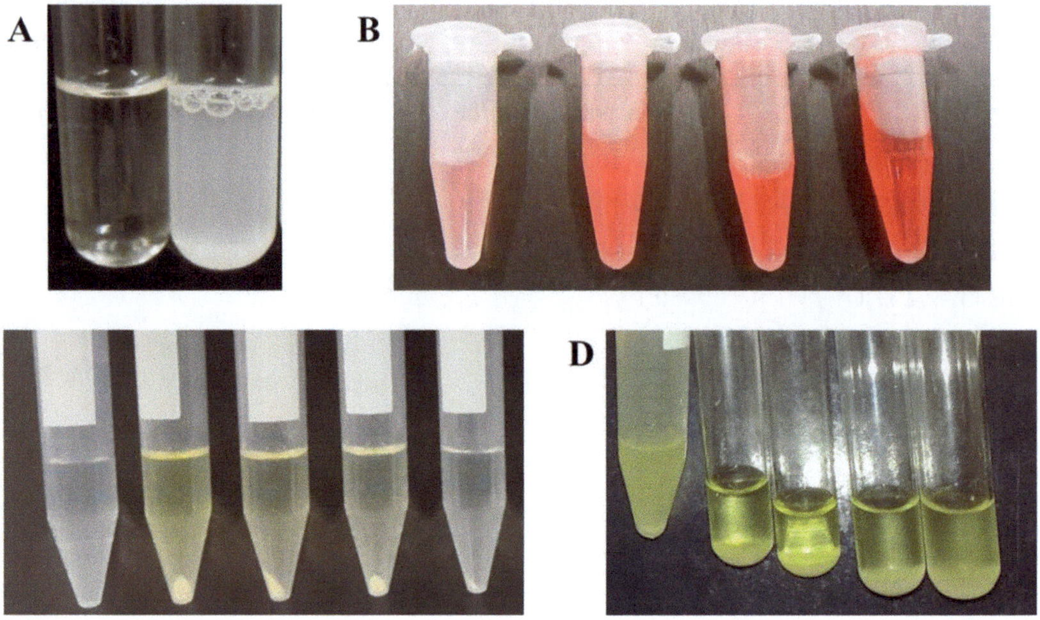

Fig. 1 Schematic representation of the procedure to select riboflavin-overproducing derivative strains. Isolates able to grow in a CDRFM (**a**) were submitted to the exposure of increasing concentration of roseoflavin (**b**). Roseoflavin-resistant isolates inoculated in CDRFM showed different ability to produce riboflavin (**c**), a phenotypic trait that should be monitored for about 100 generations (**d**)

2. Sediment at 5,000 rpm for 3 min at room temperature by using a microfuge. Wash twice by resuspension in sterile saline solution and sedimentation as above, in order to remove residual vitamin B_2.
3. Resuspend the pellet in the same volume of sterile saline solution, and use the cell suspension to inoculate CDRFM (1:1000 v/v). Incubate at the optimal growth conditions until reaching the early stationary phase.
4. Repeat **steps 2** and **3** two more times in CDRFM in order to remove any trace of residual vitamin B_2 (*see* **Note 6**).
5. Strains showing prototrophy for vitamin B_2 (Fig. 1a) are cryopreserved in CDRFM supplemented with sterile glycerol (20% v/v) and stored at −80 °C.

3.2 Selection of Spontaneous Roseoflavin-Resistant Strains

1. Inoculate the culture (1:1000 v/v) from a cryopreserved stock into CDRFM supplemented with roseoflavin at a concentration of 10 mg/L. Incubate at the optimal growth conditions until reaching the early stationary phase.
2. If you observe growth, repeat the assay with subsequent inocula into CDRFM supplemented with increasing concentration of roseoflavin (i.e., 50, 100, and 200 mg/L) (Fig. 1b) (*see* **Note 7**).

3. Prepare serial decimal dilution in sterile saline solution from the culture submitted to the exposure to 200 mg/L of roseoflavin. Spread 100 μL onto the surface of riboflavin-free medium agar plates and incubate at the optimal growth conditions for 48 h (*see* **Note 8**).
4. Randomly, pick up single colonies with a sterile loop. Inoculate the roseoflavin-resistant isolates into CDRFM. Incubate at the optimal growth conditions until reaching the mid-exponential phase.
5. A qualitative selection of the best riboflavin-overproducer derivative isolates can be made after 48-h of growth by the visual observation of the yellow intensity (Fig. 1c) (*see* **Note 5**).
6. To evaluate the stability of the riboflavin-overproducing phenotype, subculture (inoculum 1:1000 v/v) the derivative strains in CDRFM for 24 h, and repeat this step five consecutive times (Fig. 1d) (*see* **Note 9**).
7. To preserve the strains of interest, recover the bacterial biomass by centrifugation in a microfuge (5,000 rpm for 3 min at room temperature) and remove the supernatant. Resuspend the pellet in fresh CDRFM. Add sterile glycerol (20% v/v) and store the stock at −80 °C.

3.3 Direct Quantification of Riboflavin by Fluorescence in Culture Supernatants

3.3.1 Calibration Curve

1. Dissolve riboflavin in CDRFM at 10 mg/L and dilute the sample to obtain solutions with the following concentrations: 0, 0.25, 0.50, 1.25, 2.50, 5.00 mg/L (*see* **Note 10**).
2. Transfer an aliquot of 200 μL of each solution into a well of Nunc™ 96-Well Optical-Bottom Plates with Polymer Base. Carry out triplicate repetition for each concentration (*see* **Note 11**).
3. Insert the plates in a Varioskan Flash System® reader and set the instrument to make a reading of fluorescence ($\lambda_{ex} = 450$ nm; $\lambda_{em} = 530$ nm).
4. At the end of the reading, build a table in Microsoft Office Excel with the average of the three measurements for every riboflavin concentration and then build a scatter chart with trend line, square R-value, and the corresponding equation.

3.3.2 Quantification of Riboflavin Produced by the LAB Strains

1. Recover bacterial cells from MRS overnight cultures by centrifugation in a microfuge (5,000 rpm for 3 min at room temperature).
2. Wash twice and sedimentat the cells as in **step 1**. Then, resuspend the pellet in the same volume of CDRFM.
3. Use this bacterial suspension to inoculate (1:100 v/v) fresh CDRFM and incubate the culture for 24 h at the optimal growth conditions.

4. Sediment bacteria by centrifugation in a microfuge (11,000 × *g*, 10 min at room temperature).
5. Gently remove 0.2-mL aliquots of culture supernatants to measure fluorescence as described above. Carry out the assay in triplicate (*see* **Note 12**).
6. Determine the riboflavin concentration by interpolation of fluorescence values in the calibration curve.

3.4 Simultaneous Measurement of Cell Growth and Fluorescence

1. Inoculate the strain from a cryopreserved stock culture in MRS broth (1:1000 v/v) and incubate at 37 °C (or the optimal temperature for the strain) until it reaches an OD_{600} ranging between 0.5 and 0.6. Stop the microbial growth by immersion in ice (*see* **Note 13**).
2. Take an aliquot (1 mL) of culture and transfer it into a sterile Eppendorf tube. Centrifuge by using a microfuge (11,000 × *g* for 5 min at room temperature), then carefully remove the supernatant and resuspend the microbial pellet in PBS pH 7.4. Repeat this operation twice.
3. Resuspend the pellet in CDRFM in order to achieve an OD_{480} ranging between 1.0 and 1.2.
4. Dilute the suspension in fresh CDRFM (1:100 v/v) and transfer an aliquot (200 μL) into a well of a Nunc™ 96-Well Optical-Bottom Plates with Polymer Base. Each sample should be analyzed at least in triplicate.
5. Insert the Optical-Bottom Plates in a Varioskan Flash System® reader and set the instrument to make a reading of absorbance ($\lambda = 480$ nm) and fluorescence ($\lambda_{ex} = 450$ nm; $\lambda_{em} = 530$ nm) every 30 min for 24 h (Fig. 2).
6. The Varioskan Flash System® settings is as follows:
   (a) Temperature = 37 °C or the optimal temperature for growth of the strain (Wait until instrument has reached the target temperature);
   (b) Shaking (step duration = 10 s; speed = 600 rpm; diameter = 1 mm);
   (c) Photometric measurement (measurement time = 100 ms; wavelength = 480 nm; bandwidth = 5 nm);
   (d) Fluorometric measurement (measurement time = 100 ms; bandwidth = 5 nm; excitation = 450 nm; emission = 530 nm);
   (e) Pause (waiting time = 30 min);
   (f) Repeat shaking, photometric measurement, fluorometric measurement, and pause for 48 times to make a 24-h growth curve.

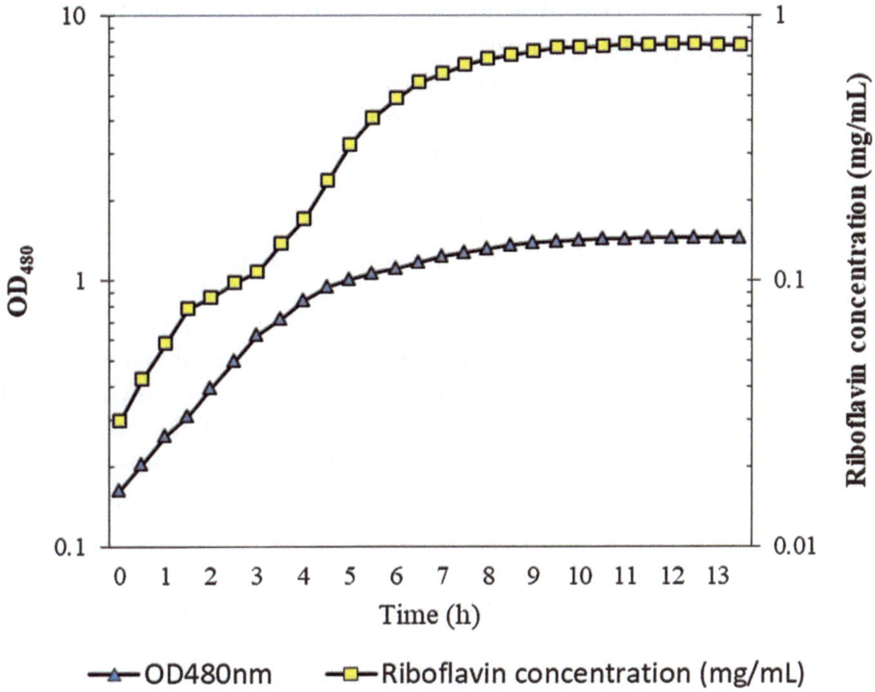

Fig. 2 Example of simultaneous and real-time detection of bacterial growth (triangle) and riboflavin quantification (square)

7. The growth rate at exponential phase (μ) is determined applying the following equation:

$$\ln\left(\frac{OD_t}{OD_0}\right) = \mu(t - t_0)$$

where OD_t and OD_0 represent, respectively, the optical density of the culture at the times t and t_0, and μ is the slope of the linear regression in the graph of the experimental values of ln (OD_t/OD_0) during the exponential growth phase versus the incubation time.

8. Specific riboflavin production can be determined as the ratio between direct fluorescence quantification and the biomass (estimated from the OD_{480}) (*see* **Note 14**).

4 Notes

1. The chemically defined riboflavin-free medium (CDRFM) is prepared as described by Terrade et al. [25] and modified according to Russo et al. [9]. Although different commercial riboflavin-free media are available, we suggest to use the

CDRFM, whose composition is here reported (Table 1). This medium provides the following advantages:

(a) It has been optimized for the growth of wine lactic bacteria and successfully tested with 22 wine strains of the genera *Oenococcus* as well as lactobacilli and *Pediococcus* [25], that are LAB known for their fastidious nutritional requirements.

(b) Unlike other riboflavin-free commercial media, after sterilization by filtration the medium is colorless. This feature allows for easy qualitative discrimination of the riboflavin production only based on the yellow intensity. Moreover, compared to other riboflavin-free media, the CDRFM shows the lower basal level of fluorescence that could interfere with vitamin B_2 quantification.

2. Take care to dissolve each compound under the condition (neutral, basic or acid) indicated in Table 1, to obtain their complete solubilization.
3. Sterilization by autoclaving could destroy thermolabile molecules. Moreover, heat causes Maillard reactions with consequent browning of the medium.
4. Light-sensitive molecules (mainly some vitamins) are lost from exposure to light.
5. Optimal growth conditions should be determined depending on the cultured microorganism (i.e., temperature, aeration, incubation time).
6. According to Terrade and Mira de Orduña [26] for the determination of auxotrophies, at least three subcultures in deficient media should be made, reducing intra- and extracellular nutrient carry over by low inoculation rates and washing cells twice between transfers.
7. At the higher concentration of roseoflavin, it could be difficult to detect growth by the turbidity of the medium due to the change of the color from light pink to red. Therefore, it could be useful to centrifuge the culture in a microfuge (5,000 rpm for 3 min at room temperature), in order to detect the formation of bacterial pellet.
8. Different colonies could be derivative strains carrying different mutations or deletions in the riboswitch element, thus resulting in a different capability to produce riboflavin, as reported by Burgess et al. [8]. Therefore, a screening of a high number of colonies based on the intensity of the yellow color after 48 h of growth could be useful to select derivatives with different phenotypic traits.
9. The suggested time of 120 h of growth is considered to correspond to approximately 100 generations for *L. plantarum*. To

evaluate the stability of the riboflavin-overproducing phenotype of other LAB species, the incubation period should be modified according to the specific generation time.

10. The concentration of riboflavin solutions used to determine the calibration curve should range among the expected amounts of microbial vitamin B_2 synthesized by the LAB strain.
11. Each sample should be analyzed at least in triplicate to increase the reliability of the results.
12. Riboflavin is degraded into various photoproducts on exposure to light [27]. Therefore, take care to work under dark conditions in order to avoid underestimation of the riboflavin concentration.
13. The following protocol has been optimized for *Lactiplantibacillus plantarum* strains.
14. Since riboflavin accumulates in the medium, determining the specific riboflavin concentration does not allow normalizing the vitamin production in cultures that have reached different OD. In fact, the specific fluorescence increases with the OD of the culture (instead of remaining constant).

Acknowledgments

We thank Dr. Stephen Elson for critical reading of the manuscript. Pasquale Russo is the beneficiary of a grant by MIUR in the framework of "AIM: Attraction and International Mobility" (PON R&I2014-2020) (practice code D74I18000190001). This work was supported by the Spanish Ministry of Science, Innovation and Universities (grant RTI2018-097114-B-I00).

References

1. Capozzi V, Russo P, Dueñas MT et al (2012) Lactic acid bacteria producing B-group vitamins: a great potential for functional cereals products. Appl Microbiol Biotechnol 96:1383–1394. https://doi.org/10.1007/s00253-012-4440-2
2. LeBlanc JG, Milani C, de Giori GS et al (2013) Bacteria as vitamin suppliers to their host: a gut microbiota perspective. Curr Opin Biotechnol 24:160–168. https://doi.org/10.1016/j.copbio.2012.08.005
3. Powers HJ (2003) Riboflavin (vitamin B2) and health. Am J Clin Nutr 77:1352–1360. https://doi.org/10.1093/ajcn/77.6.1352
4. Thakur K, Tomar SK, Singh AK et al (2017) Riboflavin and health: a review of recent human research. Crit Rev Food Sci Nutr 57:3650–3660. https://doi.org/10.1080/10408398.2016.1145104
5. Thakur K, Tomar SK, De S (2016) Lactic acid bacteria as a cell factory for riboflavin production. Microb Biotechnol 9:441–451. https://doi.org/10.1111/1751-7915.12335
6. Abbas CA, Sibirny AA (2011) Genetic control of biosynthesis and transport of riboflavin and flavin nucleotides and construction of robust biotechnological producers. Microbiol Mol Biol Rev 75:321. https://doi.org/10.1128/MMBR.00030-10
7. Burgess C, O' Connell-Motherway M, Sybesma W et al (2004) Riboflavin production in *Lactococcus lactis*: potential for in situ

production of vitamin-enriched foods? Appl Environ Microbiol 70:5769–5777. https://doi.org/10.1128/AEM.70.10.5769-5777.2004

8. Burgess CM, Smid EJ, Rutten G, Sinderen D (2006) A general method for selection of riboflavin-overproducing food grade micro-organisms. Microb Cell Factories 5:24. https://doi.org/10.1186/1475-2859-5-24
9. Russo P, Capozzi V, Arena MP et al (2014) Riboflavin-overproducing strains of *Lactobacillus fermentum* for riboflavin-enriched bread. Appl Microbiol Biotechnol 98:3691–3700. https://doi.org/10.1007/s00253-013-5484-7
10. Mohedano ML, Hernández-Recio S, Yépez A et al (2019) Real-time detection of riboflavin production by *Lactobacillus plantarum* strains and tracking of their gastrointestinal survival and functionality in vitro and in vivo using mCherry labeling. Front Microbiol 10:1748. https://doi.org/10.3389/fmicb.2019.01748
11. Capozzi V, Menga V, Digesu AM et al (2011) Biotechnological production of vitamin B2-enriched bread and pasta. J Agric Food Chem 59:8013–8020. https://doi.org/10.1021/jf201519h
12. Russo P, de Chiara MLV, Capozzi V et al (2016) *Lactobacillus plantarum* strains for multifunctional oat-based foods. LWT Food Sci Technol 68:288–294. https://doi.org/10.1016/j.lwt.2015.12.040
13. Yépez A, Russo P, Spano G et al (2019) In situ riboflavin fortification of different kefir-like cereal-based beverages using selected Andean LAB strains. Food Microbiol 77:61–68. https://doi.org/10.1016/j.fm.2018.08.008
14. Juarez del Valle M, Laiño JE, Savoy de Giori G, LeBlanc JG (2014) Riboflavin producing lactic acid bacteria as a biotechnological strategy to obtain bio-enriched soymilk. Food Res Int 62:1015–1019. https://doi.org/10.1016/j.foodres.2014.05.029
15. Arena MP, Russo P, Capozzi V et al (2014) Probiotic abilities of riboflavin-overproducing *Lactobacillus* strains: a novel promising application of probiotics. Appl Microbiol Biotechnol 98:7569–7581. https://doi.org/10.1007/s00253-014-5837-x
16. Russo P, Iturria I, Mohedano ML et al (2015) Zebrafish gut colonization by mCherry-labelled lactic acid bacteria. Appl Microbiol Biotechnol 99:3479–3490. https://doi.org/10.1007/s00253-014-6351-x
17. Juarez del Valle M, Laiño JE, de Moreno de LeBlanc A et al (2016) Soyamilk fermented with riboflavin-producing *Lactobacillus plantarum* CRL 2130 reverts and prevents ariboflavinosis in murine models. Br J Nutr 116:1229–1235. https://doi.org/10.1017/S0007114516003378
18. Levit R, de Giori GS, de Moreno de LeBlanc A, LeBlanc JG (2017) Evaluation of the effect of soymilk fermented by a riboflavin-producing *Lactobacillus plantarum* strain in a murine model of colitis. Benef Microbes 8:65–72. https://doi.org/10.3920/BM2016.0063
19. Levit R, Savoy de Giori G, de Moreno de LeBlanc A, LeBlanc JG (2018) Effect of riboflavin-producing bacteria against chemically induced colitis in mice. J Appl Microbiol 124:232–240. https://doi.org/10.1111/jam.13622
20. Russo P, Peña N, de Chiara MLV et al (2015) Probiotic lactic acid bacteria for the production of multifunctional fresh-cut cantaloupe. Food Res Int 77:762–772. https://doi.org/10.1016/j.foodres.2015.08.033
21. Russo P, de Chiara MLV, Vernile A et al (2014) Fresh-cut pineapple as a new carrier of probiotic lactic acid bacteria. Biomed Res Int 2014:309183. https://doi.org/10.1155/2014/309183
22. Jakobsen J (2008) Optimisation of the determination of thiamin, 2-(1-hydroxyethyl)thiamin, and riboflavin in food samples by use of HPLC. Food Chem 106:1209–1217. https://doi.org/10.1016/j.foodchem.2007.06.008
23. Fatima Z, Jin X, Zou Y et al (2019) Recent trends in analytical methods for water-soluble vitamins. J Chromatogr A 1606:360245. https://doi.org/10.1016/j.chroma.2019.05.025
24. Thakur K, Tomar SK, Brahma B, De S (2016) Screening of riboflavin-producing lactobacilli by a polymerase-chain-reaction-based approach and microbiological assay. J Agric Food Chem 64:1950–1956. https://doi.org/10.1021/acs.jafc.5b06165
25. Terrade N, Noël R, Couillaud R, de Orduña RM (2009) A new chemically defined medium for wine lactic acid bacteria. Food Res Int 42:363–367. https://doi.org/10.1016/j.foodres.2008.12.011
26. Terrade N, Mira de Orduña R (2009) Determination of the essential nutrient requirements of wine-related bacteria from the genera *Oenococcus* and *Lactobacillus*. Int J Food Microbiol 133:8–13. https://doi.org/10.1016/j.ijfoodmicro.2009.03.020
27. Sheraz MA, Kazi SH, Ahmed S et al (2014) Photo, thermal and chemical degradation of riboflavin. Beilstein J Org Chem 10:1999–2012. https://doi.org/10.3762/bjoc.10.208

Chapter 2

Recent Advances in Construction of the Efficient Producers of Riboflavin and Flavin Nucleotides (FMN, FAD) in the Yeast *Candida famata*

Dariya V. Fedorovych, Kostyantyn V. Dmytruk, and Andriy A. Sibirny

Abstract

The approaches used by the authors to design the *Candida famata* strains capable to overproduce riboflavin, flavin mononucleotide (FMN), and flavin adenine dinucleotide (FAD) are described. The metabolic engineering approaches include overexpression of *SEF1* gene encoding positive regulator of riboflavin biosynthesis, *IMH3* (coding for IMP dehydrogenase) orthologs from another species of flavinogenic yeast *Debaryomyces hansenii*, and the homologous genes *RIB1* and *RIB7* encoding GTP cyclohydrolase II and riboflavin synthase, the first and the last enzymes of riboflavin biosynthesis pathway, respectively. Overexpression of the above mentioned genes in the genetically stable riboflavin overproducer AF-4 obtained by classical selection resulted in fourfold increase of riboflavin production in shake flask experiments.

Overexpression of engineered enzymes phosphoribosyl pyrophosphate synthetase and phosphoribosyl pyrophosphate amidotransferase catalyzing the initial steps of purine nucleotide biosynthesis enhances riboflavin synthesis in the flavinogenic yeast *C. famata* even more.

Recombinant strains of *C. famata* containing *FMN1* gene from *D. hansenii* encoding riboflavin kinase under control of the strong constitutive *TEF1* promoter were constructed. Overexpression of the *FMN1* gene in the riboflavin-producing mutant led to the 30-fold increase of the riboflavin kinase activity and 400-fold increase of FMN production in the resulting recombinant strains which reached maximally 318.2 mg/L.

FAD overproducing strains of *C. famata* were also constructed. This was achieved by overexpression of *FAD1* gene from *D. hansenii* in *C. famata* FMN overproducing strain. The 7- to 15-fold increase in FAD synthetase activity as compared to the wild-type strain and FAD accumulation into cultural medium were observed. The maximal FAD titer 451.5 mg/L was achieved.

Key words Riboflavin, Flavin mononucleotide, Flavin adenine dinucleotide, Yeast *Candida famata*, Metabolic engineering, Riboflavin, FMN and FAD overproducers

1 Introduction

Riboflavin (vitamin B_2) is one of the most important vitamins required for human and animal nutrition. This vitamin is metabolic precursor of flavin nucleotides involved as coenzymes in numerous

Maria Barile (ed.), *Flavins and Flavoproteins: Methods and Protocols*, Methods in Molecular Biology, vol. 2280,
https://doi.org/10.1007/978-1-0716-1286-6_2,

enzymatic reactions, mostly of oxidative metabolism. Riboflavin deficiency causes retardation of growth, skin, nervous, and eye diseases. This compound is manufactured for use as a vitamin in human and animal nutrition and as a food colorant, for example, for soft drinks and yogurt. Approximately 10% of manufactured riboflavin is used in medicine as a component of multivitamin mixtures and in the treatment of a number of skin, vision, and nervous system diseases, including migraine and malaria.

The current world annual production of riboflavin is about 9000 tons which corresponds in price equivalent to 250 million US$. It was produced by both synthetic and fermentation processes. Currently, the production of riboflavin is exclusively done using microbial fermentation [1]. Bacteria *Bacillus subtilis*, the yeast *Candida famata*, and the filamentous fungus *Ashbya gossypii* were mainly used for the biosynthesis of riboflavin via large-scale production though currently production using yeast *C. famata* is discontinued [2–6]. Riboflavin-overproducing strains of *B. subtilis* and *A. gossypii* were constructed using metabolic engineering approaches. Riboflavin production in *B. subtilis* was elevated by the increasing the copy numbers of the riboflavin operon and of the *ribA* gene, coding for GTP cyclohydrolase II, the first enzyme of riboflavin biosynthesis as well as selection of engineered strains resistant to purine and riboflavin analogs by protein engineering of the enzyme to improve its kinetic characteristics and other approaches [4, 5, 7]. In *A. gossypii*, riboflavin production was improved by disruption of *VMA1* encoding vacuolar ATPase leading to the complete excretion of riboflavin into media [8] and due to enhanced supplementation of purine precursors through the de novo purine biosynthetic pathway [9] and phosphoribosyl pyrophosphate synthetase [10] as well as deregulation of genes involved in the biosynthesis of purines and glycine by C-terminal deletion in the *BAS1* gene encoding a Myb-related transcription factor [11], see also mentioned reviews [2, 4, 5]. The main problem for the all fermentative processes is low genetic stability and productivity of overproducers.

The process of fermentation with the use of yeast producers has a range of advantages (simple media and cheap renewable carbohydrates as a carbon source, resistance against phage infection, and others). The flavinogenic potential of the yeast *C. famata* is very high, and some mutants of this species are the most flavinogenic organism known [2, 4, 5]. Despite limited knowledge concerning mechanisms of iron-dependent regulation of riboflavin biosynthesis, *C. famata* overproducers were selected by conventional mutagenesis. *C. famata* dep8 strain, an industrial yeast riboflavin producer, was selected by conventional mutagenesis because molecular tools for *C. famata* have only recently been developed. This organism accumulates high amounts of riboflavin (near 21 g/L in large bioreactor) and secretes it into the cultural medium

[5]. The industrial production of riboflavin using mutant strains of the yeast *C. famata* by ADM Company (USA) was stopped due to lack of profitability. One of the main reasons for the termination of this process was low stability of the riboflavin producer used [5]. A stable riboflavin-overproducing strain of *C. famata* were selected by us using a random mutagenesis based on the selection of mutants resistant to different antimetabolites [12, 13]. The conventional mutagenesis involved consecutive selection for resistance to riboflavin structural analog 7-methyl-8-trifluoromethyl-10-(1′-D-ribityl)isoalloxazine), 8-azaguanine, 6-azauracil, 2-diazo-5-oxo-L-norleucine, and guanosine as well as screening for yellow colonies at high pH. Flavinogenic activity of selected AF-4 strain was, however, lower than that of industrial riboflavin producer *C. famata* ATCC 20849 (dep8). However, both these strains produce rather small amounts of riboflavin, far from the theoretical maximum—only 4% of carbon substrate, glucose, is converted to riboflavin while the remaining 96% is used for biomass building or is evolved as CO_2. Therefore, rational approaches of metabolic engineering were focused on the improvement of riboflavin synthesis in stable AF-4 strain *C. famata*. This strain was used as a parental strain for metabolic engineering of more efficient riboflavin overproducers and for the construction of strains capable of FMN and FAD oversynthesis [14, 15].

During last decade, methods of transformation [16] and insertional mutagenesis [17] have been developed and structural genes of riboflavin synthesis of *C. famata*, namely *RIB1*, *RIB2*, *RIB5*, *RIB6*, and *RIB7*, and *D. hansenii RIB3* were isolated and identified [18]. *C. famata* transcription factor Sef1 has been identified as a positive regulator of riboflavin biosynthesis [17]. Point or knockout mutations of *C. famata* and *Pichia guilliermondii SEF1* convert phenotype of these flavinogenic species to that of majority of yeasts, where iron does not regulate riboflavin synthesis. Expression of additional copy of *SEF1 D. hansenii* in strain *C. famata* overproducer of riboflavin dep8 resulted in simultaneous increasing of riboflavin productivity and stability of this feature [19].

Manipulation with the purine nucleotide biosynthesis pathway has been successfully used to increase metabolic flow through the pathway in the flavinogenic fungus *A. gossypii*, resulting in the enhancement of the riboflavin production [9]. GTP is the immediate precursor for riboflavin synthesis. It is synthesized through the de novo purine nucleotide biosynthetic pathway, which starts with the formation of phosphoribosyl pyrophosphate (PRPP). The first two reactions in the purine nucleotide biosynthetic pathway catalyzed by PRPP synthetase and PRPP amidotransferase are tightly regulated by transcription repression and feedback inhibition mechanisms. It was shown that the upregulation of the purine biosynthetic genes (*pur* operon) and modification of PRPP amidotransferase which led to alleviation of its feedback inhibition resulted to threefold increase of riboflavin production in the industrial riboflavin producing strain of *B. subtilis* [20, 21]. Protein engineering of PRPP synthetase or PRPP amidotransferase leading to insensitivity to feedback inhibition of the enzymes and subsequent overexpression of the engineered genes in *A. gossypii* led to tenfold or twofold increases, respectively, in riboflavin production [9, 10].

In contrast to riboflavin biosynthesis, which occurs only in plants, fungi, and most of prokaryotes, entirely all organisms, including animals, are able to synthesize flavin nucleotides FMN and FAD (quoted to [5, 22]. The flavin coenzyme systems are involved in a wide range of biochemical processes, particularly in mitochondrial electron transport, photosynthesis, fatty acid oxidation, metabolism of vitamins B_6, B_{12}, and folates. Flavin nucleotides are used in pharmacy and food industry [5, 23].

Chemically synthesized FMN is abundant on the market (3% of total amount of produced riboflavin) and cheap; however, it contains a lot of impurities of flavin nature which could act as riboflavin antagonists [5]. There are many microorganisms that can overproduce riboflavin but they unable to oversynthesize FMN or FAD. The exception is *Eremothecium ashbyi*, which overproduces FAD and excretes it into medium. Various approaches were applied for the microbial production of flavin nucleotides. Using the FAD synthetase–overproducing recombinant bacterial cells, the production FAD from FMN or riboflavin and ATP, and the specific production of FMN from riboflavin and ATP were developed [24–28]. However, the main drawback of proposed approaches is a production of the mixture of FMN and FAD and requirement for addition of high concentrations of ATP and riboflavin into the cultural medium for phosphorylation reactions. Apparently, there is an intrinsic drawback in the development of bacterial organisms for production of flavin nucleotides because FMN is a corepressor of *RFN* riboswitches, which prevents accumulation of large amounts of FMN and FAD during de novo biosynthesis [5].

Eukaryotes generally use two different enzymes for FMN and FAD production, whereas most prokaryotes depend on a single bifunctional enzyme, the riboflavin kinase/FAD synthetase [29]. In eukaryotes, riboflavin is converted into catalytically active cofactors via the sequential actions of ATP:riboflavin phosphotransferase (RF kinase, EC 2.7.1.26), which phosphorylates the vitamin into FMN and ATP:FMN adenylyl transferase (FAD synthetase, EC 2.7.7.2), which adenylates FMN to FAD. Both enzymes have been purified from yeast and rat tissues, and biochemically characterized and genes coding for these enzymes were identified [5].

2 Construction of the Efficient Riboflavin Overproducers in the Yeast *C. famata* Using Metabolic Engineering Approaches

2.1 Overexpression of SEF1 Gene (Encodes Transcription Factor) into C. famata Riboflavin-Overproducing Strains

It has been shown that putative transcriptional factor Sef1 is involved in positive regulation of riboflavin synthesis in the *C. famata* [17]. One might expect that the introduction of additional copies of *SEF1* into the genome of *C. famata* riboflavin-overproducing strain AF-4 would further increase riboflavin production. There are several successful examples of enhancement of gene dosages of positive regulators bestowing a substantial increase in the synthesis of desired products [30, 31]. To this end, an additional copy of *D. hansenii SEF1* gene was integrated into the genome of AF-4 strain, which resulted in 1.7-fold improvement of riboflavin production as compared to the parental strain. As *SEF1* gene sequence of *C. famata* was unavailable, the *SEF1* ortholog from closely related flavinogenic yeast *D. hansenii* CBS767 complementing insertion mutation *sef1::LEU2* of *C. famata* [18] was used for this purpose. To introduce the additional copy of *SEF1* from *D. hansenii*, strain AF-4 was transformed with integrative plasmid pTDhSEF1 (Fig. 1). Isolated transformants bearing an additional copy of the *SEF1* gene accumulated enhanced amounts of riboflavin as compared to the parental strain. Riboflavin synthesis of AF-4/SEF1 strain is represented in Table 1. To introduce an additional copy of *SEF1* gene from *D. hansenii* additional selective markers was required. A gene *IMH3* encoding IMP dehydrogenase, a rate-limiting enzyme in the de novo synthesis of GTP, an immediate riboflavin precursor [32] was used. *IMH3* gene serves as a dominant selective marker conferring resistance to mycophenolic acid (MPA) [33]. Consequently, introduction of the *IMH3* gene should simultaneously bestow resistance to mycophenolic acid, elevate activity of IMP dehydrogenase, and increase riboflavin synthesis in yeast transformants. Strains AF-4/IMH3, AF-4/SEF1/IMH3, and AF-4/2xSEF1/IMH3 were obtained via transformation of AF-4 and AF-4/SEF1 strains with integrative plasmids pDhIMH3 and pTDhSEF1/DhIMH3, respectively (Fig. 1). Results of riboflavin production by the constructed strains are shown in Table 1. The AF-4/IMH3 strain showed a 1.6-fold improvement in the production of riboflavin relative to the parental strain AF-4. The AF-4/SEF1 and AF-4/SEF1/IMH3 strains showed more significant 1.9-fold improvement in the production of riboflavin, whereas the AF-4/2xSEF1/IMH3 strain showed even much higher riboflavin overproduction, with 3.2-fold enhancement as compared to the strain AF-4. Constructed strains also produced much more riboflavin compared to industrial strain *C. famata* ATCC 20849 (dep8) ([34], Table 1).

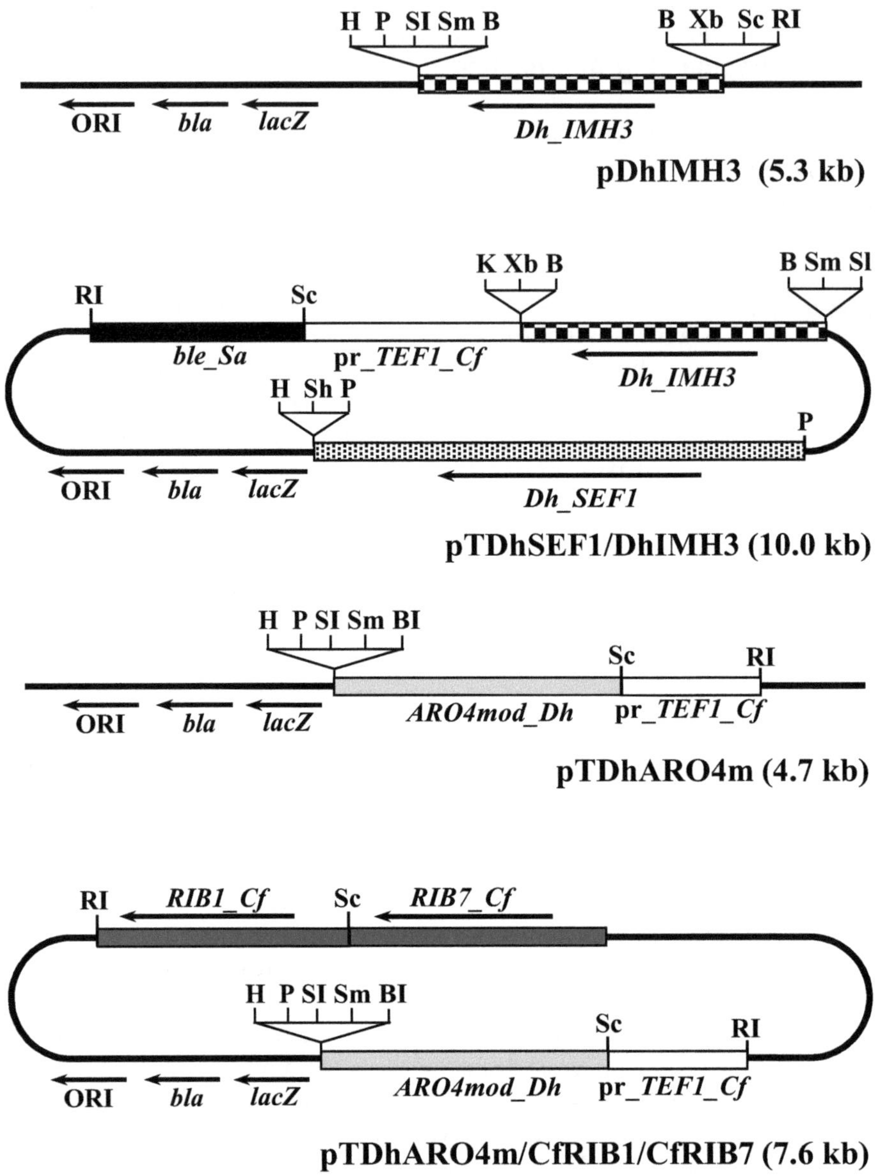

Fig. 1 Schemes of the plasmids used in this study: pDhIMH3, pTDhSEF1/DhIMH3, pTDhARO4m, and pTDhARO4m/CfRIB1/CfRIB7. The *C. famata* DNA fragment harboring the *TEF1* promoter is shown as a white box. The *Staphylococcus aureus* DNA fragment carrying the *ble* gene—black box. DNA fragments containing the *DhIMH3*, *DhSEF1*, modified *DhARO4* and *CfRIB1* with *CfRIB7* genes are shown as tessellated, dotted, light-gray, and gray boxes, respectively. pUC57 sequence—thin line. Origin of replication ORI, ampicillin resistance gene (*bla*) and *lacZ*—arrows. Restriction sites: H, *Hind*III; Sh, *Sph*I; P, *Pst*I; Sl, *Sal*I; Xb, *Xba*I; B, *Bam*HI; Sm, *Sma*I; K, *Kpn*I; Sc, *Sac*I; RI, *Eco*RI [13, 15]

Table 1
Riboflavin production, riboflavin-specific production rate, and riboflavin yield of *C. famata* transformants and control strains (YPD medium on the third day of cultivation) [34]

Strain	Riboflavin [mg/L]	Riboflavin-specific production rate [mg/g of cells/h]	Riboflavin yield [mg/g of cells]	Riboflavin yield [mg/g consumed glucose]
dep8	284 ± 14	0.74 ± 0.04	53.5 ± 2.6	14.2 ± 0.7
AF-4	250 ± 12	0.67 ± 0.03	48.1 ± 2.3	12.5 ± 0.6
AF-4/IMH3	410 ± 20	1.12 ± 0.05	80.4 ± 3.9	20.5 ± 1.0
AF-4/SEF1	468 ± 23	1.28 ± 0.06	91.8 ± 4.5	23.4 ± 1.2
AF-4/RIB1/RIB7	345 ± 17	0.94 ± 0.04	67.6 ± 3.3	17.3 ± 0.9
AF-4/SEF1/IMH3	472 ± 23	1.34 ± 0.06	96.3 ± 4.7	23.6 ± 1.2
AF-4/2xSEF1/ IMH3	792 ± 38	2.34 ± 0.11	168.5 ± 8.1	40.2 ± 1.9
AF-4/2xSEF1/ IMH3 /RIB1/ RIB7	1026 ± 50	3.17 ± 0.15	228.0 ± 11.1	53.2 ± 2.6
BRP	1463 ± 21	3.38 ± 0.19	240.1 ± 10.0	73.2 ± 4.8
BRP/PRS3m/ ADE4m	2855 ± 35	5.39 ± 0.31	390.2 ± 20.4	147.8 ± 8.2

2.2 Overexpression of Limiting Structural Genes of Riboflavin Synthesis for Selection of C. famata Strains with Improved Riboflavin Production

A significant increase in riboflavin production was reached through manipulation of structural genes involved in riboflavin synthesis. To introduce additional copies of riboflavin structural genes *RIB1* and *RIB7* coding for enzymes of the first and the final reactions of riboflavin synthesis, the modified version of the gene *D. hansenii ARO4* (coding for 3-deoxy-D-arabino-heptulosonate-7-phosphate (DAHP) synthase) which catalyzes the first step in aromatic amino acid biosynthesis and is insensitive to feedback inhibition by tyrosine driven by the *CfTEF1* promoter was used as additional dominant selective marker conferring yeast resistance to fluorophenylalanine [35, 36]. The integrative plasmids pTDhARO4m and pTDhARO4m/CfRIB1/CfRIB7 (Fig. 1) were used for transformations of the corresponding AF-4 and AF-4/2xSEF1/IMH3 strains. The AF-4/2xSEF1/IMH3/RIB1/RIB7 strain showed even much higher riboflavin overproduction, with 4.1-fold enhancement as compared to the strain AF-4 (Table 1). Amplification of the first and last genes of riboflavin synthesis pathway resulted in 1.4- and 1.3-fold enhancement for AF-4/RIB1/RIB7 and AF-4/2xSEF1/IMH3/RIB1/RIB7, as compared to the corresponding parental strains AF-4 and AF-4/2xSEF1/IMH3, respectively (Table 1).

2.3 Deregulation of Purine Nucleotide Synthesis De Novo for Activation of Riboflavin Synthesis

To study the effects of increased metabolic flux through the de novo purine nucleotide on riboflavin synthesis in *C. famata* a simultaneous overexpression of initial enzymes of purine nucleotide synthesis in the previously constructed advanced stable riboflavin producer AF-4/SEF1/RIB1/RIB7 [13] was performed. To this end, the genes *D. hansenii PRS3* and *ADE4* encoding PRPP synthetase and PRPP amidotransferase were cloned and subjected to site-directed mutagenesis. Closely related flavinogenic yeast *D. hansenii* CBS767 as a source of target genes was chosen since genome sequence of *C. famata* is unavailable. The highly conserved amino acid residues involved in feedback inhibition reported for Prs3 and Ade4 of *A. gossypii* [9, 10] were substituted in the corresponding enzymes of *D. hansenii* expecting alleviation of feedback inhibition in a similar way. The strictly conserved residues leucine 132 and histidine 195, were substituted with isoleucine 132 and glutamine 195 in PRPP synthetase to avoid the inhibitory effect of ADP. The conserved amino acid residues aspartic acid 315, lysine 338, and alanine 422 that are involved in feedback inhibition of the PRPP amidotransferase were substituted with valine 315, glutamine 338, and tryptophan 422 in order to overcome inhibition by ATP and GTP. The engineered genes of *PRS3m*, *ADE4m* were cloned under the control of the strong constitutive promoter of *C. famata TEF1* gene by overlap PCR.

The AF-4/SEF1/RIB1/RIB7 [37] designated as BRP was used as the parental strain for overexpression of the engineered PRPP synthetase and PRPP amidotransferase encoded by the modified *PRS3m* and *ADE4m* genes, respectively. To achieve this goal, the constructed plasmids harboring *PRS3m*, *ADE4m*, or both *PRS3m* and *ADE4m* genes under the control of the constitutive *TEF1* promoter were used (Fig. 2). Corresponding integrative plasmids pPRS3m/IMH3, pADE4m/IMH3, and pPRS3m/ADE4m/IMH3 were used for transformation of strain BRP.

The overexpression of the *PRS3* and *ADE4* genes in obtained transformants was confirmed by qRT-PCR. It was found that the expression profile of the *PRS3* and *ADE4* was 13.8 and 6.5 fold increased as compared to that of *D. hansenii*. The expression of tested genes was not detectible in BRP. The riboflavin production by the BRP strain, amounted to 1.46 g/L. Strain BRP/PRS3m/ADE4m demonstrated twofold increase in riboflavin production that amounted to 2.85 g/L in the YPD medium. The riboflavin yield by the BRP/PRS3m/ADE4m amounted to 0.39 g/g of cells. The strain had a 1.6-fold increase in riboflavin yield per g of cells when compared to that of the parental BRP strain ([37], Table 1). Obtained results indicate that the simultaneous overexpression of initial enzymes of purine nucleotide synthesis in the previously constructed stable riboflavin producer AF-4/SEF1/RIB1/RIB7 activate riboflavin biosynthesis in *C. famata*. Constructed strain BRP/PRS3m/ADE4m produced 2.85 g/L of riboflavin in flask

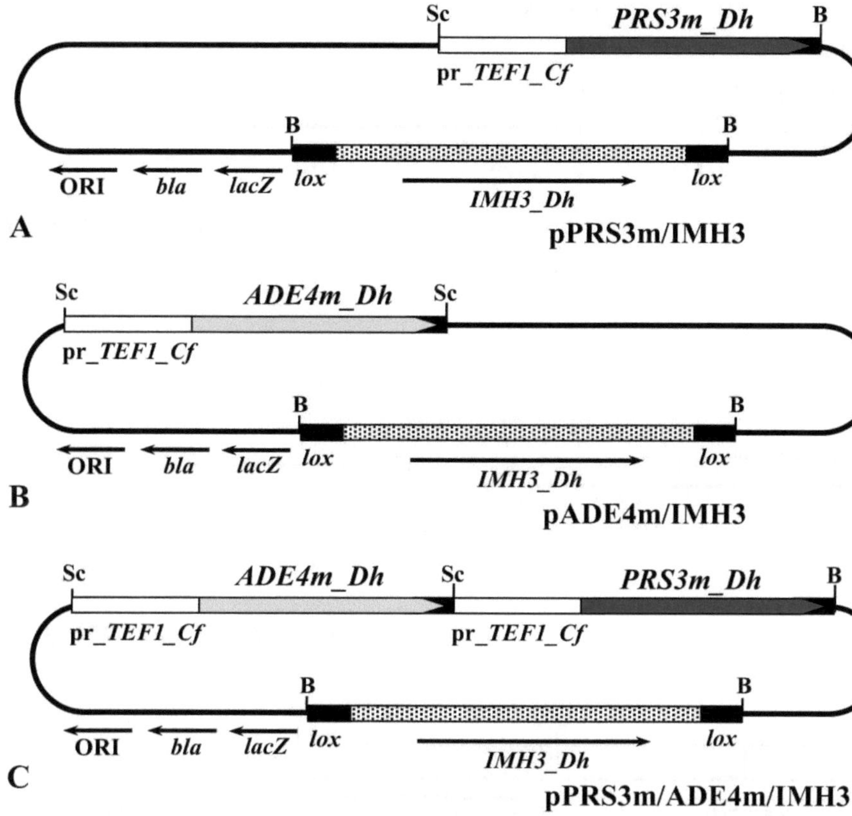

Fig. 2 Schemes of the plasmids: pPRS3m/IMH3, pADE4m/IMH3, and pPRS3m/ADE4m/IMH3. The *C. famata TEF1* promoter is shown as white box, modified *D. hansenii PRS3* gene is shown as dark-gray arrow, modified *D. hansenii ADE4* gene is shown as light-gray arrow, *D. hansenii IMH3* gene is shown as dotted box, thin line—bacterial vector pUC57, origin of replication ORI, ampicillin resistance gene (*bla*) and *lacZ*—arrows. Sites of restriction endonucleases: B—BamHI, Sc—SacI [37]

batch culture, which is close to the highest amount described in literature for recombinant *B. subtilis* reaching 4.2 g/L of riboflavin in shake flask cultivation medium [41] and much higher than was described in scientific and patent literature for yeast and fungi riboflavin overproducers, including *A. gossypii* [9].

3 Construction of the Recombinant Strains of the Flavinogenic Yeast *C. famata* Able to Produce FMN

To construct strains with higher riboflavin kinase activity, *FMN1* gene, encoding riboflavin kinase, has been cloned from *D. hansenii* CBS767 and overexpressed in *C. famata* wild type strain L2105 under the control of its own strong promoter *TEF1* (Fig. 3). Most of the recombinant strains possessed six- to eightfold increased activity of this enzyme and accumulation of very small amounts (no more 1.5 mg/L) of FMN in cultural medium [38]. 25–34% of

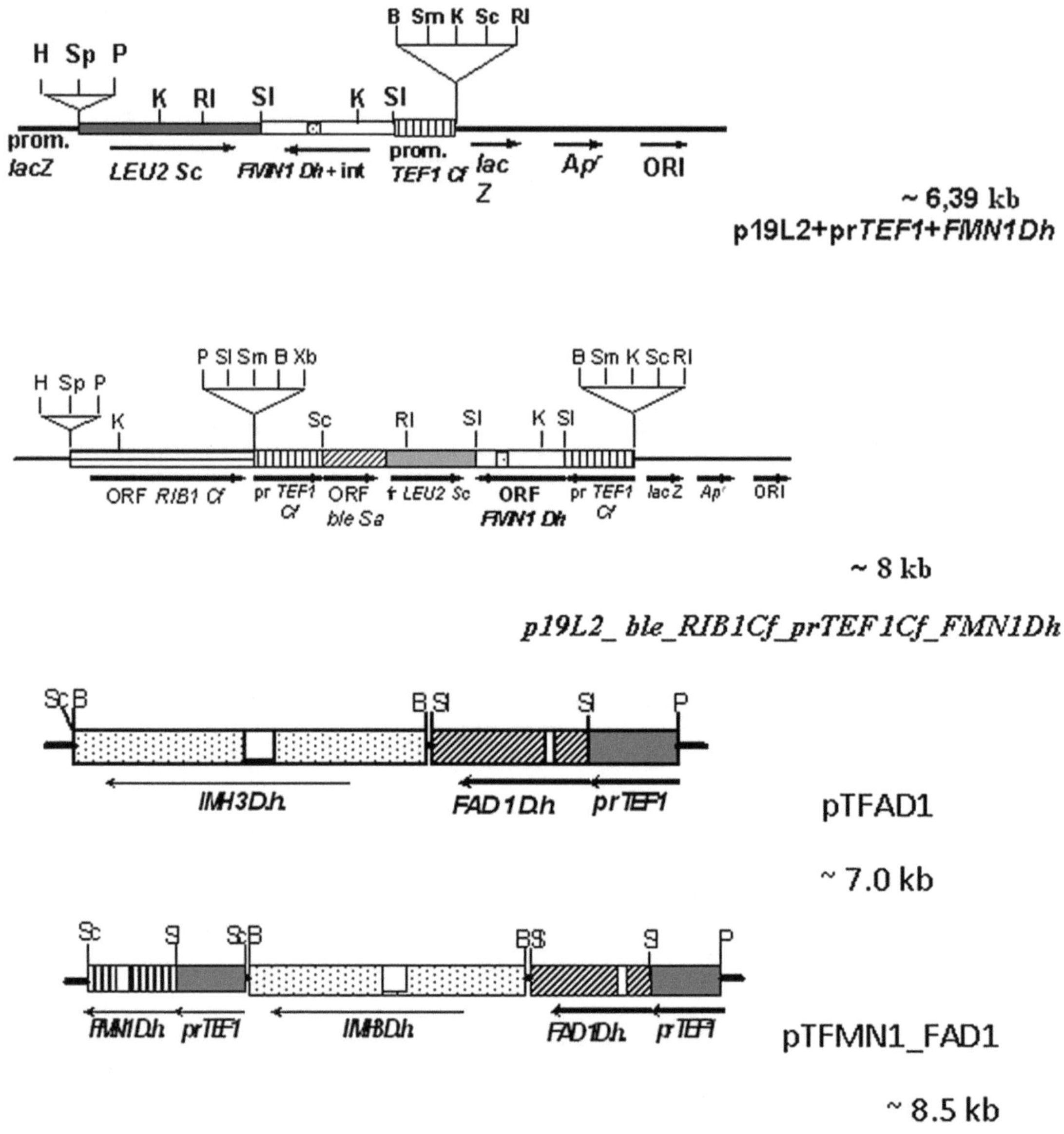

Fig. 3 Linear schemes of plasmid p19L2 + *prTEF1* + *FMN1Dh*, p19L2_ble_*RIB1Cf_prTEF1_FMN1Dh*, pTFAD1, and pTFMN1_FAD1. The fragment containing the *LEU2* gene of *S. cerevisiae* is indicated by a gray bar; the fragment containing the ARS of *C. famata* is indicated by a black bar; a fragment containing the *RIB1* promoters of *C. famata* and *TEF1* of *C. famata*, respectively, indicated by vertical strokes; fragment containing the open reading frame *FMN1 C. albicans*, marked with oblique strokes; the fragment containing the open reading frame *FMN1 D. hansenii*, indicated by a white strip; *D. hansenii* DNA fragment with *FAD1* gene (box with slanting hatches); *D. hansenii* DNA fragment with *FMN1* gene (box with vertical hatches); *FMN1* intron *D. hansenii*, dotted; the thin black line indicates the bacterial part of the vector. Restriction Sites: H, HindIII; Sp, SphI; P, PstI; K, KpnI; RI, EcoRI; SI, SalI; Xb, XbaI; B, BamHI; Sm, SmaI; Sc, SacI [13–15]

the total flavin content in culture medium was FMN, while the FMN content of wild-type strain L20105 did not exceed 10%. To express the additional copy of *FMN1* gene under the *TEF1* promoter in *C. famata*, the plasmid p19L2_ble_RIB1Cf_pr-*TEF1_FMN1Dh* (Fig. 3) was introduced into obtained earlier recombinant strain #18, which already contains the recombinant *FMN1*. The multicopy integration of recombinant *FMN1* gene into genome of the strains was confirmed by Southern hybridization. Resulting strains had higher copy numbers of the recombinant gene compared to *C. famata* recipient strain #18. The strains that contain six to eight copies of *FMN1* gene were characterized with 250-fold increase in riboflavin kinase (RF kinase) activity and 40-fold increase in FMN content in medium relative to the wild-type strain [14]. The strains also showed 3.5-fold increase in FMN production and 70-fold increase in RF kinase activity as compared with the recipient strain *C. famata* #18. However, despite the high RF kinase activity, the accumulation of FMN was still low and a large amount of riboflavin was detected in the cultural medium.

All strains described above are capable to overproduce riboflavin only under conditions of iron depletion. It is very inconvenient for developing commercial strains, because of impossibility to support the iron-deficient conditions during industrial production. In addition, the amounts of riboflavin synthesized as substrate by the strains are insufficient for RF kinase–catalyzed reaction. Therefore, as recipient the *C. famata* strain AF-4 that has deregulated riboflavin synthesis pathway and high level of riboflavin production was used [13]. Stable recombinant strains derived from AF-4 that contained the *FMN1* gene under the *TEF1* promoter were constructed. The strains showed 30-fold increase in RF kinase activity as compared to that of the wild-type strain (Fig. 4). The high-density cultivation (final biomass, 5 g dry weight per L) of

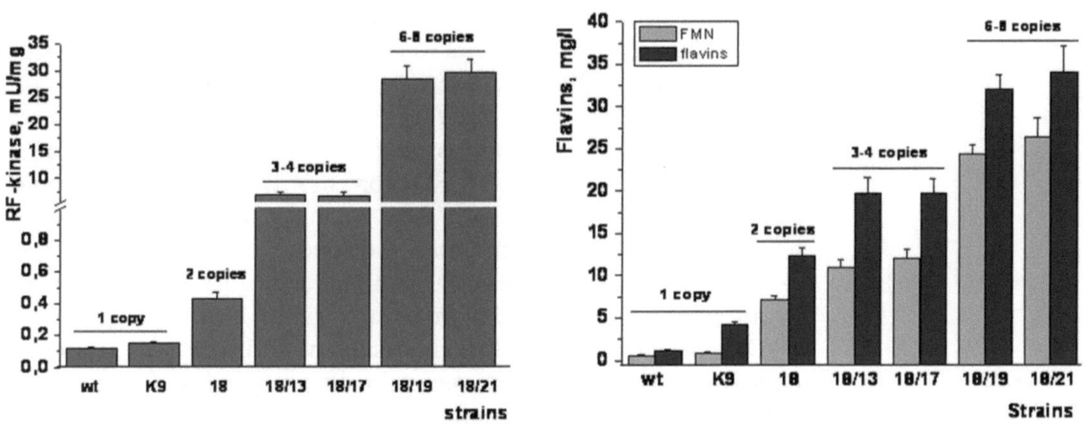

Fig. 4 RF kinase activity of transformants (**a**) and flavin accumulation in cultural medium during incubation (**b**) of *C. famata* recombinant strains containing the *FMN1* gene under the *TEF1* promoter. wt—*C. famata* VKM Y-9, K9—*C. famata* transformant (leu^{+}) (1 copy of *FMN1* gene), 18—*C. famata* transformant (p19L2_*prTEF1_FMN1Dh*) (2 copies of *FMN1* gene), 18/13-18/17—*C. famata* transformants (p19L2_*prTEF1_FMN1Dh* and p19L2_ble_*RIB1Cf_prTEF1_FMN1Dh*) (3–4 copies of *FMN1* gene), 18/19–18/21—*C. famata* transformants (p19L2_*prTEF1_FMN1Dh* and p19L2_ble_*RIB1Cf_prTEF1_FMN1Dh*) (6–8 copies of *FMN1* gene) [14, 37]

transformants in modified Burkholder medium resulted in 120-fold increase in FMN production reaching 72 mg/L. It was near 60% of total flavin amount in the recombinant strains [14]. The FMN production by different clones of the selected recombinant strain varied from 45 to 120 mg/L. One of such clones was accumulated 200–250 mg/L FMN under the high-density cultivation condition (final biomass, 5 g dry weight per L). The statistical optimization of medium constituents was employed to enhance FMN production of this recombinant strain [39, 40]. A final FMN concentration of 318.2 mg/L was obtained.

4 Construction of FAD Overproducing Strains of *C. famata* by Overexpression of FAD Synthetase

For construction of *C. famata* strains with the ability for FAD overproduction, the sequence of *D. hansenii* CBS767 was used. As the recipient, the *C. famata* strain obtained earlier, characterized by high level of FMN production, was used [14]. The amount of FMN (substrate of FAD synthetase) overproduced in this strain could be enough for FAD overproduction.

To express the *FAD1* gene under the *TEF1* promoter in *C. famata*, two different plasmids were constructed [15]. The first plasmid, pTFAD1, contained a *FAD1* gene under the *TEF1* promoter (Fig. 3). The second plasmid pTFMN1_FAD1 contained an additional copy of *D. hansenii FMN1* gene under the *TEF1* promoter (Fig. 3). Both integrative plasmids were introduced to a FMN-producing strain, T-OP 13-76. The reason for expression of the additional copy of *FMN1* gene was more production of FMN as a substrate for adenylyltransferase reaction.

Colonies of transformants were selected by resistance to mycophenolic acid (20 mg/L) as the plasmids pTFAD1 and pTFMN1_FAD1 contain the *IMH3* gene conferring the resistance to this antibiotic. Stable strains were isolated among selected transformants (designated as T-FD for strains, contained plasmid pTFAD1 and T-FD-FM for strains, contained plasmid pTFMN1_FAD1). The integration of two copies of *FAD1* expression module into genome of the strains was confirmed by Southern hybridization. All recombinant strains had 6- to 17-fold higher FAD synthetase activity than the recipient and wild-type strains of *C. famata* [15]. Strains containing additional copies of *FAD1* gene accumulated in cultural medium 3.9–9.6 mg/L of FAD. That is more than tenfold increase when compared to the recipient strain. The FAD production in strains containing both *FAD1* and *FMN1* genes accumulated 30–60 mg/L after 48 h of cultivation. The total extracellular flavin content (riboflavin+FMN+FAD) did not differ in recombinant and recipient strains (400 mg/L). However, FAD

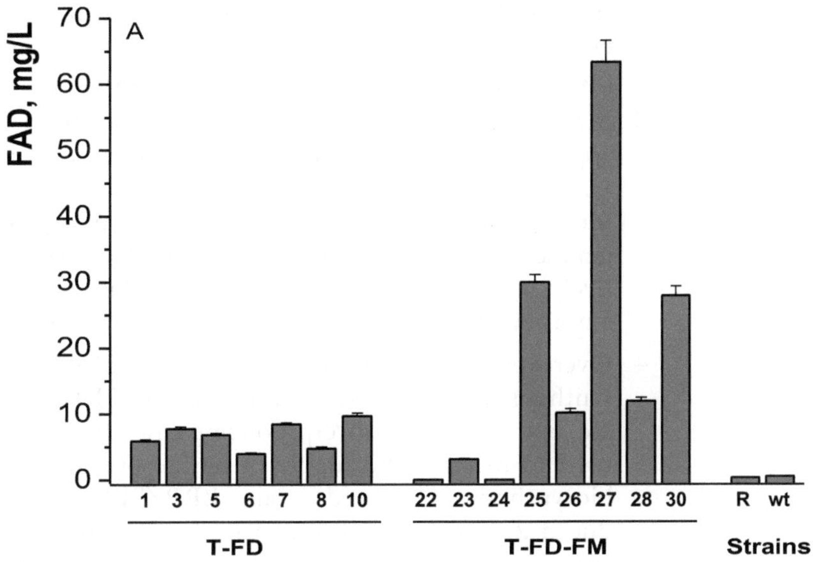

Fig. 5 FAD accumulation in culture medium of *C. famata* recombinant strains: T-FD—strains, contained plasmid pTFAD1 (containing the *FAD1* gene under the *TEF1* promoter); T-FD-FM—strains, contained plasmid pTFMN1_FAD1 (containing the *FAD1* gene under the *TEF1* promoter and the *FMN1* gene under the *TEF1* promoter); R—recipient strain *C. famata* T-FM 13-76 (containing the *FMN1* gene under the *TEF1* promoter); wt—*C. famata* VKM Y-9 [15]

percentage varied from 1 to 2.5% in strains belonging to group T-FD and up to 18% in strains with the highest FAD content. The effectiveness of FAD production under different growth conditions by one of these recombinant strains was evaluated [15]. The two-level Plackett–Burman and central composite design were used for achieving maximum FAD yield. Implementation of these optimization strategies increased FAD production to 387.0 mg/L. The maximum FAD yield of 451.5 mg/L after 40 h of incubation in batch culture in the mineral medium without addition of expensive components was achieved (Fig. 5).

5 Concluding Remarks

1. Riboflavin is an essential compound for humans and animals. Currently, it is produced on a large scale by microbial synthesis. Industrial riboflavin production is achieved through the use of recombinant strains of the gram-positive bacterium *Bacillus subtilis* and filamentous fungus *Ashbya gossypii*.
2. The process of fermentation with the use of yeast producers has a range of advantages (simple media and cheap renewable carbohydrates as a carbon sources, resistance against phage infection and other). *C. famata* riboflavin overproducers were selected by means of classic genetics methods and had been

successfully used for production of riboflavin for years. However, the efficacy of the process has to be improved for providing dependable industrial production.

3. Metabolic engineering approaches have been successfully applied for construction of riboflavin overproducer in *C. famata*. Four genes *SEF1*, *IMH3*, *RIB1*, and *RIB7* coding for the transcription factor, inosine monophosphate dehydrogenase, GTP cyclohydrolase II, and riboflavin synthase, respectively, were simultaneously overexpressed in the background of a nonreverting riboflavin-producing *C. famata* mutant AF-4.
4. Overexpression of both initial enzymes of purine nucleotide synthesis de novo, PRPP synthetase and PRPP amidotransferase, in the riboflavin-overproducing strain has been successfully used to increase the supply of purine precursor to riboflavin biosynthesis. Obtained strain BRP/PRS3m/ADE4m produced 2.85 g/L of riboflavin in flask batch culture, which is close to the highest amount described in literature for recombinant *B. subtilis* and much higher than that described in scientific and patent literature for yeast and fungi riboflavin overproducers, including *A. gossypii*.
5. The construction of stronger riboflavin-overproducing yeast strains will require further metabolic engineering. One of the approaches is to study the effects of overproduction of the second immediate riboflavin precursor, ribulose-5-phosphate, on riboflavin biosynthesis. In addition, identification of the rate-limiting enzymes of riboflavin synthesis pathway, genes responsible for riboflavin excretion, and new transcriptional factors involved in regulation of riboflavin synthesis are important for construction effective yeast producers of this vitamin.
6. *C. famata* strains overproducing FMN de novo without adding high concentrations of exogenous riboflavin and ATP were engineered. The obtained strains expressing the *FMN1* gene under the *TEF1* promoter could be used for the further construction of the improved FMN overproducers.
7. FMN production by the constructed strains was accompanied with riboflavin accumulation in the medium in spite of high activity of riboflavin kinase. One may suggest that it was caused by the hydrolytic activity of phosphatases or FMN hydrolases. Identification and disruption of the corresponding hydrolytic genes could have a positive impact on FMN accumulation by recombinant strains of *C. famata*. Further improvement of constructed strains can be achieved by overexpression of *FMN1* gene in the best overproducer of riboflavin.
8. For the first time, the yeast strains overproducing FAD were obtained. However, the FAD content did not exceed 18% of the total amount of flavins in the cultural medium. Further

enhancement of FAD production could be achieved by inactivation of the pyrophosphatase genes, involved in FAD hydrolysis and by multicopy integration of *FMN1* and *FAD1* genes in the newly isolated more efficient riboflavin-overproducing *C. famata* mutants.

Acknowledgments

This study was supported by Polish National Science Center, grant Opus UMO-2018/29/B/NZ1/01-497, by National Academy of Sciences of Ukraine (Grant 36-19) and by Ministry of Education and Science of Ukraine grant M/ 32 -2020, National Research Foundation of Ukraine grant 2020.01/0090.

References

1. Schwechheimer SK, Park EY, Revuelta JL, Becker J, Wittmann C (2016) Biotechnology of riboflavin. Appl Microbiol Biotechnol 100:2107–2119
2. Stahmann KP, Revuelta JL, Seulberger EH (2000) Three biotechnical processes using *Ashbya gossypii*, *Candida famata*, or *Bacillus subtilis* compete with chemical riboflavin production. Appl Microbiol Biotechnol 53:509–516
3. Survase SA, Bajaj IB, Singhal RS (2006) Biotechnological production of vitamins. Food Technol Biotechnol 44:381–396
4. Hohmann HP, Stahman KP (2010) Biotechnology of riboflavin production. In: Mander L, Liu HW (eds) Comprehensive natural products. II. Chemistry and biology. 7 cofactors. Elsevier, Philadelphia, PA, pp 115–135
5. Abbas CA, Sibirny AA (2011) Genetic control of biosynthesis and transport of riboflavin and flavin nucleotides and construction of robust biotechnological producers. Microbiol Mol Biol Rev 75:321–360
6. Revuelta JL, Ledesma-Amaro R, Lozano-Martinez P et al (2017) Bioproduction of riboflavin: a bright yellow history. J Ind Microbiol Biotechnol 44:659–665
7. Fischer M, Bacher A (2005) Biosynthesis of flavocoenzymes. Nat Prod Rep 22:324–350
8. Förster C, Santos MA, Ruffert S, Kramer R, Revuelta JL (1999) Physiological consequence of disruption of the *VMA1* gene in the riboflavin overproducer *Ashbya gossypii*. J Biol Chem 274:9442–9448
9. Jiménez A, Santos MA, Pompejus M, Revuelta JL (2005) Metabolic engineering of the purine pathway for riboflavin production in *Ashbya gossypii*. Appl Environ Microbiol 71:5743–5751
10. Jiménez A, Santos MA, Revuelta JL (2008) Phosphoribosyl pyrophosphate synthetase activity affects growth and riboflavin production in *Ashbya gossypii*. BMC Biotechnol 8:67
11. Mateos L, Jiménez A, Revuelta JL et al (2006) Purine biosynthesis, riboflavin production, and trophic-phase span are controlled by a Myb-related transcription factor in the fungus *Ashbya gossypii*. Appl Environ Microbiol 72:5052–5060
12. Sibirny AA, Dmytruk KV, Fedorovych DV (2010) *Candida famata* IMB Y-5034 yeast strain overproducing riboflavin (vitaminB_2). UA Patent 90741, 25 May 2010
13. Dmytruk KV, Yatsyshyn VY, Sibirna NO et al (2011) Metabolic engineering and classic selection of the yeast *Candida famata* (*Candida flareri*) for construction of strains with enhanced riboflavin production. Metab Eng 13:82–88
14. Yatsyshyn VY, Ischuk OP, Voronovsky AY et al (2009) Production of flavin mononucleotide by metabolically engineered yeast *Candida famata*. Metabol Eng 11:163–167
15. Yatsyshyn VY, Fedorovych DV, Sibirny AA (2014) Metabolic and bioprocess engineering for obtaining of the FAD overproducing strains in the yeast *Candida famata*. J Ind Microbiol Biotechnol 41:823–835
16. Voronovsky AA, Abbas CA, Fayura LR et al (2002) Development of a transformation system for the flavinogenic yeast *Candida famata*. FEMS Yeast Res 2:381–388

17. Dmytruk KV, Voronovsky AY, Sibirny AA (2006) Insertion mutagenesis of the yeast *Candida famata (Debaryomyces hansenii)* by random integration of linear DNA fragments. Curr Genet 50:183–191
18. Voronovsky AY, Abbas CA, Dmytruk KV et al (2004) Candida famata (Debaryomyces hansenii) DNA sequences containing genes involved in riboflavin synthesis. Yeast 21:1307–1316
19. Voronovsky AY, Dmytruk KV, Sibirny AA, Fedorovych DV, Yatsyshyn VY. 2008. *Candida famata* IMB Y-5033 yeast strain—stable producer of riboflavin (vitamin B_2). UA Patent 90741, 25 May 2010
20. Shi SB, Shen Z, Chen X, Chen T, Zhao XM (2009) Increased production of riboflavin by metabolic engineering of the purine pathway in *Bacillus subtilis.* Biochem Eng J 46:28–33
21. Shi T, Wang YC, Wang ZW et al (2014) Deregulation of purine pathway in *Bacillus subtilis* and its use in riboflavin biosynthesis. Microb Cell Factories 13:101–116
22. Yatsyshyn VY, Fedorovych DV, Sibirny AA (2009) The microbial synthesis of flavin nucleotides: a review. Appl Biochem Microbiol 45:133–142
23. Shimizu S (2001) Vitamins and related compounds: microbial production. In: Reed G, Rehm H-J (eds) Biotechnology, vol 10. Wiley VCH, Weinheim, pp 320–340
24. Hagihara T, Fujio T, Aisaka K (1995) Cloning of FAD synthetase gene from *Corynebacterium ammoniagenes* and its application to FAD and FMN production. Appl Microbiol Biotechnol 42:724–729
25. Nakagawa S, Hagihara T, Fujio T, Aisaka K (1995) Metaphosphate-dependent phosphorylation of riboflavin to FMN by *Corynebacterium ammoniagenes.* Appl Microbiol Biotechnol 43:325–329
26. Efimov I, Kuusk V, Zhang X, McIntire WS (1998) Proposed steady-state kinetic mechanism for *Corynebacterium ammoniagenes* FAD synthetase produced by *Escherichia coli.* Biochemistry 37:9716–9723
27. Mack M, van Loon AP, Hohmann HP (1998) Regulation of riboflavin biosynthesis in *Bacillus subtilis* is affected by the activity of the flavokinase/flavin adenine dinucleotide synthetase encoded by *ribC.* J Bacteriol 180:950–955
28. Kitatsuji K, Ishino S, Teshiba S, Arimoto M (1996) Method of producing flavine nucleotides. US Patent 5,514,574
29. Frago S, Velázquez-Campoy A, Medina M (2009) The puzzle of ligand binding to *Corynebacterium ammoniagenes* FAD synthetase. J Biol Chem 284:6610–6619
30. des Etages SA, Falvey DA, Reece RJ, Brandriss MC (1996) Functional analysis of the *PUT3* transcriptional activator of the proline utilization pathway in *Saccharomyces cerevisiae.* Genetics 142:1069–1082
31. Kuscer E, Coates N, Challis I et al (2007) Roles of *rapH* and *rapG* in positive regulation of rapamycin biosynthesis in *Streptomyces hygroscopicus.* J Bacteriol 189:4756–4763
32. Hyle JW, Shaw RJ, Reines D (2003) Functional distinctions between IMP dehydrogenase genes in providing mycophenolate resistance and guanine prototrophy to yeast. J Biol Chem 278:28470–28478
33. Köhler GA, Gong X, Bentink S et al (2005) The functional basis of mycophenolic acid resistance in *Candida albicans* IMP dehydrogenase. J Biol Chem 280:11295–11113
34. Dmytruk K, Lyzak O, Yatsyshyn V, Kluz M et al (2014) Construction and fed-batch cultivation of *Candida famata* with enhanced riboflavin production. J Biotechnol 172:11–17
35. Fukuda K, Watanabe M, Asano K et al (1991) A mutated *ARO4* gene for feedback-resistant DAHP synthase which causes both o-fluoro-DL-phenylalanine resistance and beta-phenethyl-alcohol overproduction in *Saccharomyces cerevisiae.* Curr Genet 20:453–456
36. Cebollero E, Gonzalez R (2004) Comparison of two alternative dominant selectable markers for wine yeast transformation. Appl Environ Microbiol 70:7018–7023
37. Dmytruk KV, Ruchala J, Fedorovych DV et al (2020) Overexpression of two engineered enzymes involved in the initial steps of purine nucleotide biosynthesis enhances riboflavin synthesis in the flavinogenic yeast Candida famata. Biotechnol J 15(7):e1900468. https://doi.org/10.1002/biot.201900468
38. Ischuk O P, Yatsyshyn VY, Dmytruk K V et al (2006) Construction of the flavinogenic yeast *Candida famata* strains with high riboflavin kinase activity using gene engineering. Ukr Biochim J 78:53–59
39. Sibirny AA, Yatsyshyn VY, Fedorovych DV et al (2010) Yeast strain *Candida famata* IMB Y-5028 producing flavin mononucleotide (5′FMN). UA Patent no. 90754 2010
40. Yatsyshyn VY, Fedorovych DV, Sibirny AA (2010) Medium optimization for production of flavin mononucleotide by the recombinant strain of the yeast *Candida famata* strain using statistical designs. Biochem Eng J 49:52–60
41. Wang G, Shi T, Chen T et al (2018) Integrated whole-genome and transcriptome sequence analysis reveals the genetic characteristics of a riboflavin-overproducing *Bacillus subtilis.* Metab Eng 48:138–149

Chapter 3

Overexpression of Riboflavin Excretase Enhances Riboflavin Production in the Yeast *Candida famata*

Andriy O. Tsyrulnyk, Dariya V. Fedorovych, Kostyantyn V. Dmytruk, and Andriy A. Sibirny

Abstract

Many microorganisms are capable of riboflavin oversynthesis and accumulation in a medium, suggesting that they efficiently excrete riboflavin. The mechanisms of riboflavin efflux in microorganisms remain elusive. *Candida famata* are representatives of a group of so-called flavinogenic yeast species that overproduce riboflavin (vitamin B_2) in response to iron limitation. The riboflavin overproducers of this yeast species have been obtained by classical mutagenesis and metabolic engineering. Overproduced riboflavin accumulates in the cultural medium rather than in the cells suggesting existence of the special mechanisms involved in riboflavin excretion. The appropriate protein and gene have not been identified in yeasts till recently. At the same time, the gene *BCRP* (breast cancer resistance protein) has been identified in mammal mammary glands. Several homologs of the mammal *BCRP* gene encoding putative riboflavin efflux protein (excretase) were identified in the flavinogenic yeasts *Debaryomyces hansenii* and *C. famata*. Here we evaluate the yeast homologs of *BCRP* with respect to improvement of a riboflavin production by *C. famata*. The closest homologs from *D. hansenii* or *C. famata* were expressed under the control of *TEF1* promoter of these yeasts in the wild-type and riboflavin-overproducing strains of *C. famata*. Resulted transformants overexpressed the corresponding genes (designated as *DhRFE* and *CfRFE*) and produced 1.4- to 6-fold more riboflavin as compared to the corresponding parental strains. They also were characterized by overexpression of *RIB1* and *RIB6* genes which encode the first and the last structural enzymes of riboflavin synthesis and exhibited elevated specific activity of GTP cyclohydrolase II. Thus, overexpression of yeast homolog of mammal gene *BCRP* may be useful to increase the riboflavin yield in a riboflavin production process using a recombinant overproducing *C. famata* strain or other flavinogenic microorganisms.

Key words *Candida famata* yeast, Riboflavin biosynthesis, *BCRP* (breast cancer resistance protein) gene, Riboflavin excretion

1 Introduction

Currently, riboflavin is produced on a large scale by microbial synthesis. The microbiological production has replaced chemical synthesis [1, 2]. Industrial riboflavin production is achieved through the use of recombinant strains of the gram-positive bacterium *Bacillus subtilis* and filamentous fungus *Ashbya gossypii* [3, 4].

Maria Barile (ed.), *Flavins and Flavoproteins: Methods and Protocols*, Methods in Molecular Biology, vol. 2280,
https://doi.org/10.1007/978-1-0716-1286-6_3,

Formerly, the flavinogenic yeast *Candida famata* was also used for industrial riboflavin production [5, 6]; however, this process was terminated due to low genetic stability of the producer [7]. The available schemes for selection of riboflavin producers are based mainly on overexpression of the genes involved in riboflavin biosynthesis and of its purine precursor. Despite significant increase in riboflavin accumulation, further strain improvement is needed to compete with current industrial producers, *B. subtilis* and *A. gossypii*. Enhanced export of riboflavin may also be useful to increase the riboflavin yield. Heterologous expression of the codon optimized *ribM* gene from bacterium *Streptomyces davaonensis* (*S. davawensis*), encoding a putative facilitator of riboflavin transport, in the riboflavin-producing strain of *B. subtilis*, improved riboflavin production, though the increase was relatively small [8]. The reason of activation was not elucidated.

There are two rather old publications on riboflavin excretion in yeasts, in *Saccharomyces cerevisiae* [9] and in the flavinogenic yeast *Pichia guilliermondii* [10]. No gene(s) encoding riboflavin excretion was identified till now in these organisms so far. This is not the case in animals as the mammalian mammary glands contain specific riboflavin efflux protein encoding by *BCRP* gene [11]. Several genes encoding putative riboflavin excretase (RF excretase) were identified in *P. guilliermondii* genome by searching for homology to human BCRP [Boretsky, personal communication]. We found that the flavinogenic yeasts *Debaryomyces hansenii* and *C. famata* contain genes homologous to the mammalian *BCRP* gene. Since BCRP protein actively secrets riboflavin into milk, we suggested that its yeast homolog is responsible for riboflavin excretion from the yeast cell and that its overexpression could activate overall riboflavin production in riboflavin-overproducing strains of *C. famata*. To check this hypothesis, the homolog of mammal *BCRP* from the flavinogenic yeast *D. hansenii* and *C. famata* were overexpressed in the wild-type and riboflavin-overproducing strains of *C. famata*. The resulted strains showed 1.4- to 6-fold increase in riboflavin production in shake flask relative to the corresponding parental strains.

2 Materials and Methods

2.1 Strains and Growth Conditions

Candida famata VKM Y-9 (All-Russian Collection of Microorganisms, Pushchino, Russia), L20105 (*leu2*) [12], riboflavin overproducers AF-4 [13] and AF-4/SEF1/RIB1/RIB7 (designated as BRP from the *B*est *R*iboflavin *P*roducer) [14] strains, and *Debaryomyces hansenii* CBS767 (Centraalbureau voor Schimmelcultures, Utrecht, the Netherlands) were used throughout this work. Yeast cells were cultured at 30 °C in modified Burkholder medium [12], Burkholder medium, supplemented with 0.2% of yeast extract or YPD media (0.5% yeast extract, 1% peptone, and 2% glucose).

The *Escherichia coli* DH5α strain (Φ80d*lacZ*ΔM15, *recA*1, *endA*1, *gyrA*96, *thi*-1, *hsdR*17(r_K^-, m_K^+), *supE*44, *relA*1, *deoR*, Δ(*lacZYA-argF*)U169) was used as a host for plasmid propagation. The DH5α strain was grown at 37 °C in LB medium as described previously [15]. The transformed *E. coli* cells were maintained on a medium containing 100 mg/l of ampicillin.

2.2 Biochemical Analyses

Cell biomass was determined turbidimetrically with a Helios Gamma spectrophotometer (OD, 590 nm; cuvette, 10 mm) with gravimetric calibration. Riboflavin concentration was determined by measuring fluorescence (Turner Quantech FM 109510-33 fluorometer, excitation maximum = 440 nm, emission maximum = 535 nm). The flavin content of the cells was determined after extraction with 5% trichloroacetic acid. Protein content [16] and GTP-cyclohydrolase II activity were measured in cell-free extracts that were prepared by yeast cells vortexing with glass beads in Eppendorf microtubes at 4 °C for 10 min with 20 mM tris-HCl buffer pH 8.2 after dialysis [17]. The GTP-cyclohydrolase II assay mixture contained Tris–HCl buffer (pH 8.2) 20 mM, $MgCl_2$ 2 mM, GTP 0.5 mM and dithiothreitol 2 mM. The reaction was initiated with the addition of cell extract (0.5 mg of protein).

2.3 Molecular Biology Techniques

The plasmid DNA isolation, restriction, ligation, electrophoresis in agarose gel, electrotransformation, and PCR were carried out using standard methods [15]. Genomic DNA of *D. hansenii* and *C. famata* was isolated using the NucleoSpin® Tissue Kit. Restriction endonucleases and DNA ligase were used following the manufacturer's specifications. Plasmid isolation from *E. coli* was performed with the ZymoPURE Plasmid Miniprep. PCR amplification of the fragments of interest was performed using Phusion High-Fidelity DNA Polymerase according to the manufacturer specification. PCRs were performed in a GeneAmp PCR System 9700 thermocycler.

2.4 Cloning BCRP Gene Homolog from Genome of D. hansenii [18]

Flavinogenic yeast *D. hansenii* CBS767 was used for isolation of BCRP homolog. This yeast is closely related to *C. famata* [19] and its genome database is available (mycocosm.jgi.doe.gov/Debha1/Debha1.home.html). Blast search of *Homo sapiens* Bcrp1 (NP_001335914.1) against *D. hansenii* database revealed 14 putative transporters of ABC superfamily. However only DEHA2C03784g (29% identities, 49% positives, 5% gaps) showed the highest similarity to *H. sapiens* Bcrp1 according to location and number of predicted transmembrane domains. Corresponding *D. hansenii* gene was designated as *RFE1* (*RiboFlavin Excretase*). Promoter and terminator of *TEF1* gene of *D. hansenii* encoding translation elongation factor 1 were amplified with primers Ko427 (AAA AAG CTT GCT CCC CCC TAC CAA GCC TAC) and Ko428 (GAC TTT CCC AAA TAC GTT TAA GGA GCC TGC

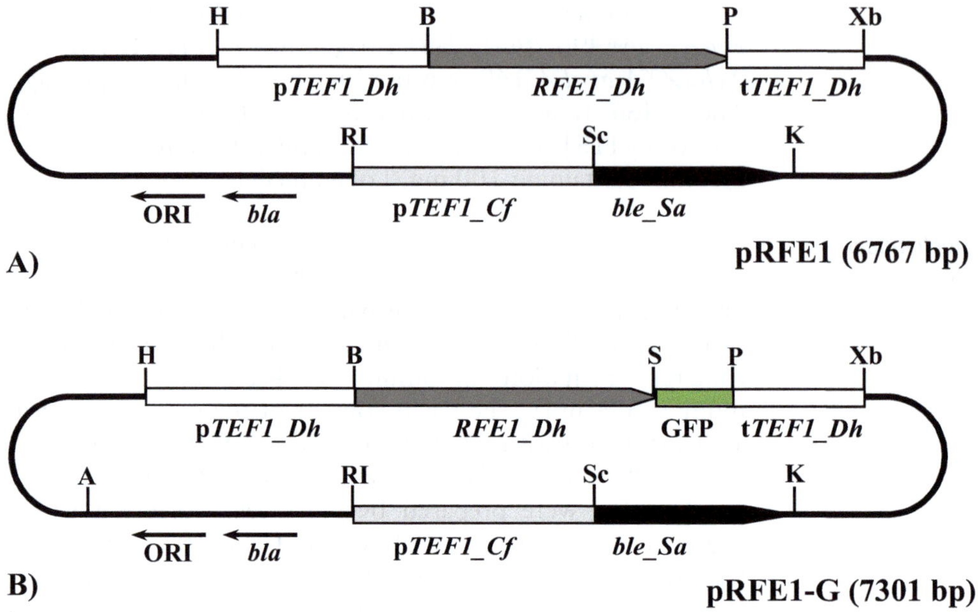

Fig. 1 The circular schemes of plasmids pRFE1 (**a**) and pRFE1-G (**b**). *D. hansenii TEF1* promoter and terminator are shown as white boxes, *D. hansenii RFE1* gene is shown as dark gray box, *C. famata TEF1* promoter is shown as gray box, GFP is sown as green box, gene *ble Staphylococcus aureus*—black box, thin line—bacterial vector pUC57, origin of replication ORI, ampicillin resistance gene (*bla*)—arrows. Sites of restriction endonucleases: H, HindIII; B, BamHI; P, PstI; Xb, XbaI; K, KpnI; S, SalI; Sc, SacI; RI, EcoRI; A, AatII

AGG TCG ACG CGG CCG CGG ATC AAA AAG CTT GCT CCC CCC TAC CAA GCC TAC CTT TGA TTA ATT AAT ATA TAA AAT ATA TGT TTG TGG) and Ko429 (CCA CAA ACA TAT ATT TTA TAT ATT AAT TAA TCA AAG GAT CCG CGG CCG CGT CGA CCT GCA GGC TCC TTA AAC GTA TTT GGG AAA GTC) and Ko430 AAA TCT AGA GAT TAT TGA CTC GAG ATG TTG CGC CG) from the genomic DNA of *D. hansenii* CBS767, respectively. Both fragments were combined by overleap PCR with primers Ko427/Ko430, HindIII/XbaI double-digested and cloned into corresponding sites of plasmid pTb [20]. The resulted recombinant plasmid was named pTTb. Gene DEHA2C03784g (*RFE1*) homolog of mammal BCRP was amplified using genomic DNA of *D. hansenii* CBS767 as a template and primers Ko817 (CGC GGA TCC ATG ATA TCT ATA AGT AAC CCA ATG) and Ko818(AAA CTG CAG TCA CTT TCT CAA CTT TAA ACC AAC). Obtained fragment was BamHI/PstI-digested and subcloned into BamHI/PstI-linearized plasmid pTTb. As a result, recombinant plasmid pRFE1 was constructed (Fig. 1a).

2.5 Transformants' Selection

The XmaI-linearized plasmid pRFE1 harboring *RFE1* expression module was used for transformation of the riboflavin-overproducing strain of *C. famata* AF-4/SEF1/RIB1/RIB7 strain [14]. Cell transformation was performed using electroporation method, with the following parameters: $R = 200\ \Omega$; $C = 25\ \mu F$; $V = 2.5$ kV. Yeast transformants were selected on YPD supplemented with phleomycin (20 mg/l). Transformation frequency marker was around 50 transformants/mg of DNA. The obtained transformants were stabilized by cultivation under nonselective conditions over the course of 10–12 generations with subsequent transfer onto selective medium with phleomycin. The presence of the corresponding plasmids in the stabilized transformants was confirmed by PCR.

2.6 Study of Flavinogenic Activity of Transformants

Initial characterization of riboflavin production was performed in the recombinant strains of *C. famata*, expressing *RFE1* on 72 h of growth in YPD medium. Tested recombinant strains revealed some heterogeneity in riboflavin yield, which can be explained by different copy number of expression module as well as impact of random integration of plasmid to the genome of BRP strain (Fig. 2).

A study of flavinogenic activity of one of the best transformants (BRP/RFE1) was conducted in more detail. Growth and riboflavin production by BRP and BRP/RFE1 strains were estimated after 72 h of growth in the Burkholder medium, Burkholder medium supplemented with 0.2% yeast extract and YPD. The biomass accumulation by constructed strains was similar to that observed in recipient strain BRP. Growth of both strains was very weak in the synthetic Burkholder medium without yeast extract.

The constructed strain BRP/RFE1 revealed 70% and 90% increase in riboflavin yield on Burkholder medium supplemented with 0.5% yeast extract and YPD as compared to that of the parental strain BRP (Table 1). Representative profiles of riboflavin

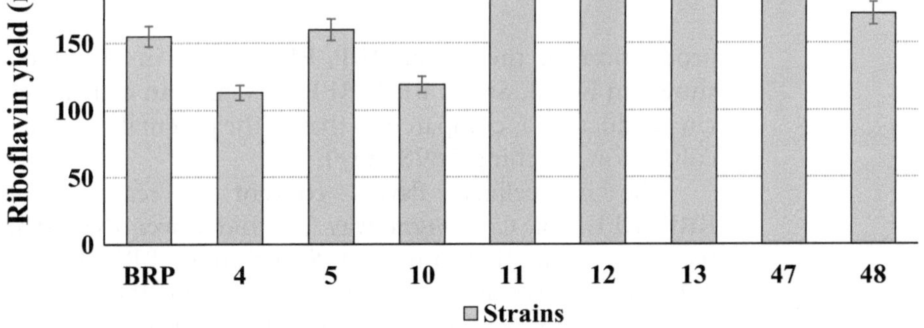

Fig. 2 Riboflavin yields of recombinant strains of *C. famata* with gene *RFE D. hansenii* in the YPD medium

Table 1
Cell biomass, riboflavin production, and riboflavin yield of recombinant strains BRP and BRP/RFE1 of *C. famata* in different media

Medium	Strains	Biomass (g/l)	Riboflavin (mg/l)	Riboflavin yield (mg/g of cells)
YPD	BRP	6.24 ± 0.31	1227.20 ± 61.36	196.66 ± 9.81
	BRP/RFE1	4.34 ± 0.13	1698.5 ± 67.94	391.36 ± 15.65
Burkholder medium	BRP	0.29 ± 0.01	81.42 ± 2.44	280.76 ± 8.42
	BRP/RFE1	0.39 ± 0.02	147.12 ± 8.46	377.23 ± 22.03
Burkholder medium + yeast extract	BRP	5.75 ± 0.32	610.25 ± 20.41	106.12 ± 3.50
	BRP/RFE1	4.75 ± 0.19	915.30 ± 42.10	192.69 ± 8.86

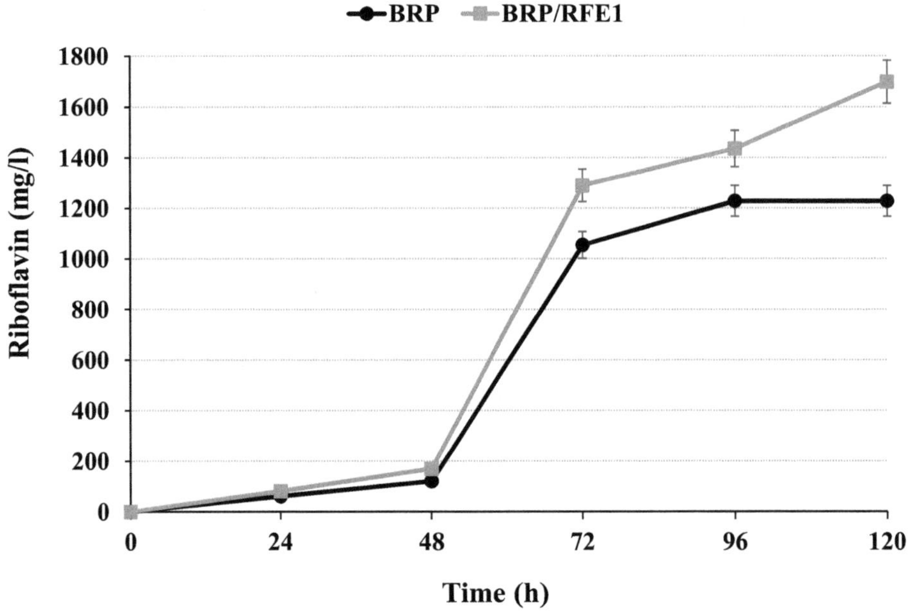

Fig. 3 Time profiles of riboflavin concentration of *C. famata* BRP (circles) and BRP/RFE1 (squares) on YPD medium in flasks

production for the strains BRP, BRP/RFE1 in YPD medium are shown in Fig. 3. Strain BRP/RFE1 revealed an increase in riboflavin production as compared to that of the parental strain at 120 h of cultivation, reaching 1698 mg/l.

The intracellular flavins content in recombinant strain BRP/RFE1 of *C. famata* was 1.5-fold lower than that of the recipient strain BRP (Table 2). Surprisingly, BRP/RFE1 showed 2.2-fold increase of the specific activity of GTP-cyclohydrolase II (the first enzyme of riboflavin biosynthesis pathway) as compared to that of parental strain (Table 2).

Table 2
GTP-cyclohydrolase II activity and content of flavins in the cells of recombinant strains BRP, BRP/RFE, and wild type *C. famata* VKMY-9

Strains	GTP-cyclohydrolase II activity, E/mg of protein $\times 10^{-5}$	Intracellular flavins, mg/g dry cells
VKMY-9	3.70 ± 0.11	1.02 ± 0.04
BRP	37.45 ± 1.31	2.15 ± 0.08
BRP/RFE1	82.62 ± 4.13	1.45 ± 0.05

2.7 Study the Level of Expression of the CfRIB1, CfRIB6, and DhRFE1

Expression of the *CfRIB1*, *CfRIB6* [7], and *DhRFE1* genes was estimated by qRT-PCR. Total RNA was extracted from yeast cells using the GeneMATRIX Universal RNA Purification Kit with DNaseI. The qRT-PCR was performed by 7500 Fast Real-Time PCR System with SG OneStep qRT-PCR kit using corresponding pairs of primer RIB1f/RIB1r (GAG AAT GGG TCA TCT ATT AGA G/AGT AAT CAA CAT CTG CTC TAT TTG), RIB6f/RIB6r (TTT GAT GAT GAG ATT GGA TGA TTG/AAA TTA CTG GTA AAA GAA AGG CC), RFE1f/RFE1r (GTT CAT TGC CTC TGT TTT CCC/TCG CAG TCA AAT ATA CGT TGT TC) and ACT1f/ACT1r (TAA GTG TGA TGT CGA TGT CAG/TTT GAG ATC CAC ATT TGT TGG AA), RNA as a template and ROX reference passive dye following the manufacturer's instructions as previously described [21]. Sequences of the genes *RFE1* and *ACT1* were taken from *D. hansenii* genome sequence database (mycocosm.jgi.doe.gov/Debha1/Debha1.home.html).

The expression level of *RIB1* and *RIB6* (genes coding for enzymes of first reactions of riboflavin pathway) was estimated in BRP/RFE1, parental strain BRP and the wild-type strain of *C. famata*. It was found that BRP strain is characterized by moderate increase in expression of *RIB6* gene and strong 50-fold activation of *RIB1* gene relative to the wild-type strain (gene *RIB1* was overexpressed during BRP strain construction) [14]. It was also unexpectedly found that BRP/RFE1 showed further increased expression of both *RIB1* (1.8 times) and especially of *RIB6* (5.1 times) genes relative to that in the parental strain BRP (Table 3).

2.8 Localization of Riboflavin Excretase in Cells of Wild-Type Strain with Overexpressed Gene Rfe1

To label Rfe1, the codon-optimized yeast enhanced green fluorescent protein (GFP) was amplified with primers Ko1043 (GAA AGG TCG ACT CTA AAG GTG AAG AAT TAT TCA CTG G)/Ko1044 (AGC CTG CAG TCA ATT GAA CAG AAC CAT CTT CAA TGT TG) from pGFP [22]. Primers Ko1045 (TTC AAT TGA CTG CAG GCT CCT TAA ACG TAT T) and Ko1046 (CTT TAG AGT CGA CCT TTC TCA ACT TTA AAC CAA CGT AAC) were

Table 3
The relative quantification (RQ) *of RFE1*, *RIB1*, and *RIB6* genes expression. Strains were cultivated on YNB medium supplemented with yeast extract at 28 °C, 220 rpm

	RQ		
Strain	***RFE1***	***RIB1***	***RIB6***
VKM Y-9	–	1 ± 0.123	1 ± 0.152
CBS767	1 ± 0.098	–	–
BRP	–	50.08 ± 8.172	2.13 ± 1.115
BRP/RFE1	3.5 ± 0.682	91.03 ± 12.610	11.32 ± 3.888

The mRNA quantification was normalized to *ACT1* mRNA; – not detected

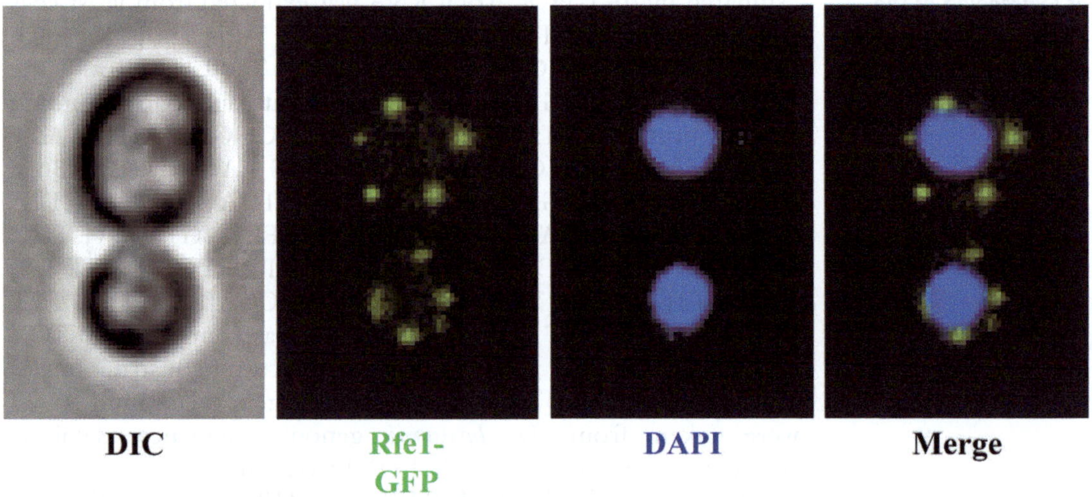

Fig. 4 Fluorescence images of the *C. famata* cells of the L20105/RFE1-*GFP strain*, grown in the liquid YNB medium with glucose for 18 h. Cells were stained with DAPI and observed with a fluorescence microscope. *DIC* differential interference contrast

used to amplify pRFE1 plasmid. Both, GFP and pRFE1 were digested by SalI/PstI and ligated to create pRFE1-G (Fig. 1b).

To visualize Rfe1 and confirm its membrane localization, *RFE1-GFP fusion* gene was generated by *fusing* the codon-optimized yeast enhanced green fluorescent protein gene to the C terminus of *D. hansenii RFE1* gene. *C. famata* L20105 was transformed by the AatII-linearized plasmid pRFE1-G (Fig. 4).

Selection of plasmid-containing transformants was performed on a YPD supplemented with phleomycin. The presence of the plasmid in the genome of transformants L20105/RFE1-*GFP* was confirmed by PCR with pair of primers Ko1043/Ko1044 (data not shown).

```
Dh_Rfe1     1 MISISNPMKVNSFEWSEISLYLPSKRNEEEVCLLDNVSGLMKPGEIMALMGPSGSGKTTL
Cf_Rfe1     1 MLSVNNPIKVRCFEWSDISLYLSSKKNEEDVCLLENVSGLMITGEIMALMGPSGSGKTTL

Dh_Rfe1    61 LNRLSNRSNPKSSKQTGEILINKEVATLAELKEVSNYVEQEDSLIGSLTVKETVEFSAKE
Cf_Rfe1    61 LNALSNRSNVKSSRYSGNVLINKKVATPAELKEVSNYVEQEDSLIGSLTVKETVEFSAKE

Dh_Rfe1   121 ANIPKRFRGDLVDGIIRLLGLENQKNLKIGTPLLKGISGGQKRRTSIASQVLSKPQILFL
Cf_Rfe1   121 ANIPKRYRGELVNEIISLLGLKNQKDLKIGTPLLKGISGGQKRRTSIASQVLGRPQILFL

Dh_Rfe1   181 DEPTSGLDSVASREVINTLKKIAISEKIIVIASIHQPSTSTFQLFDKVLFLSKGKPIYNS
Cf_Rfe1   181 DEPTSGLDSVASREVINTLKKIAISERIIVIASIHQPSTFTFQLFDKVMFLSKGKPIYNG

Dh_Rfe1   241 KVSEIPAYFESIHYGIPQYHNPSEYILDLINTDFSNNLASDEESTIGDKEMIVQDLVNKW
Cf_Rfe1   241 RVSEIPDYFKSIHYEIPTYHNPSEFILDLINTDFSNSTPSDEESVIGDKEMIVRDLVNKW

Dh_Rfe1   301 RKVEERQKASQVDEDYTNLQSEKSSQMYENEYLLPCIRFKNILVRESIRTGILLQRLLIK
Cf_Rfe1   301 RKVEERQKAQQADESYSNFQSEKSLQIDKNESLLPCIRFKNILVRESNRTGILLQILLIK

Dh_Rfe1   361 SRRDVLAYYVRIIMYLGLAILMGTVWLRLDNNQDNIQPFTNAIFFSGAFMSFMSVAYIPS
Cf_Rfe1   361 SRRDVLAYYVRIIMYLGLAILMGTVWLRLDNSQDNIQPFTNAIFFSGAFMSFMSVAYIPS

Dh_Rfe1   421 FIEDYSSYKKEKMNGDYGPFAFVFSNFIIGIPFLFVISLLFSIVTYFMCHFHDSSQGFGY
Cf_Rfe1   421 FIEDYNSYKKEKMNGNYGPFSFVLSNFIIGIPFLFIITLLFSIVTYFMCHFHDSPKSFGY

Dh_Rfe1   481 YVMWLFLDLLAAESMTVFIASVFPNFVVSLALTAFANGLWMSVGGFLVSSKILNDFWYYT
Cf_Rfe1   481 YVMWLFLDLVAAESMTVFIASLFPNFVVSLALTAFANGLWMSVGGFLVSSKILNNFWYYT

Dh_Rfe1   541 FYWINYQRYVFQGMMFNEFEQRIFDCDSNCHCLYESSLSDQCKIAGTAVLENLGYSHSDK
Cf_Rfe1   541 FYWVNYQRYVFQGMMFNEFEYRIFDCDSNCHCLYESGLSSQCKIAGTAVLENLGYSQNDR

          Dh_Rfe1   601 GLWIGILIVLIFVFRLGSYVGLKLRK
          Cf_Rfe1   601 ALWICILIVLIFVFRLGSYVALKVRK
```

Fig. 5 Alignment of putative amino acid sequences predicted from genes *DhRFE1 and CfRFE1* encoding riboflavin excretase

The presence of fluorescent labels in the L20105/RFE1-*GFP* strain was confirmed by fluorescent microscopy. Rfe1-GFP are visible as green dots in the periphery of the cells suggesting plasma membrane localization (Fig. 5). *DAPI* staining further confirmed that the protein does not localize to the nucleus.

2.9 Cloning BCRP Gene Homolog from Genome of C. famata

Blast search of *D. hansenii RFE1* (DEHA2C03784g) against roughly assembled contigs of *C. famata* VKM Y-9 revealed sequence with high homology to query. Corresponding *C. famata* gene was designated as *CfRFE1* (*Candida famata* RiboFlavin Excretase). Nucleotide align of *DhRFE1* and *CfRFE1* revealed 83% identities, 0% gaps. Align of protein sequences between *DhRfe1* and *CfRfe1* revealed 87% identities, 93% positives, 0% gaps (Fig. 5).

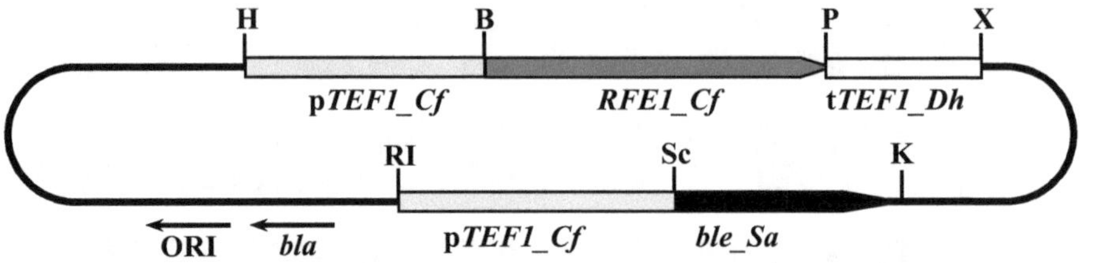

Fig. 6 The circular scheme of plasmid pCfRFE1 *C. famata TEF1* promoter and *D. hansenii TEF1* terminator are shown as gray and white boxes, *C. famata RFE1* gene is shown as dark gray box, gene *ble Staphylococcus aureus*—black box, thin line—bacterial vector pUC57, origin of replication ORI, ampicillin resistance gene (*bla*)—arrows. Sites of restriction endonucleases: H, HindIII; B, BamHI; P, PstI; Xb, XbaI; K, KpnI; Sc, SacI; RI, EcoRI

Promoter of *TEF1* gene of *C. famata* was amplified with primers Ko977 CCC AAG CTT AAA TTG ACT GGT CTG AAA TAA TAG and Ko978 CGC GGA TCC TTT GCT TAA TGT ATA ATA ATA GTA TA from the genomic DNA of *C. famata* VKMY-9 and cloned as HindIII/BamHI fragment in plasmid pCfTTb instead of *DhTEF1* promoter. Gene *CfRFE1* was amplified using genomic DNA of *C. famata* VKMY-9 as a template and primers Ko979 CGC GGA TCC ATG TTA TCC GTA AAT AAC CCA ATC and Ko980 AAA CTG CAG TCA TTT TCT TAC CTT TAG AGC AAC. Obtained fragment was BamHI/PstI-digested and subcloned into BamHI/PstI-linearized plasmid pCfTTb. As a result, recombinant plasmid pCfRFE1 was constructed (Fig. 6).

2.10 Selection of Transformants and Their Flavinogenic Activity

The XbaI linearized plasmid pCfRFE1 harboring *RFE1 C. famata* expression module was used for transformation of the wild type strain L20105 and riboflavin-overproducing strains of *C. famata* AF-4 [13] and AF-4/SEF1/RIB1/RIB7 *C. famata* [14]. Cell transformation was performed using electroporation method. Yeast transformants were selected on YPD supplemented with phleomycin (20 mg/l). Transformation frequency marker was around 50 transformants/mg of DNA. The obtained transformants were stabilized by cultivation under nonselective conditions over the course of 10–12 generations with subsequent transfer onto selective medium with phleomycin. The presence of the corresponding plasmids in the stabilized transformants was confirmed by PCR.

Riboflavin production was analyzed in the recombinant strains of *C. famata*, expressing *RFE1* on 72 h of growth in YPD medium (Table 4). The highest increase in riboflavin production was observed in the case of *RFE1* gene expression in the wild-type strain L20105 (6 times). Strain AF4/CfRFE1 revealed an increase in riboflavin production up to 3.9-fold as compared to that of the parental strain AF4. The constructed strain BRP/CfRFE1 revealed 2.3-fold increase in riboflavin yield relative to that of the parental strain BRP.

Table 4
Riboflavin yield of recombinant strains of *C. famata* with gene *RFE C. famata* in the YPD medium. Time of cultivation—72 h

Strains	Riboflavin yield (mg/g of cells)
L20105	0.3 ± 0.11
L20105/CfRFE1	1.80 ± 0.09
AF4	43.45 ± 1.72
AF4/CfRFE1	169.46 ± 5.93
BRP	156.62 ± 6.24
BRP/CfRFE1	362.8 ± 12.06

Thus, overexpression of the own riboflavin excretase gene *RFE1* significantly increases riboflavin production in *C. famata* as compared to expression of heterologous *RFE1* gene from *D. hansenii.*

Acknowledgments

The authors declare no conflict of interest. This study was supported by Polish National Science Center, grant Opus UMO-2018/29/B/NZ1/01-497, National Research Foundation of Ukraine grant 2020.01/0090, Ministry of Education and Science of Ukraine grant M/32-2020 and by National Academy of Sciences of Ukraine (Grant 36-19).

References

1. Acevedo-Rocha CG, Gronenberg LS, Mack M, Commichau FM, Genee HJ (2019) Microbial cell factories for the sustainable manufacturing of B vitamins. Curr Opin Biotechnol 56:18–29
2. Liu S, Hu W, Wang Z, Chen T (2020) Production of riboflavin and related cofactors by biotechnological processes. Microb Cell Factories 19:31. https://doi.org/10.1186/s12934-020-01302-7
3. Revuelta JL, Ledesma-Amaro R, Lozano-Martinez P, Diaz-Fernandez D, Buey RM, Jimenez A (2017) Bioproduction of riboflavin: a bright yellow history. J Ind Microbiol Biotechnol 44:659–665
4. Schwechheimer SK, Park EY, Revuelta JL, Becker J, Wittmann C (2016) Biotechnology of riboflavin. Appl Microbiol Biotechnol 100:2107–2119
5. Lim SH, Choi JS, Park EY (2001) Microbial production of riboflavin using riboflavin overproducers, *Ashbya gossypii, Bacillus subtilis*, and *Candida famata*: an overview. Biotechnol Bioprocess Eng 6:75–88
6. Stahmann KP, Revuelta JL, Seulberger H (2000) Three biotechnical processes using *Ashbya gossypii*, *Candida famata*, or *Bacillus subtilis* compete with chemical riboflavin production. Appl Microbiol Biotechnol 53:509–516
7. Abbas CA, Sibirny AA (2011) Genetic control of biosynthesis and transport of riboflavin and flavin nucleotides and construction of robust biotechnological producers. Microbiol Mol Biol Rev 75:321–360
8. Hemberger S, Pedrolli DB, Stolz J, Vogl C, Lehmann M, Mack M (2011) RibM from *Streptomyces davawensis* is a riboflavin/

roseoflavin transporter and may be useful for the optimization of riboflavin production strains. BMC Biotechnol 11:119. https://doi.org/10.1186/1472-6750-11-119

9. Perl M, Kearney EB, Singer TP (1976) Transport of riboflavin into yeast cells. J Biol Chem 251(11):3221–3228
10. Sibirnyi AA, Shavlovskii GM, Ksheminskaia GP, Orlovskaia AG (1978) Effect of glucose and its derivatives on systems of riboflavin uptake and excretion in the yeast *Pichia guilliermondii*. Biokhimiya 43 (8):1414–1422
11. van Herwaarden AE, Jonker JW, Wagenaar E, Brinkhuis RF, Schellens JH, Beijnen JH, Schinkel AH (2003) The breast cancer resistance protein (Bcrpl/Abcg2) restricts exposure to the dietary carcinogen 2-amino-1-methyl-6-phenylimidazo(4,5-b)pyridine. Cancer Res 63:6447–6452
12. Voronovsky AA, Abbas CA, Fayura LR, Kshanovska BV, Dmytruk KV, Sybirna KA, Sibirny AA (2002) Development of a transformation system for the flavinogenic yeast. FEMS Yeast Res 2:381–388
13. Dmytruk KV, Yatsyshyn VY, Sybirna NO, Fedorovych DV, Sibirny AA (2011) Metabolic engineering and classic selection of the yeast *Candida famata* (*Candida flareri*) for construction of strains with enhanced riboflavin production. Metab Eng 13:82–88
14. Dmytruk K, Lyzak O, Yatsyshyn V, Kluz M, Sibirny V, Puchalski C, Sibirny A (2014) Construction and fed-batch cultivation of *Candida famata* with enhanced riboflavin production. J Biotechnol 172:11–17
15. Sambrook J, Fritsh EF, Maniatis T (1989) Molecular cloning: a laboratory manual, 2nd edn. Cold Spring Harbor, New York, NY
16. Lowry OH, Rosebrough NJ, Farr AL, Randall RJ (1951) Protein measurement with the Folin phenol reagent. J Biol Chem 193:265–275
17. Shavlovsky GM, Logvinenko EM, Benndorf R (1980) First reaction of riboflavin biosynthesis—catalysis by a guanosine triphosphate cyclohydrolase from yeast. Arch Microbiol 124:255–259
18. Tsyrulnyk AO, Andreieva YA, Ruchala J, Fayura LR, Dmytruk KV, Fedorovych DV, Sibirny AA (2020) Expression of yeast homolog of the mammal *BCRP* gene coding for riboflavin efflux protein activates vitamin B_2 production in the flavinogenic yeast *Candida famata*. Yeast 37:467–473
19. Voronovsky AY, Abbas CA, Dmytruk KV, Ishchuk OP, Kshanovska BV, Sybirna KA, Gaillardin C, Sibirny AA (2004) Candida famata (Debaryomyces hansenii) DNA sequences containing genes involved in riboflavin synthesis. Yeast 21(15):1307–1316. https://doi.org/10.1002/yea.1182
20. Dmytruk KV, Voronovsky AY, Sibirny AA (2006) Insertion mutagenesis of the yeast *Candida famata* (*Debaryomyces hansenii*) by random integration of linear DNA fragments. Curr Genet 50:183–191
21. Ruchala J, Kurylenko OO, Soontorngun N, Dmytruk KV, Sibirny AA (2017) Transcriptional activator Cat8 is involved in regulation of xylose alcoholic fermentation in the thermotolerant yeast *Ogataea (Hansenula) polymorpha*. Microb Cell Factories 16:36
22. Barelle CJ, Manson CL, MacCallum DM, Odds FC, Gow NA, Brown AJ (2004) GFP as a quantitative reporter of gene regulation in *Candida albicans*. Yeast 21(4):333–340

Part II

Measuring Riboflavin Transport and Flavin Cofactor Metabolism with Recombinant Human Proteins/Cell Fractions

Chapter 4

Functional Study of the Human Riboflavin Transporter 2 Using Proteoliposomes System

Lara Console, Maria Tolomeo, and Cesare Indiveri

Abstract

Riboflavin is essential for cell viability. The biologically active forms of riboflavin, FMN and FAD, participate in many biochemical redox reactions including the metabolism of carbohydrates, amino acids, and lipids. Differently from bacteria, fungi, and plants which synthesize riboflavin, higher organisms have lost the ability to synthesize the vitamin and must absorb it from food and intestinal microflora production. The riboflavin flux through cell membranes occurs via specific transporters belonging to the SLC52 family. Three members of this family have been identified so far which show poor homology with the riboflavin transporters of *Saccharomyces cerevisiae* or bacteria. Alterations of RFVTs are causative of severe diseases. Indeed, under pathological stress, humans are susceptible of developing riboflavin deficiency. Such a deficiency in pregnancy induces fetus abnormalities, and has been indicated as a risk factor for anemia, cancer, cardiovascular diseases, and neurodegeneration. Moreover, inherited diseases are also of interest; the most well-described is the Brown–Vialetto–van Laere syndrome, a rare neurological disorder characterized by infancy onset sensorineural deafness and pontobulbar palsy. Numerous polymorphisms of Slc52a2 and Slc52a3 genes associated with this syndrome have been discovered. In spite of their important metabolic role and their relevance to human health, the riboflavin transporters are still poorly characterized. Bacterial overexpression, purification, and protein reconstitution in liposomes represent an up-to-date methodology for obtaining functional data information. The methodology for reconstituting the RFVT2 into proteoliposomes and performing transport assay is described. These methods will be suitable for investigating the functional defects of the variants of RFVTs associated with human pathologies.

Key words *E. coli* protein expression, Protein purification, Liposomes, Reconstitution, Transport

1 Introduction

The existence of mammalian transport proteins devoted to the cellular uptake of riboflavin was postulated since the middle of 1960, but the molecular identity of these transporters remained unknown for many years. The delay in their identification is due to the scarce homology with the riboflavin transporters of yeast and bacteria. Three human riboflavin transporters have been cloned and characterized. They have been classified into the Solute carrier family 52 [1]. Mutations of member 2 (SLC52A2 also known as

Maria Barile (ed.), *Flavins and Flavoproteins: Methods and Protocols*, Methods in Molecular Biology, vol. 2280, https://doi.org/10.1007/978-1-0716-1286-6_4,

RFVT2) and 3 (SLC52A3 also known as RFVT3) of this family are causative of a severe metabolic disorder called Brown–Vialetto–van Laere syndrome [2–8]. Moreover, since mammals have lost the ability to synthesize riboflavin, these three transporters play a fundamental role in maintaining the flavoproteome homeostasis. Indeed, riboflavin is the precursor of FMN and FAD which participate in various biochemical reactions occurring in the metabolism of carbohydrates, amino acids, and lipids [9]. Despite the importance of these transporters for human health, they are poorly characterized. Almost all the information available is obtained using an intact eukaryotic cell system in which specific transporters were expressed. This experimental model allowed the characterization of a large number of transporters but has some technical limitations. Indeed, in intact cells it is hard to modify the membrane lipid composition and is almost impossible to control the experimental conditions in the internal compartment (cytosol). Furthermore, the transport measurement could be affected by enzyme activities that modify the chemical structure of the transported compounds or by the interference of other transporters. Thus, an alternative experimental strategy is needed. A suitable model for transport studies is the proteoliposome system [10]. It was already used to study the functional properties of many transporters located both in the plasma membrane and in the membrane of cell organelles such as mitochondria [11] and lysosomes [12]. Thanks to this model, it is possible to control the experimental conditions in the internal compartment, which is essential to determine kinetic parameters such as the Km for internal substrates. The reconstitution of the purified recombinant transport protein abolished the interferences due to enzymes or other transporters. This technique allows the reconstitution of one single transport protein molecule per proteoliposome, thus prolonging the time course of the substrate uptake and leading to better resolution of the initial transport rate for kinetic measurements. Another advantage is the possibility to easily manipulate the lipid composition of the membrane.

The protocol for performing the bacterial overexpression, purification, reconstitution in proteoliposomes and transport assay of riboflavin transporter 2 (RFVT2) is described here [13].

2 Materials

2.1 Culture Media, Bacterial Growth, and Induction of Protein Overexpression

1. Luria-Bertani (LB) broth preparation. Dissolve 10 g tryptone, 5 g NaCl, 5 g yeast extract in 950 mL of distilled water. Adjust pH to 7.0 with 1 N NaOH. Bring the volume up to 1 L. Aliquot 500 mL of broth in 2 L flasks and close them whit aluminum foil. Autoclave on liquid cycle at 1.5 psi and 121 °C for 20 min. Store at room temperature for 1 month. For plates, 1.5% agar will be added before autoclaving. After the

temperature reaches 55 °C add the antibiotic, mix, and fill petri dishes. After drying, the plates can be stored at 4 °C for 1 month.

2. The stock solution of ampicillin is made dissolving 1 g of ampicillin powder in 10 mL of deionized sterile water.
3. Prepare a stock solution 1 M solution of isopropyl-β-D-1-thio-galactopyranoside (IPTG).

2.2 Solubilization of the E. coli *Insoluble Fraction*

1. Sonication buffer: NaCl 200 mM and 50 mM Hepes–Tris pH 7.5.
2. Stock buffer pH 8.0: 200 mM Tris–HCl pH 8.0. Use this solution to prepare all dilute solutions. The stock solution can be stored at 0–4 °C for several weeks.
3. *E. coli* pellet washing buffer: Tris–HCl 0.1 M pH 8.
4. Solubilisation buffer: 3 M urea, 0.8% Sarkosyl, 200 mM NaCl, 10 mM Tris–HCl pH 8.0.

2.3 Purification of Recombinant Human SLC52A2

1. Stock buffer pH 7.0: 200 mM Tris–HCl pH 7.0. Use this solution to prepare all the diluted solution. The stock buffer pH 7.0 can be stored at 0–4 °C for several weeks.
2. Pack 500 μL of His-Select resin slurry in a glass column and wash 3 times with no gas water to eliminate the storage buffer. Immediately before use wash the resin with 8 ml of the equilibration Buffer (*see* **Note 1**).
3. Equilibration Buffer: Tris–HCl 20 mM pH 8, 200 mM NaCl, 10% Glycerol, 0.1% Sarkosyl.
4. Washing buffer: Tris–HCl 20 mM pH 8, 200 mM NaCl, 10% Glycerol, 0.1% C12E8.
5. Imidazole buffer: Tris–HCl 20 mM pH 8, 200 mM NaCl, 10% Glycerol, 0.1% C12E8, 50 mM Imidazole.

2.4 Reconstitution of Purified Recombinant Human SLC52A2 into Proteoliposomes

1. Amberlite resin preparation: swell Amberlite XAD-4 incubating 1 volume of resin with 2 volumes of methanol for about 30 min at room temperature. Mix with a glass stick each 15 min. Remove the supernatant methanol and replace it with fresh ones (*see* **Note 2**). Wash the resin with water to remove the methanol 5 times (*see* **Note 3**). Leave the resin in water and store the resin at 0–4 °C for 1 month.
2. Liposomes preparation: dissolving 1 g phospholipid from egg yolk (Sigma) in a final volume of 10 mL water, under stirring at 4 °C until particles are still present. Sonicate 2 mL of the lipid suspension in a glass tube for 2 min. Set a sonicator with a 3 mm diameter microtip at 40 W output power to perform in pulse sonication (1 s sonication, 1 s intermission). After sonication liposomes can be stored at 0–4 °C for 24 h.

2.5 Transport Assay

1. Sephadex resin preparation. Swell Sephadex G-75 with water overnight. After swelling eliminate air by vacuum pump for 15 min. The suspension can be stored at 0–4 °C for several months by adding Azide.
2. Sephadex G-75 packing in glass Columns: Fill glass columns (Econo-Columns from Bio-Rad or similar) of 0.7 cm internal diameter up to 15 cm height with Sephadex G-75 resin. The columns can be reused several times and stored the columns at 0–4 °C. Immediately before use equilibrat with a buffer containing 20 mM Tris–HCl at pH 7.0 and 30 mM NaCl.
3. Sephadex G-75 packing in the Pierce column. Fill Glass Columns-Reusable from Pierce of 0.6 cm internal diameter up to 8 cm height with Sephadex G-75 resin Sephadex resin. The columns can be reused and stored at 0–4 °C. Immediately before use equilibrated with a buffer containing 50 mM NaCl.
4. Labeled riboflavin preparation. Dilute the stock solution of radiolabeled riboflavin in bidistilled sterile water until obtaining the desired concentration (1.0 mCi/mL). The concentrations of labeled riboflavin may vary depending on the experiment.

3 Methods

The protocol for performing the study of the transport mediated by the human Riboflavin transporter 2 (RFVT2) consists of four main steps: (1) production of human recombinant RFVT2 in *E. coli*; (2) purification by affinity chromatography; (3) protein reconstitution into proteoliposomes; (4) transport assay. The *E. coli* optimized cDNA of human RFVT2 is cloned into the pH6EX3 vector between EcoRI and XhoI restriction sites. This vector allows for the insertion of a 6His-tag at the N-terminus of RFVT2. The recombinant protein expression is achieved transforming the *E. coli* strain Rosetta (DE3) with the hRFVT2–pH6EX3 construct. After plate selection, colonies, carrying the recombinant plasmid, are inoculated in the LB liquid medium. The protein expression is induced by the addition of 0.4 mM IPTG to the bacterial culture [13]. After 4 h, cells are centrifuged and the pellet is suspended in an adequate volume of sonication buffer. Cells are disrupted by mild sonication at 4 °C. The lysate is centrifuged to collect the bacterial insoluble fraction which contains the recombinant hRFVT2. An aliquot of the insoluble fraction is solubilized and applied onto a column containing the Ni^{2+}-chelating resin. Indeed, the His-tag at the N-terminus of the recombinant hRFVT2 is able to bind the Ni^{2+} resin. The column is washed with the purification washing buffer. The elution of pure hRFVT2 is performed adding imidazole buffer [14]. The first step to achieved reconstitution of the hRFVT2 into proteoliposome is to mix the purified recombinant transporter with phospholipids and detergent for mixed micelle formation. The mixture is then incubated on Amberlite XAD-4 columns to remove the detergent [15].

The nonincorporated components of the reconstitution mixture are removed by size exclusion chromatography on Sephadex G-75 filled column. Then, proteoliposomes can be used for the transport assay. Transport assay starts by adding the [^{3}H]riboflavin to the proteoliposomes and it is stopped by passing the samples through columns containing Sephadex G-75 to remove the radioactive substrate which is not entered into proteoliposome. The transport has to be measured also in liposomes without incorporated hRFVT2. This measurement represents the unspecific permeability of the proteoliposomes toward a radiolabeled substrate that will be subtracted from the [^{3}H]riboflavin taken up by the samples (*see* **Note 4**). After stopping the transport at defined times, the external radioactivity is removed by Sephadex G-75 chromatography and the radioactivity entrapped inside the vesicles is counted [13].

The described method represents a suitable strategy to perform kinetic studies, to discover the effect of different types of molecules as inhibitors and effectors (specific reagents, nucleotides, ions, phospholipids).

3.1 Culture Media, Bacterial Growth, and Induction of Protein Overexpression

1. Incubate 150 μL of Chemically Competent *E. coli* Rosetta (DE3) cells with 90 ng hRFVT2–pH6EX3 construct for 30 min in ice.
2. Transfer cells at 42 °C for 90 s and add 150 of sterile LB medium.
3. Incubate the cells at 37 °C for 60 min under continuous stirring and then spread the cells on a plate containing ampicillin.
4. Incubate overnight at 37 °C (*see* **Note 5**).
5. Store the plate at 4–10 °C until the evening and then pick a colony and put it in 100 mL of LB medium plus ampicillin. Incubate overnight at 37 °C under stirring.
6. The day after, dilute the preculture (1:10) in 1 L of fresh LB medium and incubate at 37 °C until reaching the OD of 0.6 (*see* **Note 6**).
7. Induce the expression of hRFVT2 adding 0.4 mM IPTG (final concentration).

8. Incubate cells for 4 h at 28 °C under continuous stirring and then collect them by centrifugation.
9. Suspend the pellet in the sonication buffer.
10. Disrupted cells by mild sonication at 4 °C (*see* **Note 7**).

3.2 Solubilization of the E. coli Insoluble Fraction

1. Centrifuge the bacterial lysate at 12,000 × *g* for 10 min at 4 °C;
2. Solubiliz the resulting pellet with a buffer made of 3 M urea, 0.8% Sarkosyl, 200 mM NaCl, 10 mM Tris–HCl pH 8.0.
3. Centrifuge at 12,000 × *g* for 10 min at 4 °C.

3.3 Purification of Recombinant Human SLC52A2

1. Apply the supernatant onto a His-select column preconditioned with 7 mL of a buffer containing 0.1% Sarkosyl, 200 mM NaCl, 10 mM Tris–HCl pH 8.0.
2. Wash the column with 7 mL of a buffer containing 0.1% C12E8, 200 mM NaCl, 10 mM Tris–HCl pH 8.0.
3. Perform the elution adding 3 mL of the same buffer supplemented with 50 mM imidazole.
4. Collect the column eluate in fractions of 0.5 mL.
5. Analyze each fraction by SDS PAGE to identify the fraction contending the purified hRFVT2 (*see* Fig. 1).

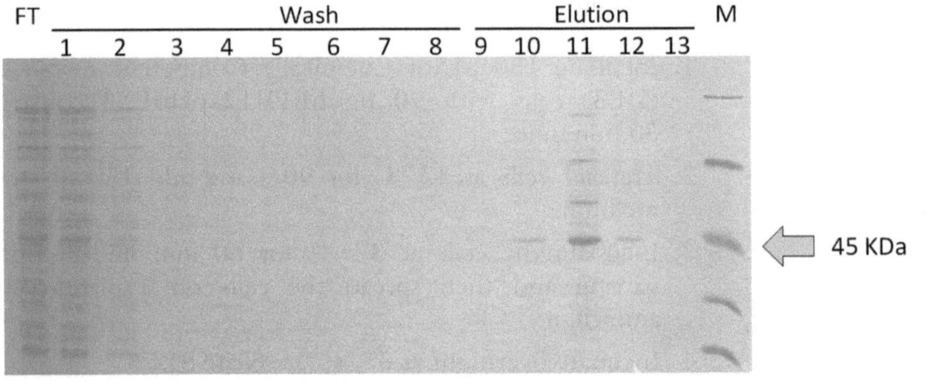

Fig. 1 Identification of the elution fractions containing the recombinant human RFVT2 by SDS-PAGE. 15 μL of each fraction were separated by SDS–PAGE on 12% polyacrylamide gel and stained with Coomassie Blue. FT is pass-through fraction; lanes 1–8, fractions corresponding to washing step; lanes 9–13, elution with imidazole buffer. Fractions 10–12 contain the purified RFVT2 which corresponds to the 45 kDa band of the marker

3.4 Reconstitution of Purified Recombinant Human SLC52A2 into Proteoliposomes

1. Remove the water from Amberlite resin.
2. Put 0.5 g of resin in a 2 mL tube.
3. Prepare the initial reconstitution mixture in 1.5 mL tube adding the different components in the following order: 400 μL of purified hRFVT2, 60 μL of 10% C12E8, 100 μL of 10% egg yolk phospholipids in the form of sonicated liposomes, 70 μL of 200 mM Tris–HCl at 7.0 in a final volume of 700 μL and vortex for 30 s.
4. Incubate the reconstitution mixture with the Amberlite XAD-4 under rotatory stirring at room temperature for 90 min [16].
5. After the incubation time, discharge the Amberlite XAD-4 and store the active proteoliposomes in ice for a maximum of 4 h.

3.5 Transport Assay (See Note 8)

1. Preequilibrate a Sephadex G-75 column with a buffer containing 20 mM Tris–HCl at pH 7.0 and 30 mM NaCl.
2. Apply 550 μL proteoliposomes onto a Sephadex G-75 column.
3. Collect about 600 μL of eluate (*see* **Note 9**).
4. Prepare five tubes with 100 μL of eluted liposomes.
5. Started the transport assay by adding 10 μL the [^{3}H]riboflavin to each tube containing proteoliposome and mix.
6. Stop the transport after 5, 10, 30, 60, and 90 min by passing the samples through Pierce columns preequilibrated with 50 mM NaCl to remove the radioactive substrate which is outside of proteoliposome.
7. Collect the fraction (about 1 mL) corresponding to proteoliposomes in a vial of the appropriate size to be analyzed in a β-counter (*see* **Note 10**).
8. After addition of at least 4 mL of scintillation cocktail (Pico-Fluor 40 from PerkinElmer) vortex the vial and count the radioactivity.
9. To determine the total radioactivity put in a separate vial (total radioactivity) 10 μL of the labeled riboflavin plus 4 mL of scintillation cocktail and analyzed in a β-counter.
10. The same procedure has to be performed also for the control sample containing only liposome.
11. Calculate the specific transport using the following equation: transport (nmol/mg protein) = (cpm sample − cpm control)/(SR × mg protein) where transport is the specific transport at each time considered (*see* **step 6**), cpm sample and cpm control are the cpm value measured by β-counter, SR is the specific radioactivity (cpm/nmol) calculated as: total radioactivity (*see* **step 9**)/nmoles of riboflavin per sample and mg protein is the amount of RFVT2 in each 100 μL of sample (*see* **Notes 11** and

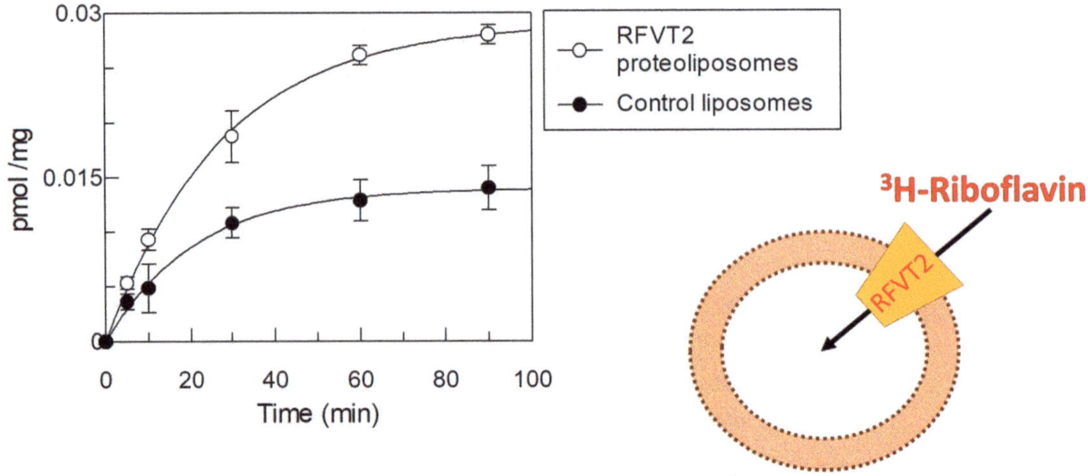

Fig. 2 [^{3}H]Riboflavin uptake into proteoliposomes reconstituted whit recombinant hRFVT2 in comparison with [^{3}H]Riboflavin diffusion trough liposome. The transport measurement was started adding 0.1 μM [^{3}H]Riboflavin and stopped at the indicated times. The values are means ± S.D. from three experiments

12). Experimental data expressed as nmol/mg protein obtained above are interpolated by nonlinear regression analysis using the first order rate equation (*see* Fig. 2).

4 Notes

1. The nickel resin packed in the column must never dry out.
2. Repeat the procedure until the supernatant methanol remains limpid.
3. An interval of 15 min between the washings is advised.
4. Liposomes, used as a control in the transport assay, are produced using the same procedure of RFVT2-proteoliposome. Elution buffer has been added to the mixture instead of purified RFVT2.
5. After overnight incubation at 37 °C colonies appear on the plate.
6. Split the 100 mL of preculture in two 2 L flasks containing 500 mL of LB plus ampicillin.
7. Set a sonicator with a 6 mm diameter tip at 35 W output power to perform pulse sonication (1 s sonication, 1 s intermission) and sonicate for 15 min. During sonication the sample have to be kept cool using ice.
8. Tips, tubes, and all the materials which directly touch radiolabeled riboflavin have to be collected in special containers and

treated according to local legislation on the use and disposal of radioactive substances.

9. Collect the drops which appear turbid.
10. Apply 100 μL of the reaction mix onto each Pierce column, wait for the complete adsorption and add 100 μL, 200 μL, 200 μL, and 400 μL of 50 mM NaCl. Wait for the complete adsorption and than apply 1000 μL, of 50 mM NaCl. Collect them in a vial.
11. The amount of RFVT2 in each sample can be derived from the quantity of the protein added to the initial reconstitution mixture (*see* **step 3** of Subheading 3.4).
12. It is important to know that Pierce column are reusable. These have to accurately washed using 10 mL of 50 mM NaCl. The solution eluted from the columns during this washing step contains radioactivity (^{3}H); it must be collected in special containers and treated according to local legislation on the use and disposal of radioactive waste.

Acknowledgments

This work was funded by a grant from Cure RTD (Year 2019) http://curertd.org/news/new/.

References

1. Yonezawa A, Inui K (2013) Novel riboflavin transporter family RFVT/SLC52: identification, nomenclature, functional characterization and genetic diseases of RFVT/SLC52. Mol Asp Med 34(2–3):693–701
2. Foley AR, Menezes MP, Pandraud A, Gonzalez MA, Al-Odaib A, Abrams AJ, Sugano K, Yonezawa A, Manzur AY, Burns J, Hughes I, McCullagh BG, Jungbluth H, Lim MJ, Lin JP, Megarbane A, Urtizberea JA, Shah AH, Antony J, Webster R, Broomfield A, Ng J, Mathew AA, O'Byrne JJ, Forman E, Scoto M, Prasad M, O'Brien K, Olpin S, Oppenheim M, Hargreaves I, Land JM, Wang MX, Carpenter K, Horvath R, Straub V, Lek M, Gold W, Farrell MO, Brandner S, Phadke R, Matsubara K, McGarvey ML, Scherer SS, Baxter PS, King MD, Clayton P, Rahman S, Reilly MM, Ouvrier RA, Christodoulou J, Zuchner S, Muntoni F, Houlden H (2014) Treatable childhood neuronopathy caused by mutations in riboflavin transporter RFVT2. Brain 137(Pt 1):44–56
3. Ciccolella M, Corti S, Catteruccia M, Petrini S, Tozzi G, Rizza T, Carrozzo R, Nizzardo M, Bordoni A, Ronchi D, D'Amico A, Rizzo C, Comi GP, Bertini E (2013) Riboflavin transporter 3 involvement in infantile Brown-Vialetto-Van Laere disease: two novel mutations. J Med Genet 50(2):104–107
4. Manole A, Jaunmuktane Z, Hargreaves I, Ludtmann MHR, Salpietro V, Bello OD, Pope S, Pandraud A, Horga A, Scalco RS, Li A, Ashokkumar B, Lourenco CM, Heales S, Horvath R, Chinnery PF, Toro C, Singleton AB, Jacques TS, Abramov AY, Muntoni F, Hanna MG, Reilly MM, Revesz T, Kullmann DM, Jepson JEC, Houlden H (2017) Clinical, pathological and functional characterization of riboflavin-responsive neuropathy. Brain 140(11):2820–2837
5. Haack TB, Makowski C, Yao Y, Graf E, Hempel M, Wieland T, Tauer U, Ahting U, Mayr JA, Freisinger P, Yoshimatsu H, Inui K, Strom TM, Meitinger T, Yonezawa A, Prokisch H (2012) Impaired riboflavin transport due to missense mutations in SLC52A2 causes Brown-Vialetto-Van Laere syndrome. J Inherit Metab Dis 35(6):943–948
6. Udhayabanu T, Subramanian VS, Teafatiller T, Gowda VK, Raghavan VS, Varalakshmi P, Said

HM, Ashokkumar B (2016) SLC52A2 [p.P141T] and SLC52A3 [p.N21S] causing Brown-Vialetto-Van Laere Syndrome in an Indian patient: first genetically proven case with mutations in two riboflavin transporters. Clin Chim Acta 462:210–214

7. Woodcock IR, Menezes MP, Coleman L, Yaplito-Lee J, Peters H, White SM, Stapleton R, Phelan DG, Chong B, Lunke S, Stark Z, Pitt J, Ryan MM, Robertson C, Yiu EM (2018) Genetic, radiologic, and clinical variability in Brown-Vialetto-van Laere syndrome. Semin Pediatr Neurol 26:2–9
8. Bosch AM, Abeling NG, Ijlst L, Knoester H, van der Pol WL, Stroomer AE, Wanders RJ, Visser G, Wijburg FA, Duran M, Waterham HR (2011) Brown-Vialetto-Van Laere and Fazio Londe syndrome is associated with a riboflavin transporter defect mimicking mild MADD: a new inborn error of metabolism with potential treatment. J Inherit Metab Dis 34(1):159–164
9. Barile M, Giancaspero TA, Leone P, Galluccio M, Indiveri C (2016) Riboflavin transport and metabolism in humans. J Inherit Metab Dis 39(4):545–557
10. Scalise M, Pochini L, Giangregorio N, Tonazzi A, Indiveri C (2013) Proteoliposomes as tool for assaying membrane transporter functions and interactions with xenobiotics. Pharmaceutics 5(3):472–497
11. Indiveri C, Iacobazzi V, Tonazzi A, Giangregorio N, Infantino V, Convertini P, Console L, Palmieri F (2011) The mitochondrial carnitine/acylcarnitine carrier: function, structure and physiopathology. Mol Asp Med 32(4–6):223–233
12. Rebsamen M, Pochini L, Stasyk T, de Araujo ME, Galluccio M, Kandasamy RK, Snijder B, Fauster A, Rudashevskaya EL, Bruckner M, Scorzoni S, Filipek PA, Huber KV, Bigenzahn JW, Heinz LX, Kraft C, Bennett KL, Indiveri C, Huber LA, Superti-Furga G (2015) SLC38A9 is a component of the lysosomal amino acid sensing machinery that controls mTORC1. Nature 519 (7544):477–481
13. Console L, Tolomeo M, Colella M, Barile M, Indiveri C (2019) Reconstitution in proteoliposomes of the recombinant human riboflavin transporter 2 (SLC52A2) overexpressed in E. coli. Int J Mol Sci 20(18):4416
14. Galluccio M, Pochini L, Amelio L, Accardi R, Tommasino M, Indiveri C (2009) Overexpression in E. coli and purification of the human OCTN1 transport protein. Protein Expr Purif 68(2):215–220
15. Kramer R, Heberger C (1986) Functional reconstitution of carrier proteins by removal of detergent with a hydrophobic ion exchange column. Biochim Biophys Acta 863 (2):289–296
16. Spagnoletta A, De Palma A, Prezioso G, Scalera V (2008) A micro-batchwise technique method for rapid reconstitution of functionally active mitochondrial ADP/ATP carrier from Jerusalem artichoke (Helianthus tuberosus L.) tubers. J Biochem Biophys Methods 70 (6):954–957

Chapter 5

Heterologous Overexpression of Human FAD Synthase Isoforms 1 and 2

Michele Galluccio and Cesare Indiveri

Abstract

The study of human FAD synthase enzymes requires a recombinant strategy to produce large amount of purified proteins in a soluble form. *E. coli* was exploited to this aim. To achieve the production of FAD synthase in a large scale, *E. coli* strains, plasmids (promoter, tags), growth temperature, inducer concentration, medium composition, and osmotic pressure were optimized. To date there is no universal protocol for protein expression, but for each protein a specific combination of "expression parameters" can be selected in order to maximize the results. An experimental protocol for the expression of two isoforms of the human FAD synthase was set up. The final procedures are based on the use of *E. coli* Rosetta(DE3) strain. Two different plasmids were used to obtain optimal amount of the two protein isoforms. In both cases, following the addition of the IPTG inducer, the growth temperature was lowered to increase the solubility of the recombinant protein. The detailed procedures for FAD synthase isoform 1 and isoform 2 overproduction are described in this protocol.

Key words Protein overexpression, FAD, *E. coli*, Enzyme

1 Introduction

Flavine Adenine Dinucleotide (FAD) is the cofactor of hundreds of flavoenzymes, involved in bioenergetics, production of/defense from ROS, protein folding, and many other processes. In humans, dietary riboflavin (vitamin B2) is converted by riboflavin kinase (RFK, EC 2.7.1.26) to FMN which, in turn, is converted to FAD by ATP: FMN adenylyl transferase (FMNAT, EC 2.7.7.2) commonly known as FAD synthase. Different transcripts deriving from alternative splicing of the FLAD1 gene (MIM: 610595) are described in humans. They code for proteins differing not only in the subcellular localization, but also in the catalytic activity. Indeed, the longest protein isoform, namely hFADS1, characterized by 587 amino acids, is targeted to mitochondria by its mitochondrial targeting sequence [1]; on the contrary, a slightly shorter isoform (hFADS2, 490 amino acids) deriving form a downstream ATG, is

Maria Barile (ed.), *Flavins and Flavoproteins: Methods and Protocols*, Methods in Molecular Biology, vol. 2280, https://doi.org/10.1007/978-1-0716-1286-6_5,

cytosolic. Both FADS isoforms contain a N-terminal molybdopterin binding (MPTb) domain, also known as FADHy, which has FAD hydrolase activity [2], and a C-terminal 3-phosphoadenosine 5-phosphosulfate (PAPS) reductase domain, competent in FAD synthesis and pyrophosphorolysis, also known as FADSy domain. The relevance of the FAD synthase in the human pathophysiology in confirmed by the fact that mutation of the FLAD1 gene are involved in riboflavin-responsive and nonresponsive cases of Multiple Acyl CoA dehydrogenase deficiency (MADD, OMIM 231680) [3]. In this scenario, it is crucial to produce huge amounts of the recombinant FAD synthase enzymes for characterizing their activity. Moreover, the possibility to reproduce "in vitro," by site-directed mutagenesis, the naturally occurring mutations discovered in patients, can reveal the mechanism underlying the pathology giving the opportunity to hypothesize a cure. Due to its easy handling, low cultivation costs, fast growth, low complexity, and high expression yield, *E. coli* represents, to date, the most used expression host. However, several parameters need to be investigated and optimized for achieving the goal. Firstly, the target cDNA can be amplified starting from commercial sources, such as cDNA library or recombinant constructs supplied from different vendors. Alternatively, total mRNA can be extracted from human cell culture and reverse transcribed in order to obtain a total cDNA which can be used as template in a PCR reaction aimed to amplify the specific gene. More expensive, but in some case crucial for the heterologous production of the desired protein, is the codon optimization of the corresponding cDNA according to the expression host [4]. This last strategy is devoted to increase the Codon Adaptation Index (CAI) of the gene, thus facilitating the translation and preventing ribosome stalling due to the lack in *E. coli* of tRNAs specific for codons more frequently used by humans. The use of engineered *E. coli* strains, such as Rosetta(DE3), RosettaGami (DE3), and BL21 codon plus RIL which supply additional tRNAs that are rare in *E. coli*, may also improve the efficiency of expression of human genes. A further component that must be considered for its crucial role in protein production, is the expression vector. Indeed, different plasmids have different promoters and tags which profoundly affect the overexpression of the protein of interest. A strong promoter should give an high yield of the target protein, but in case of toxic protein, reducing the promoter strength could be useful. In this case, the promoter should be tightly regulated for preventing "toxic leakage." Also the specific tag ad its position with respect to the target protein (N-terminus versus C-terminus) can be crucial allowing not only identification by western blotting assay but also purification by affinity chromatography. One of the most used for its versatility, is the 6-His tag which can allow purification either under native or denaturing conditions. Finally, the medium composition, the inducer concentration, the time and temperature of growth should also be optimized. Normally, the reduction of the growth temperature after the addition of the inducer, as well as the reduction of the inducer concentration, could increase the percentage of soluble protein with respect to the total. Tuning the described parameters, different FAD synthase isoforms or "module" have been successfully overexpressed in *E. coli* [5–8]. The protocols for the expression of two isoforms of the human FAD synthase are reported.

2 Materials

2.1 Solutions

All the solutions must be prepared in bidistilled water.

2.2 Composition of LB Broth

1% tryptone, 0.5% yeast extract, 0.5% NaCl.

2.3 Preparation of LB Broth

1. Weigh out 10 g of tryptone, 5 g of yeast extract, and 5 g of NaCl.
2. Add 900 mL of H_2O and dissolve the powders by magnetic stirrer.
3. Adjust pH to 7.0 with NaOH, and reach final volume of 1 L in a graduate cylinder.
4. Do not fill the container more than 65% to prevent leakage during sterilization.
5. Partially close the container and sterilize by autoclaving for 20 min at 1 ATM and 120 °C.

2.4 Antibiotics

1. Ampicillin stock solution (1000×): 100 mg/mL in H_2O. Dissolve ampicillin powder in bidistilled H_2O and filter sterilize through a 0.22 μM filter. Store at −20 °C.
2. Chloramphenicol stock solution (1000×): 34 mg/mL in ethanol. Dissolve chloramphenicol powder in ethanol. Store at −20 °C.

2.5 LB Agar Plates

Agar concentration should be 1.5%.

1. Transfer LB Broth at pH 7.0 before sterilization in a suitable container.
2. Do not fill the container more than 65% to prevent leakage.
3. Add 1.5 g of agar every 100 mL of LB broth.
4. Partially close the container and sterilize by autoclaving as in Subheading 2.3.
5. Transfer ~30 mL in a petri dish close to the Bunsen until solidification.

6. After solidification, store at 4 °C (*see* **Note 1**).

In case of selective LB agar plates:

1. Before autoclaving, add a magnet to the container.
2. After sterilization hermetically seal the container and allow to cool under agitation to ~60 °C.
3. Open the container near to the Bunsen, add the desired antibiotic and seal the container.
4. Carefully mix and transfer ~30 mL in a petri dish close to the Bunsen until solidification.

2.6 IPTG Solution (1 M Stock Solution)

1. Dissolve 2.38 g of isopropyl-β-D-thiogalactopyranoside (IPTG) in 8 mL of bidistilled water.
2. Adjust the final volume to 10 mL.
3. Sterilize by filtration through 0.22 μM filter and store at −20 °C.

2.7 TAE 50× (Stock Solution)

Components to be mixed per liter of solution:

1. Weigh out 242 g of Tris base in a suitable container.
2. Add 100 mL of 0.5 mM EDTA pH 8.0.
3. Add 57.1 mL of glacial acetic acid and reach final volume of 1 L with bidistilled water.

2.8 Competent Cells

To prepare fresh competent TG1 cells by calcium chloride treatment, the day before transformation (*see* **Note 2**):

1. Inoculate the cells in a 50 mL conic tube containing 5 mL of sterile LB broth overnight at 37 °C under agitation (~180 rpm).
2. The day after dilute the culture 1:10 in an Erlenmeyer 250 mL flask containing 50 mL of fresh sterile LB broth and continue the growth under agitation (~180 rpm).
3. Follow cell growth by measuring optical density (OD) at 600 nm wavelength.
4. When OD value reach 0.6–0.8 (normally 30–45 min after dilution) transfer the culture in sterile conditions in a conic centrifuge tube and centrifuge at 4 °C for 20 min at 4000 × *g*.
5. Trash the supernatant and resuspend (by gently shaking the tube) the bacterial pellet in a small volume (~2–3 mL) of 50 mM sterile calcium chloride, then reach the final volume of 40 mL with 50 mM calcium chloride.
6. Gently invert the tube and leave it on ice for 30 min.
7. Centrifuge at 4 °C for 20 min at 4000 × *g*, trash the supernatant and resuspend by gently shaking the bacterial pellet in a

small volume (3–5 mL) of cold sterile 50 mM calcium chloride and leave it on ice for at least 2 h (*see* **Note 3**).

8. After 2 h on ice, TG1 cells are ready for transformation (100 μL of competent cells) or can be stocked at −80 °C after dilution 1:1 in 50% sterile glycerol (*see* **Notes 4** and **5**).

3 Methods

3.1 Amplification of hFADS1 cDNA

1. Extract total RNA from primary human fibroblasts by RNeasy Mini Kit and reverse-transcribe with oligo dT primers by RevertAid RT Reverse Transcription Kit, according to the manufacturer's instructions (*see* **Note 6**).
2. Amplify the cDNA corresponding to the hFADS1 encoding sequence (GenBank NM_025207.5) using following forward and reverse primers 5′-GCGGATCC ATGGGTTGG GATTTGGGAAC-3′ and 5′-CCCAAGCTT TCATGTGCGG GAGTTCCGCTC -3′, containing the *Bam*HI and *Hind*III restriction sites, respectively.
3. Use the following PCR protocol:
   (a) Stage 1 (1 cycle): 96 °C for 5 min.
   (b) Stage 2 (35 cycles): Step 1, 96 °C for 1 min; step 2, 60 °C for 30 s, Step 3, 72 °C for 90 s.
   (c) Stage 3 (1 cycle): 72 °C for 10 min.
   (d) Stage 4 (1 cycle): hold at 4 °C (*see* **Note 7**).
4. Load PCR product (~50 μL) on 0.8% agarose gel and run for 40 min at 75 V in TAE 1× buffer.
5. Cut the gel slice and extract the desired product (~1800 bp) by ISOLATE II PCR and Gel Kit according to manufacturer instructions (*see* **Note 8**).

3.2 Cloning of hFADS1 cDNA

3.2.1 Digestion

1. Digest in two separate 1.5 mL microcentrifuge tubes, 16 μL (~0.8 μg) of the hFADS1 cDNA purified as described in Subheading 3.1, **step 5**, and 1 μg (~1 μL) of the plasmid pET-21a (+), by double digestion using 10 U of *Bam*HI and 20 U of *Hind*III restriction enzymes, respectively, for 5 h at 37 °C in the presence of the *Bam*HI 1× buffer (*see* **Note 9**).
2. Only for plasmid: add, directly to the digestion mixture, 1 U of the FastAP Thermosensitive Alkaline Phosphatase in the presence of the supplied buffer and incubate at 37 °C for 15 min.
3. Purify both digestion mixtures using the ISOLATE II PCR and Gel Kit and elute each sample with 20 μL of the supplied elution buffer.

3.2.2 Quantification

Quantify the double digested DNAs by using a spectrophotometer or by UV-fluorescence after agarose gel electrophoresis.

3.2.3 Ligation

Mix ~100 ng of double digested pET-21a(+) DNA with ~300 ng of double digested hFADS1 cDNA and 2 U of T4 DNA ligase in the presence of the supplied buffer and incubate at 22 °C for at least 2 h (*see* **Note 10**).

3.2.4 Transformation

All the manipulation should be performed under sterility.

1. Mix 100 μL of fresh or 170 μL of stocked competent cells with 200 ng (approx. half ligation mix, 10 μL) in a sterile 1.5 mL microcentrifuge tube, by gently pipetting up and down three times; transfer the same volume of competent cells in an empty tube, as negative control.
2. Incubate the mixtures on ice for 30 min.
3. Transfer the tubes in a thermoblock at 42 °C for 90 s.
4. Complete the heat shock by leaving the tubes for 1–2 min on ice.
5. Add 100 μL or 170 μL of LB broth, respectively, and mix by gently pipetting up and down three times.
6. Incubate at 37 °C for 1 h, under agitation at ~200 rpm.
7. Plate on LB-AGAR plates, ampicillin added, and incubate at 37 °C for 12–16 h.

3.3 Screening of Positive Clones

Extract DNA from several colonies by QIAprep Spin Miniprep Kit (Qiagen) and verify the cloning by restriction digestion. Alternatively, starting directly from agar plates, colony PCR procedure could be used.

3.4 Expression of T7-Tagged hFADS1 Protein

E. coli Rosetta(DE3) cells must be transformed as described in Subheading 3.2.4, with hFADS1-pET-21a(+) construct, which encodes a T7 tagged fusion protein carrying the N-terminal sequence MASMTGGQQMGR. This strain supply tRNAs for the codons AUA, AGG, AGA, CUA, CCC, and GGA on compatible chloramphenicol-resistant plasmids. Since these codons are rare in *E. coli* but frequent in eukaryotic genes, Rosetta cells could overcome this problem reducing ribosomal stalling.

1. Inoculate a colony of *E. coli* Rosetta(DE3) cells carrying the recombinant plasmid in a 50 mL conic tube containing 10 mL of LB medium, supplemented with 100 μg/mL ampicillin and 34 μg/mL chloramphenicol, and culture overnight at 37 °C with rotary shaking (~200 rpm).
2. The day after, dilute 1:20 splitting two 5 mL aliquot of the cell culture to 0.1 L of fresh LB medium supplemented with 100 μg/mL ampicillin and 34 μg/mL chloramphenicol.

3. Follow cell growth by measuring optical density (OD) at 600 nm wavelength.
4. At mid-logarithmic phase of the growth (normally at OD ~ 0.8–1) induce the protein synthesis by adding in one of the flasks 1 mM IPTG, to induce the expression of the recombinant protein (*see* **Note 11**).
5. After induction, growth temperature should be lowered to 30 °C in order to increase the solubility of the recombinant protein (*see* **Note 12**).
6. To monitor the production of the protein, aliquots of the cell culture can be collected at specific time intervals (i.e., every 2 h).
7. After 8 h of induction harvest the culture by centrifugation at 4000 × *g* for 10 min at 4 °C (*see* **Note 13**).
8. The bacterial pellets can be stored at −20 °C or analyzed.

3.5 Protein Verification

1. Resuspend the pellets in a buffer containing 300 mM NaCl, 30 mM Hepes, pH 7.4 with NaOH, supplemented with 0.2 mL of Protease Inhibitor Cocktail and 0.5 mM PMSF (*see* **Note 14**).
2. Disrupt the cells by mild sonication for 5 min at 40 W in a VCX130 PB sonifier.
3. Centrifuge the cell lysates at 12,000 × *g* for 10 min at 4 °C to separate soluble and insoluble cell fractions.
4. Analyze the pellets and the supernatant by SDS-PAGE, to verify the presence of the overexpressed protein.

3.6 Cloning of hFADS2 cDNA in pET-21a(+)

1. Starting from total cDNA obtained as described in Subheading 3.1, **step 1**, amplify the cDNA corresponding to the hFADS2 encoding sequence (GenBank NM_201398.3) using following forward and reverse primers 5′-CCGGAATTC ATGACATC TAGGGCCTCTGAACTT-3′ and 5′-CCGCTCGAG TCATG TGCGGGAGTTCCGCTCCTCT -3′, containing the *Eco*RI and *Xho*I sites, respectively.
2. Use the following PCR protocol:
   (a) Stage 1 (1 cycle): 96 °C for 5 min.
   (b) Stage 2 (35 cycles): Step 1, 96 °C for 1 min; step 2, 65 °C for 30 s, Step 3, 72 °C for 90 s (*see* **Note 15**).
   (c) Stage 3 (1 cycle): 72 °C for 10 min.
   (d) Stage 4 (1 cycle): hold at 4 °C.

   (note Use Taq with proofreading activity to prevent random mutagenesis.)

3. Load PCR product (~50 μL) on 0.8% agarose gel and run for 40 min at 75 V in TAE 1× buffer.
4. Cut the gel slice and extract the desired product (~1500 bp) by ISOLATE II PCR and Gel Kit according to manufacturer instructions (note: to increase sample concentration, elution volume could be reduced to 20 μL).

3.6.1 Digestion

1. Digest in two separate microcentrifuge tubes, 16 μL (~0.8 μg) of the hFADS2 cDNA purified as described in Subheading 3.1, **step 5**, and 1 μg (~1 μL) of the plasmid pET-21a(+) by double digestion using 10 U of *Eco*RI and of *Xho*I restriction enzymes, for 5 h at 37 °C in the presence of the supplied Tango buffer 2× (*see* **Note 16**).
2. Only for plasmid: add, directly in the digestion mixture, 1 U of the FastAP Thermosensitive Alkaline Phosphatase in the presence of the supplied buffer and incubate the tube at 37 °C for 15 min.
3. Purify both digestion mixture using the ISOLATE II PCR and Gel Kit and elute each sample with 20 μL of elution buffer.

3.6.2 Quantification

Quantify the double digested DNAs by using a spectrophotometer or by UV-fluorescence after agarose gel electrophoresis.

3.6.3 Ligation

Mix ~100 ng of double digested pET-21a(+) DNA with ~300 ng of double digested hFADS2 cDNA and 2 U of T4 DNA ligase in the presence of the supplied buffer and incubate the tube at 22 °C for at least 2 h (*see* **Note 17**).

3.6.4 Transformation

All the manipulation should be performed under sterility.

1. Mix 100 μL of fresh or 170 μL of stocked TG1 competent cells with 200 ng (approx. half ligation mix) in a sterile microcentrifuge 1.5 mL tube, by gently pipetting up and down three times; transfer the same volume of competent cells in an empty tube, as control.
2. Leave the mixtures on ice for 30 min.
3. Transfer the tubes in a thermoblock at 42 °C for 90 s.
4. Complete the thermal shock by leaving the tubes for 1–2 min on ice.
5. Add 100 μL or 170 μL of LB broth, respectively and mix by gently pipetting up and down three times.
6. Incubate at 37 °C for 1 h under agitation at ~200 rpm.
7. Plate on LB-AGAR plates ampicillin added and incubate at 37 °C for 12–16 h.

3.7 Screening of Positive Clones

Extract DNA from several colonies by QIAprep Spin Miniprep Kit and verify the cloning by restriction digestion. Alternatively, starting directly from agar plates, colony PCR could be used.

3.8 Expression of T7-Tagged hFADS2 Protein

Rosetta(DE3) cells must be transformed as described in Subheading 3.2.4, with hFADS2-pET-21a(+) construct, which encodes a T7 tagged fusion protein carrying an N-terminal sequence MAS MTGGQQMGRGSEF.

1. Inoculate a colony of *E. coli* Rosetta(DE3) cells carrying the recombinant plasmid in a 50 mL conic tube containing 10 mL of LB medium, supplemented with 100 μg/mL ampicillin and 34 μg/mL chloramphenicol, and culture overnight at 37 °C with rotary shaking (~200 rpm).
2. The day after, dilute 1:20 splitting two 5 mL aliquot of the cell culture to 0.1 L of fresh LB medium supplemented with 100 μg/mL ampicillin and 34 μg/mL chloramphenicol.
3. Follow cell growth by measuring optical density (OD) at 600 nm wavelength.
4. At mid-logarithmic phase of the growth (normally at OD ~0.8–1) induce the protein synthesis by adding in one of the flasks 0.4 mM IPTG, to induce the expression of the recombinant protein (*see* **Note 18**).
5. After induction, growth temperature should be lowered to 30 °C in order to increase the solubility of the recombinant protein (*see* **Note 12**).
6. To monitor the production of the protein, aliquots of the cell culture can be collected at specific time intervals (i.e., every 2 h) and harvested by centrifugation at 4000 × *g* for 10 min at 4 °C (*see* **Note 13**).
7. The bacterial pellets can be stored at −20 °C or analyzed.

3.9 Protein Verification

1. Resuspend the pellets in a buffer containing 300 mM NaCl, 30 mM Hepes, pH 7.4 with NaOH, supplemented with 0.2 mL of Protease Inhibitor Cocktail and 0.5 mM PMSF (*see* **Note 14**).
2. Disrupt the cells by mild sonication for 5 min at 40 W in a VCX130 PB sonifier.
3. Centrifuge the cell lysates at 12,000 × *g* for 10 min at 4 °C to separate soluble and insoluble cell fractions.
4. Analyze the pellets and the supernatant by SDS-PAGE, to verify the presence of the overexpressed protein.

3.10 Cloning of hFADS2 cDNA in pH6EX3 Vector

The use of pH6EX3 plasmid is aimed to exploit the strong tac promoter which mix lac and trp consensus sequences. Moreover, the presence of a 6His tag may be used to purify the recombinant protein by Nickel-chelating affinity chromatography.

1. Starting from total cDNA obtained as described in Subheading 3.1, **step 1**, amplify the cDNA corresponding to the hFADS2 encoding sequence (GenBank NM_201398.3) using following forward and reverse primers, 5′-CCGGAATTCAATGACATC TAGGGCCTCT-3′ and 5′-GACCCTCGAGTCATGTGCGG GAGTT-3′, containing the *Eco*RI and *Xho*I sites, respectively (*see* **Note 19**).
2. Use the following PCR protocol:
   (a) Stage 1 (1 cycle): 96 °C for 5 min.
   (b) Stage 2 (35 cycles): Step 1, 96 °C for 1 min; step 2, 65 °C for 30 s, Step 3, 72 °C for 90 s (*see* **Note 15**).
   (c) Stage 3 (1 cycle): 72 °C for 10 min.
   (d) Stage 4 (1 cycle): hold at 4 °C.
   (note Use Taq with proofreading activity to prevent random mutagenesis.)
3. Load PCR product (~50 μL) on 0.8% agarose gel and run for 40 min at 75 V in TAE 1× buffer.
4. Cut the gel slice and extract the desired product (~1500 bp) by ISOLATE II PCR and Gel Kit according to manufacturer instructions (note: to increase sample concentration, elution volume could be reduced to 20 μL).

3.10.1 Digestion

1. Digest in two separate microcentrifuge tubes, 16 μL (~0.8 μg) of the hFADS2 cDNA purified as described in Subheading 3.1, **step 5**, and 1 μg (~1 μL) of the plasmid pH6EX3 by double digestion using 10 U of *Eco*RI and of *Xho*I restriction enzymes, for 5 h at 37 °C in the presence of the supplied Tango buffer 2× (*see* **Note 16**).
2. Only for plasmid: add, directly in the digestion mixture, 1 U of the FastAP Thermosensitive Alkaline Phosphatase in the presence of the supplied buffer and incubate the tube at 37 °C for 15 min.
3. Purify both digestion mixture using the ISOLATE II PCR and Gel Kit and elute each sample with 20 μL of elution buffer.

3.10.2 Quantification

Quantify the double digested DNAs by using a spectrophotometer or by UV-fluorescence after agarose gel electrophoresis.

3.10.3 Ligation

Mix ~100 ng of double digested pH6EX3 DNA with ~300 ng of double digested hFADS2 cDNA and 2 U of T4 DNA ligase in the presence of the supplied buffer and incubate the tube at 22 °C for at least 2 h (*see* **Note 17**).

3.10.4 Transformation

All the manipulation should be performed under sterility.

1. Mix 100 μL of fresh or 170 μL of stocked TG1 competent cells with 200 ng (approx. half ligation mix) in a sterile microcentrifuge 1.5 mL tube, by gently pipetting up and down three times; transfer the same volume of competent cells in an empty tube, as control.
2. Leave the mixtures on ice for 30 min.
3. Transfer the tubes in a thermoblock at 42 °C for 90 s.
4. Complete the thermal shock by leaving the tubes for 1–2 min on ice.
5. Add 100 μL or 170 μL of LB broth, respectively and mix by gently pipetting up and down three times.
6. Incubate at 37 °C for 1 h under agitation at ~200 rpm.
7. Plate on LB-AGAR plates ampicillin added and incubate at 37 °C for 12–16 h.

3.11 Screening of Positive Clones

Extract DNA from several colonies by QIAprep Spin Miniprep Kit and verify the cloning by restriction digestion. Alternatively, starting directly from agar plates, colony PCR could be used.

3.12 Expression of 6His-Tagged hFADS2 Protein

1. Rosetta(DE3) cells must be transformed as described in Subheading 3.2.4, with hFADS2-pH6EX3 construct, which encodes a 6His-tagged protein carrying an N-terminal sequence MSPIHHHHHHLVPRGSEASNS.
2. Inoculate a colony of *E. coli* Rosetta(DE3) cells carrying the recombinant plasmid in a 50 mL conic tube containing 10 mL of LB medium, supplemented with 100 μg/mL ampicillin and 34 μg/mL chloramphenicol, and culture overnight at 37 °C with rotary shaking (~200 rpm).
3. The day after, dilute 1:20 splitting two 5 mL aliquot of the cell culture to 0.1 L of fresh LB medium supplemented with 100 μg/mL ampicillin and 34 μg/mL chloramphenicol.
4. Follow cell growth by measuring optical density (OD) at 600 nm wavelength.
5. At mid-logarithmic phase of the growth (normally at OD ~ 0.8–1) induce the protein synthesis by adding in one of the flasks 0.4 mM IPTG, to induce the expression of the recombinant protein (*see* **Note 18**).

6. After induction, growth temperature should be lowered to 30 °C in order to increase the solubility of the recombinant protein (*see* **Note 12**).
7. To monitor the production of the protein, aliquots of the cell culture can be collected at specific time intervals (i.e., every 2 h) and harvested by centrifugation at 4000 × *g* for 10 min at 4 °C (*see* **Note 13**).
8. The bacterial pellets can be stored at −20 °C or analyzed.

3.13 Protein Verification

1. Resuspend the pellets in a buffer containing 300 mM NaCl, 30 mM Hepes, pH 7.4 with NaOH, supplemented with 0.2 mL of Protease Inhibitor Cocktail and 0.5 mM PMSF (*see* **Note 14**).
2. Disrupt the cells by mild sonication for 5 min at 40 W in a VCX130 PB sonifier.
3. Centrifuge the cell lysates at 12,000 × *g* for 10 min at 4 °C to separate soluble and insoluble cell fractions.
4. Analyze the pellets and the supernatant by SDS-PAGE, to verify the presence of the overexpressed protein.

4 Notes

1. Prewarm for 30 min at 37 °C before use.
2. All manipulation must be performed under sterility.
3. During this time, periodically gently shake the tube to resuspend the cells; reduction of resuspension volume increases the concentration of competent cells.
4. Stocked TG1 competent cells will remain competent for some years and can be thawed in a bath ice 10 min before use.
5. The same protocol can be followed also for Rosetta(DE3) strain. For Rosetta cultivation add 34 μg/mL chloramphenicol.
6. Full-length cDNA clones are commercially available and can also be purchased.
7. Use Taq with proofreading activity (Phusion™ High-Fidelity DNA Polymerase, or equivalent), to prevent random mutagenesis.
8. To increase sample concentration, elution volume could be reduced to 20 μL.
9. The final volume of the reaction should be 20 μL.
10. Ligation could be continued overnight to increase ligation efficiency; final volume of ligation mixture should be 20 μL.
11. The other flask, growing in absence of inducer represents the negative control.

12. Further temperature reduction could increase the amount of soluble protein, reducing the total expression.
13. Depending on temperature reduction, the time of induction could be consequently increased; record the OD of the cell culture at every interval.
14. The volume of resuspension buffer must be changed according to the OD of the cell culture, *see* Subheading 3.4, **step 6**.
15. Time of extension should be modified depending on Taq processivity.
16. The final volume of the reaction should be 20 μL.
17. Reaction could be continued overnight to increase ligation efficiency.
18. The other flask, grown in absence of inducer represents the negative control.
19. One extra nucleotide has been added in the sequence of the forward primer for avoiding frameshift after pH6EX3 cloning.

References

1. Torchetti EM, Brizio C, Colella M, Galluccio M, Giancaspero TA, Indiveri C, Roberti M, Barile M (2010) Mitochondrial localization of human FAD synthetase isoform 1. Mitochondrion 10 (3):263–273
2. Giancaspero TA, Galluccio M, Miccolis A, Leone P, Eberini I, Iametti S, Indiveri C, Barile M (2015) Human FAD synthase is a bi-functional enzyme with a FAD hydrolase activity in the molybdopterin binding domain. Biochem Biophys Res Commun 465(3):443–449
3. Olsen RKJ, Konarikova E, Giancaspero TA, Mosegaard S, Boczonadi V, Matakovic L, Veauville-Merllie A, Terrile C, Schwarzmayr T, Haack TB, Auranen M, Leone P, Galluccio M, Imbard A, Gutierrez-Rios P, Palmfeldt J, Graf E, Vianey-Saban C, Oppenheim M, Schiff M, Pichard S, Rigal O, Pyle A, Chinnery PF, Konstantopoulou V, Moslinger D, Feichtinger RG, Talim B, Topaloglu H, Coskun T, Gucer S, Botta A, Pegoraro E, Malena A, Vergani L, Mazza D, Zollino M, Ghezzi D, Acquaviva C, Tyni T, Boneh A, Meitinger T, Strom TM, Gregersen N, Mayr JA, Horvath R, Barile M, Prokisch H (2016) Riboflavin-responsive and -non-responsive mutations in FAD synthase cause multiple acyl-CoA dehydrogenase and combined respiratory-chain deficiency. Am J Hum Genet 98(6):1130–1145
4. Indiveri C, Galluccio M, Scalise M, Pochini L (2013) Strategies of bacterial over expression of membrane transporters relevant in human health: the successful case of the three members of OCTN subfamily. Mol Biotechnol 54 (2):724–736
5. Galluccio M, Brizio C, Torchetti EM, Ferranti P, Gianazza E, Indiveri C, Barile M (2007) Over-expression in Escherichia coli, purification and characterization of isoform 2 of human FAD synthetase. Protein Expr Purif 52(1):175–181
6. Brizio C, Galluccio M, Wait R, Torchetti EM, Bafunno V, Accardi R, Gianazza E, Indiveri C, Barile M (2006) Over-expression in Escherichia coli and characterization of two recombinant isoforms of human FAD synthetase. Biochem Biophys Res Commun 344(3):1008–1016
7. Miccolis A, Galluccio M, Giancaspero TA, Indiveri C, Barile M (2012) Bacterial over-expression and purification of the 3′phosphoadenosine 5′phosphosulfate (PAPS) reductase domain of human FAD synthase: functional characterization and homology modeling. Int J Mol Sci 13(12):16880–16898
8. Leone P, Galluccio M, Barbiroli A, Eberini I, Tolomeo M, Vrenna F, Gianazza E, Iametti S, Bonomi F, Indiveri C, Barile M (2018) Bacterial production, characterization and protein modeling of a novel monofuctional isoform of FAD synthase in humans: an emergency protein? Molecules 23(1):116

Chapter 6

Purification of Recombinant Human 6His-FAD Synthase (Isoform 2) and Quantitation of FAD/Protein Monomer Ratio by UV-Vis Spectra

Piero Leone, Stefano Quarta, Maria Tolomeo, and Maria Barile

Abstract

Here we describe a protocol for a one-step purification of a soluble form of human FAD synthase (isoform 2; hFADS2), overexpressed as a 6-His-tagged fusion protein in *Escherichia coli*, with a yield of about 15 mg from 1 L of transformed bacterial culture.

Following a desalting procedure, the protein is obtained in its FAD-bound form (about 0.8 molecules of FAD per 1 protein monomer). A simple method is also proposed here, for the rapid estimation of the [FAD]/[protein monomer] ratio, starting from the typical flavoprotein spectrum of the purified protein fraction.

The procedure described gives the protein at a quite high grade of purity (about 95%) and in its bifunctional (2.7.7.2/3.6.1.18) enzymatically active form, useful for further kinetical and molecular characterization.

Key words FAD synthase, Flavoprotein, Purification, UV-Vis spectra

1 Introduction

The protein described here is the human FAD Synthase 2, that is, an isoform among those produced by alternative splicing of *FLAD1* gene [1]. The relevance of FAD synthase in energetic metabolism was underlined years ago [2, 3], and clearly confirmed in 2016 [4], when pathological *FLAD1* gene variants were described as causative of a novel inborn error of metabolism (LSMFLAD, OMIM 255100). It causes severe respiratory chain deficiencies, sometimes responsive and sometimes not responsive to riboflavin (Rf) treatment [4].

FAD synthase (FADS, or FMN:ATP adenylyl transferase) is the last enzyme in the pathway converting Rf to FAD, the essential redox cofactor of the majority of flavoenzymes, involved not only in energetic metabolism but also in protein folding, apoptosis, chromatin remodeling and other crucial processes of cell regulation. In

Maria Barile (ed.), *Flavins and Flavoproteins: Methods and Protocols*, Methods in Molecular Biology, vol. 2280, https://doi.org/10.1007/978-1-0716-1286-6_6,

humans FADS exists in different isoforms generated by different transcript variants of the *FLAD1* gene: in our laboratory, protein isoforms 1, 2, and 6 have been over produced and characterized in some detail [1, 5–7].

Isoform 1 is a 65 kDa protein located in mitochondria; isoform 2, missing of the first 97 amino acids in N-terminus, but completely overlapping for the residual part of the sequence, is a 54 kDa polypeptide located in cytosol [1, 8]. Both isoforms consist of two domains: a molybdopterin-binding resembling (MPTb) domain in the N-terminus and a 3′-phosphoadenosine 5-′-phosphosulfate (PAPS) domain in the C-terminus.

The N-terminal domain is responsible for FAD hydrolysis, via a non-Nudix type catalysis; thus, it was recently named FADHy domain. Interestingly FADHy domain can hydrolyze also the reduced form of nicotinamide adenine dinucleotide (NADH) coenzyme. Conversely, the oxidized form (NAD) is not hydrolyzed (*see* [9] and refs. therein). This feature, already observed in cellular contests [10, 11], appear to coordinate flavin metabolism with the cellular redox state.

The C-terminal PAPS domain (or FADSy domain) of hFADS2 is per se responsible for FAD synthesis and its reverse reaction, that is, pyrophosphorolysis, as demonstrated by using an artificial truncated form of hFADS2 [12] or the novel natural isoform 6 [7].

Isoform 6 has been described, only recently, in patients with frameshift mutations of *FLAD1* gene: it is a truncated isoform of hFADS, starting from a downstream ATG codon (corresponding to Met268 in hFADS1) [4]. This leads to a 36 kDa protein containing only the FADSy domain, which can synthesize, but not hydrolyze FAD, since it lacks the FADHy domain [7]. The existence of this novel isoform, that we named “emergency protein” [7] might explain why affected individuals with biallelic *FLAD1* frameshift variants still harbor substantial FADS activity.

Most of our studies on recombinant hFADSs concerned the purified isoforms 2, produced in *E. coli* as a recombinant His-tag protein, which is still the better characterized [6].

Purified 6His-hFADS2 is a typical flavin-binding protein showing a typical flavoprotein spectrum with three major peaks at 274, 350, and 450 nm, due to the presence of flavin. Urea-unfolding experiments also demonstrated that 6His-hFADS2 binds FAD non-covalently, but very tightly, since FAD leaves the enzyme only during deep conformational modifications [6]. Steady-state kinetic studies indicate that FAD synthesis occurs via an ordered bi–bi mechanism, in which ATP binds prior to FMN, and pyrophosphate is rapidly released before FAD [13]. A very low k_{cat} value for FAD synthesis reaction indicates that FAD release to client apoflavoprotein may represent the rate-limiting step of the whole catalytic cycle [6, 13]. Experiments performed with a mutant form of hFADS6 carrying a single amino acid change (D238A) in the region of FAD binding [14] confirm both the molecular and kinetical models proposed, that is, that hFADS is quite "jealous" of the newly synthesized cofactor.

This quite surprising feature prompted us to investigate further the problem of FAD release to cognate flavoproteins during apoenzyme flavinylation, a critical step which could be controlled by a redox-switch allowed by the presence of redox-active cysteines in the FADSy domain [15].

Measuring the reconstituted activity of the apoform of D-amino acid oxidase (DAAO), used as a client apoflavoprotein, we demonstrated that hFADS2 operates not only as synthase, but also as FAD-chaperone [13]. A direct transfer of the cofactor from hFADS2 to apodimethylglycine dehydrogenase and Lysine demethylase was also demonstrated [13].

All these data give to FADS2 a central role as a component of machinery that deliver FAD cofactor to apoflavoproteins [6], strengthened by its proved relevance in mitochondrial physiopathology.

Further studies are necessary to investigate the chaperoning activity of hFADS in mammalian cells, as further investigations merit the role of FADHy domain in the cellular contest. Because of this domain hFADS2 can be considered a bifunctional enzyme, which can finely tune the level and the delivery of FAD cofactor and thus the maintenance of the whole cellular flavoproteome.

We propose here a protocol useful to one-step purification of a pure and catalytically active bifunctional hFADS2, mimicking the natural wild-type enzyme. This strategy was used for characterizing natural mutant hFADS2 from patients [4] and it can be useful to further characterize the great number of pathological variants, that are continuously emerging, since the discovery of *FLAD1* as a novel disease gene [16].

Molecular mechanisms of cofactor delivery should also be addressed further.

For this purification protocol, Rosetta *E. coli* (DE3) strains, transformed with pH6EX3 expression vector encoding for a 6His-tagged fusion protein, are used. In our lab, they are obtained from a team led by Prof. Indiveri (University of Calabria, Italy), who contributed a chapter in this book describing the protocol to opportunely transform *E. coli* strain (Chapter 5).

The purification protocol utilizes Immobilized Metal-Affinity Chromatography (IMAC) to purify recombinant 6His-tagged fusion proteins. IMAC is an affinity chromatography, based on the specific coordinate covalent bond of amino acids, particularly histidine, to metal ions, commonly copper, cobalt, and nickel, that are immobilized onto a resin. Methods used to elute the protein of interest include changing the pH, or adding a competitive molecule, such as imidazole.

In this procedure, the immobilized metal ion is nickel and the purification relies on the affinity of histidine residues for it. This affinity interaction is the result of coordination of a nitrogen of the imidazole moiety of the polyhistidine tag with a vacant coordination site on the metal.

To rapidly obtain the protein at a quite high grade of purity (about 95%, as checked by SDS-PAGE) a Chelating Sepharose Fast Flow resin has been used, charged with Ni^{2+} ions before use. A discontinuous imidazole gradient was used for elution: imidazole competes with the His-tag for binding to the metal-charged resin. At first, a low concentration of imidazole is added to interfere with the weak binding of other proteins and to elute any proteins that weakly bind. The His-tagged protein is, then, eluted with a higher concentration of imidazole.

The yield of this procedure is of about 15 mg of purified recombinant protein staring from 1 L of transformed bacterial culture. The protein is catalytically active in both its FADSy and FADHy functions, as validated by continuous and discontinuous protocols described in Chapter 7.

Further purification could be achieved by further chromatographic steps, when necessary; for example, hydroxyapatite chromatography, performed essentially as described previously [6, 17].

The protocol for the one-step purification of the recombinant human FAD synthase 2 is reported below.

2 Materials

2.1 Solutions

Prepare the following solutions using distilled water (dH_2O) and applying the equation,

$$g = \mathrm{MW(g/mol)} \cdot M\mathrm{(mol/L)} \cdot \ V\mathrm{(L)}$$

2.1.1 Preparation of LB Broth

Prepare 520 mL of LB Broth (1% peptone, 0.5% yeast extract, and 1% NaCl):

1. Weigh out 5.2 g of peptone, 2.6 g of yeast extract, and 5.2 g of NaCl.
2. Add 400 mL of distilled water (dH_2O) and dissolve the powders by magnetic stirrer.
3. Adjust pH to 7.0 with 3 M NaOH and reach final volume of 520 mL in a graduate cylinder.
4. Divide LB Broth: 500 mL in a 2 L Erlenmeyer flask (for large-scale cell culture), 10 mL in a 50 mL Erlenmeyer flask (for blank measurement, LB blank), 10 mL in a 50 mL Erlenmeyer flask (for small-scale cell culture, LB culture) (*see* **Note 1**).
5. Partially close (or close with aluminum tinfoil) the containers and sterilize by autoclaving for 20 min at 1 bar and 121 °C.

2.1.2 Antibiotics

1. Ampicillin stock solution (250×): 250 mg/mL in dH_2O. Dissolve ampicillin powder in dH_2O and sterilize by filtration through a 0.22μm filter. Aliquot and store at −20 °C.
2. Chloramphenicol stock solution (1000×): 34 mg/mL in ethanol (EtOH). Dissolve chloramphenicol powder in 100% EtOH. Aliquot and store at −20 °C.

2.1.3 IPTG Solution (1 M Stock Solution)

1. Dissolve 2.38 g of isopropyl-β-D-thiogalactopyranoside (IPTG) in 8 mL of bidistilled water.
2. Adjust the final volume to 10 mL.
3. Sterilize by filtration through 0.22μm filter. Aliquot and store at −20 °C.

2.1.4 PMSF Solution (100 mM Stock Solution)

1. Weigh out 0.1742 g of phenylmethylsulfonyl fluoride (PMSF) powder.
2. Add 100% EtOH to 10 mL and dissolve.
3. Aliquot and store at −20 °C.

2.1.5 PIC Solution

Protease inhibitor cocktail (PIC) is commercially available in dimethyl sulfoxide (DMSO) solution. Use as indicated by the manufacturer's instructions. Aliquot and store at −20 °C.

2.1.6 Buffers for Chromatography Columns and Purification

200 mM $NiSO_4$: dissolve 5.26 g of $NiSO_4$ (Nickel (II) sulfate hexahydrate) in 100 mL of dH_2O and mix. Store at 4 °C.

Buffer A: (40 mM Hepes, pH 7.4, 500 mM NaCl): dissolve 2.86 g of Hepes and 8.77 g of NaCl in 300 mL of dH_2O, mix and adjust pH with 3 M NaOH, using pH-meter. Sterilize by autoclaving for 20 min at 1 bar and 121 °C.

500 mM Imidazole: solve 3.40 g of imidazole in 100 mL of Buffer A (imidazole starting solution, ISS).

Then dilute in Buffer A to obtain:

50 mM Imidazole: dilute 2 mL of ISS in 18 mL of Buffer A.

150 mM Imidazole: dilute 6 mL of ISS in 14 mL of Buffer A.

400 mM Imidazole: dilute 16 mL of ISS in 4 mL of Buffer A.

Buffer B: (40 mM Hepes, pH 7.4, 5 mM β-mercaptoethanol): dissolve 2.86 g of Hepes in 300 mL of dH_2O. Mix and adjust pH with 3 M NaOH, using pH-meter. Sterilize by autoclaving for 20 min at 1 bar and 121 °C. Prior to use, freshly add 52.4μL of 14.3 M β-mercaptoethanol to 150 mL of 40 mM Hepes pH 7.4 sterile solution. This buffer prevents protein dimerization through disulfide bonds.

2.1.7 Regeneration Ni-Chelating Column Solutions

0.5 M NaCl, 0.05 M EDTA: dissolve 1.86 g of EDTA and 2.92 g of NaCl in 100 mL of dH_2O and mix.

0.5 M NaCl: dissolve 1.46 g of NaCl in 50 mL of dH_2O and mix.

2 M NaCl: dissolve 5.84 g of NaCl in 50 mL of dH_2O and mix.

1 M NaOH: dissolve 2.00 g of NaOH in 50 mL of dH_2O and mix.

EtOH 70%: dilute 14 mL of 100% EtOH in 6 mL of dH_2O.

EtOH 20%: dilute 4 mL of 100% EtOH in 16 mL of dH_2O.

Store solutions at 4 °C.

2.1.8 Sodium Azide

0.3% sodium azide: dissolve 1.50 g of Sodium Azide in 0.5 L of dH_2O and mix. Store at room temperature.

2.2 Chromatography Column

1. *IMAC column*: 3.5 mL Chelating Sepharose Fast Flow resin (packed according manufacturer's instructions) in a chromatography column (1–1.5 cm diameter). The column can be stored at 4 °C in the storage buffer as indicated in manufacturer's instructions at least for 1 year, depending of the use frequency (*see* **Note 2**).
2. *Gel filtration column*: prepacked PD10 desalting column containing Sephadex G25. The column can be stored at room temperature in the storage buffer as indicated in manufacturer's instructions at least for 1 year, depending on the use frequency. Do not store dry column.

3 Methods

3.1 Expression of hFADS2 Protein

Starting from LB-Agar plate with isolated colonies of transformed *E. coli* Rosetta (DE3) cells, carrying the recombinant plasmid pH6EX3 coding for 6His-hFADS2, apply the following procedure, resumed in Fig. 1:

1. Preinoculum: inoculate one colony in LB culture, supplemented with 100μg/mL ampicillin (4μL of ampicillin stock solution) and 34μg/mL chloramphenicol (10μL of chloramphenicol stock solution), and incubate together with LB blank (supplemented with antibiotics) overnight at 37 °C with rotary shaking (~200 rpm).
2. The day after, dilute (1:10) 100μL of LB culture with LB blank and measure absorbance, that is, optical density (OD) at 600 nm wavelength to verify cell growth (OD_{600} ~ 0.3–0.4) (*see* **Note 3**). Keep both LB culture and LB blank at 4 °C to stop cell growth.
3. Inoculum: about 3 h later, transfer 5 mL aliquot of the LB culture to 500 mL of fresh LB medium (for large-scale cell culture) supplemented with 100μg/mL ampicillin (200μL of

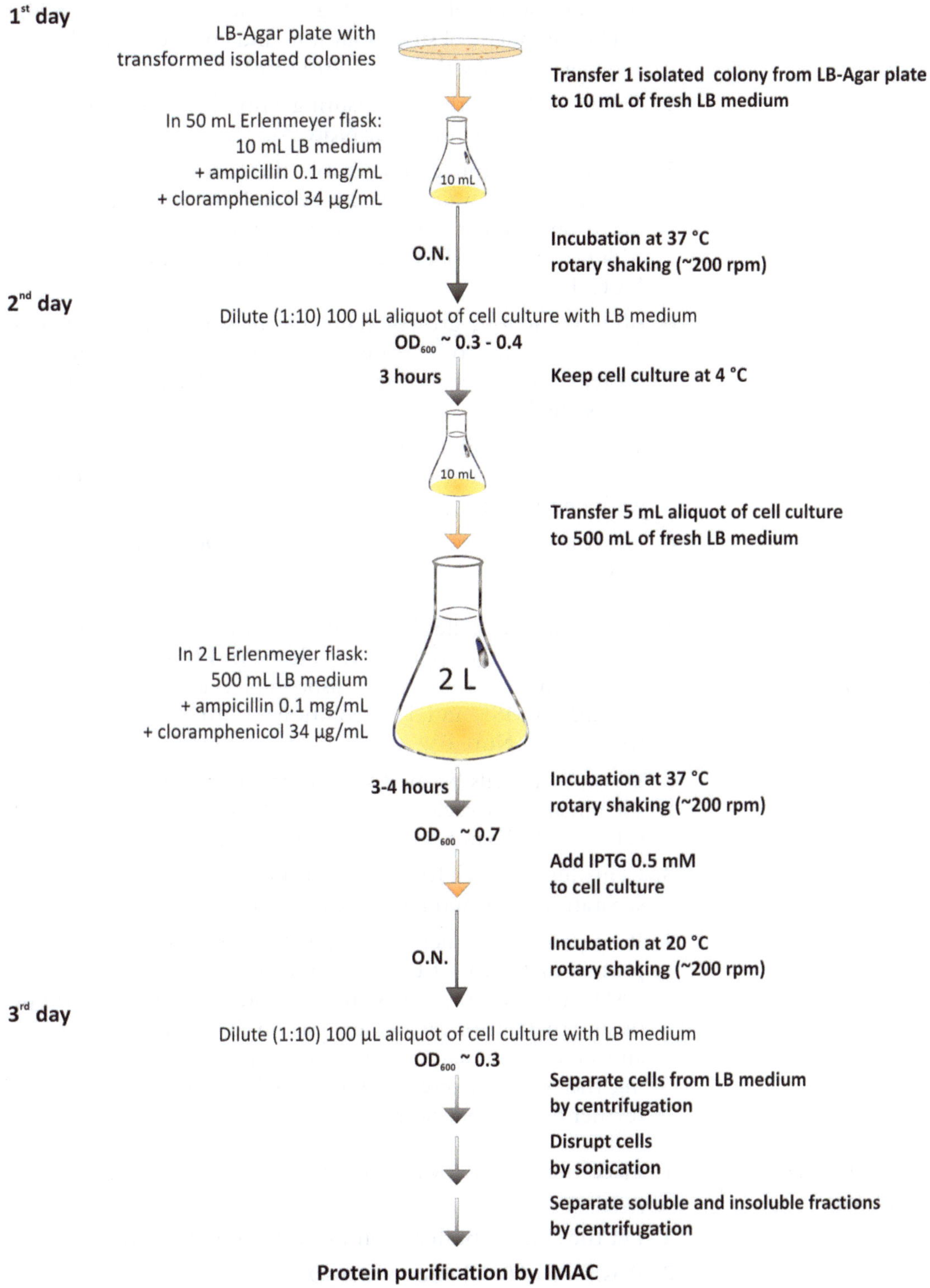

Fig. 1 Schematic representation of steps for protein purification: from cellular growth to IMAC protocol. Orange arrows indicate operative steps with cell or LB medium, gray arrows indicate incubation steps

ampicillin stock solution), 34μg/mL chloramphenicol (500μL of chloramphenicol stock solution) and culture about 3–4 h at 37 °C with rotary shaking (~200 rpm).

4. Follow cell growth by measuring optical density (OD_{600}) of not diluted cell culture (*see* **Note 3**).
5. At mid-logarithmic phase of the growth (normally at OD_{600} ~ 0.7), wait until the cell culture temperature reach the room temperature and induce the protein synthesis by adding 0.5 mM IPTG (250μL of IPTG stock solution) (*see* **Note 4**).
6. After induction, growth temperature should be lowered to 20 °C in order to increase the solubility of the recombinant protein and leave overnight with rotary shaking (~200 rpm) (*see* **Note 5**).
7. The day after, dilute (1:10) 100μL of large-scale cell culture with LB blank and measure optical density (OD) at 600 nm wavelength (OD_{600} ~ 0.3).
8. Harvest the culture by centrifugation at 3000 × *g* for 10 min at 4 °C.

3.2 Protein Extraction from E. coli Cells

1. Resuspend the pellet in 30 mL of Buffer A in a Falcon sterile tube and add the appropriate volume of PMSF stock solution (depending on the final resuspended cell volume) to reach 0.5 mM PMSF and PIC solution (1 mL/20 g of cell wet weight) (*see* **Note 6**).
2. Disrupt the cells by mild sonication on ice bath (one cycle of 10 min and one cycle of 5 min with 1 s Pulse ON and 1 s Pulse OFF, at 40 W) in a VCX130 PB sonifier.
3. Centrifuge the cell lysates at 20,000 × *g* for 30 min at 4 °C to separate soluble and insoluble cell fractions.
4. Resuspend the pellet containing the insoluble overexpressed protein in 15 mL of Buffer A and aliquot for SDS-PAGE and FADS activity assay (*see* **Note 7**). Collect one aliquot (100μL) of supernatant for further analysis on SDS-PAGE. Store pellet and supernatant aliquots at −20 °C. The supernatant containing the soluble overexpressed 6His-hFADS2 will be used for further protein purification.

3.3 Immobilizing Nickel Ions

1. Place the packed IMAC column in a ring stand in a vertical position.
2. Let the storage buffer drain from the resin by gravity flow.
3. Wash with 10 volumes of dH_2O.
4. Load 10 volumes of 200 mM $NiSO_4$ (the column becomes cyan) (Fig. 2a).

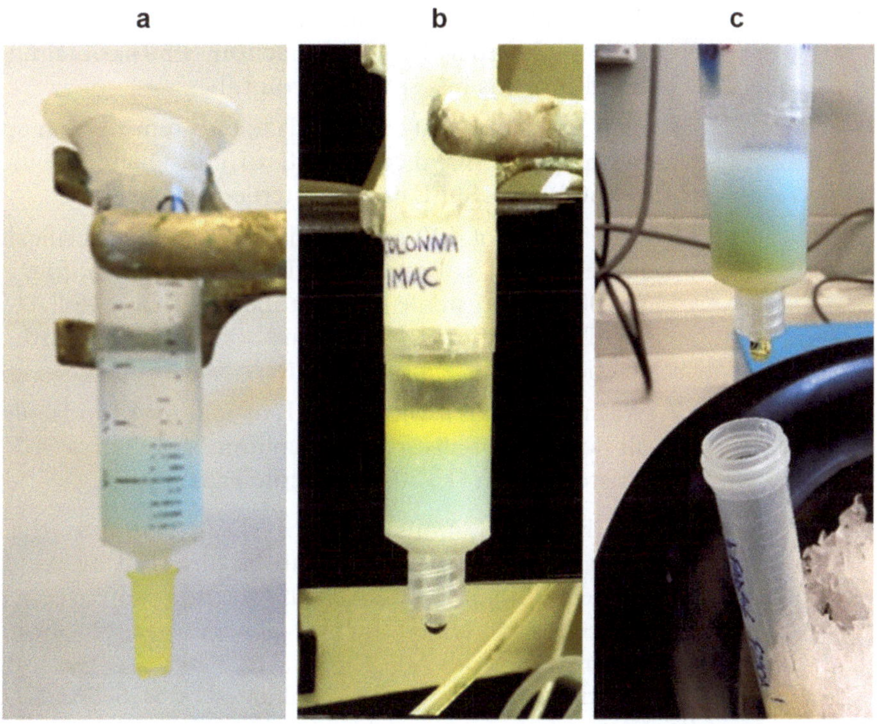

Fig. 2 In (**a**), resin for IMAC charged with Ni^{2+} ions, which give typical cyan color. In (**b**), application onto IMAC column of cellular lysate containing overexpressed 6His-hFADS2, which gives a typical yellow color (due to flavin bound protein) to the sample. In (**c**), elution of fraction containing most abundant amount of recombinant 6His-hFADS2. *See* yellow band containing 6His-hFADS2 that moves from the top to the bottom of the column (yellow droplet is visible during elution) after application of high imidazole concentration

5. Remove excess of $NiSO_4$ with 10 volumes of dH_2O.
6. Equilibrate the column with 10 volumes of Buffer A.

(For **steps 2–6**, *see* **Note 8**). Store column at 4 °C with buffer (storage or Buffer A). Do not store dry resin.

3.4 Sample Application and Elution

For each following step, *see* **Notes 8** and **9**.

1. Adjust cellular lysate supernatant by adding Buffer A to a final volume of 40 mL (starting from 500 mL of cell culture, usually you obtain about 33 mL of cellular lysate supernatant).
2. Load the lysate supernatant on the column (Fig. 2b) and collect (40 mL) the eluate fraction (Pass through 1—Pt1) in a sterile Falcon tube.
3. Wash the column with 35 mL of Buffer A to remove unbound proteins and collect (35 mL) the eluate fraction (Pass through 2—Pt2) in a sterile Falcon tube.

4. Add 35 mL of 50 mM Imidazole on the top of the column and collect (35 mL) the eluate fraction (F50) containing weakly bond proteins in a sterile Falcon tube.
5. Add 18 mL of 150 mM Imidazole to remove non-specifically bound proteins and collect (18 mL) the eluate in four fractions (F150A-D, 4.5 mL each) in sterile Falcon tubes.
6. Add 18 mL of 400 mM Imidazole to elute specifically bound proteins and collect (18 mL) the eluate in four fractions (F400A-D) in sterile Falcon tubes (Fig. 2c). Collect in order: 5 mL of fraction F400A, 3 mL of fraction F400B, 5 mL of fraction F400C and 5 mL of fraction F400D.
7. Wash the column with 4.5 mL of 500 mM Imidazole to elute all bound proteins from the column and collect (4.5 mL) the eluate fraction (F500) in a sterile Falcon tube.

In these conditions, 6His-hFADS2 elutes in the fraction F400A, as a main peak, which can be observed by eye due to its intense yellow color (Fig. 2c). In this fraction protein concentration (measured as described below) generally ranges from 1.3 to 1.8 mg/mL.

Fractionation profile can be validated by SDS-PAGE (Fig. 3) or by spectrophotometric analysis (*see* below).

F400A fraction is then loaded on a desalting column (size exclusion chromatography). When necessary, F400B fraction (normally, containing about seven to eight fold less concentrated proteins) is also desalted and analyzed further. Other fractions are normally discarded or stored at −20 °C for further analysis, for example SDS-PAGE.

3.5 Regeneration of Ni-Chelating Column

After the elution of the proteins, wash and regenerate the column (*see* **Note 8**) according to manufacturer's protocol:

1. Add 10 volumes of 0.5 M NaCl, 0.05 M EDTA in column. EDTA removes nickel from column that becomes white.
2. Add 2–3 volumes of 0.5 M NaCl to remove EDTA from the resin.
3. After 0.5 M NaCl is completely eluted from the column, close the column on the bottom, add 1 volume of 2 M NaCl and resuspend the resin to remove ionically bound proteins.
4. After complete resedimentation of resin on the base of the column, open the column and elute 2 M NaCl.
5. Add 10 volumes of 1 M NaOH to remove precipitated proteins, hydrophobically bound proteins and lipoproteins.
6. Add 4 volumes of 70% EtOH to remove lipoproteins and lipids.
7. If you need to use the column immediately, add 10 volumes of Buffer A, while if you have to keep the column, wash it with 4 volumes of 20% EtOH and store it at 4 °C in this solution.

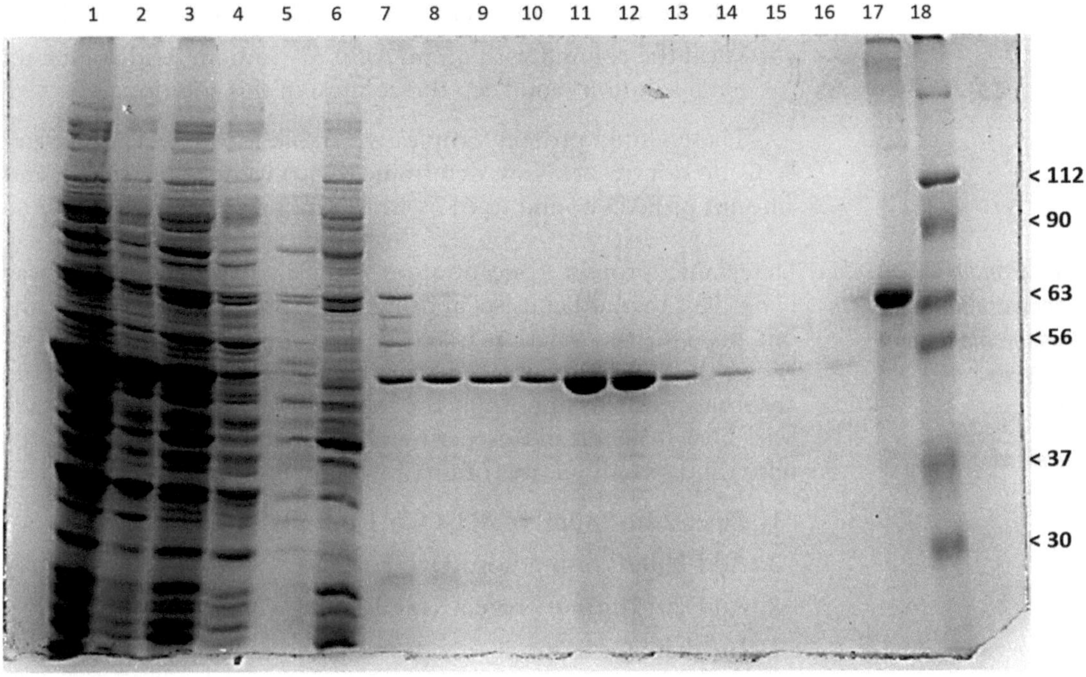

Fig. 3 Protein fractions obtained by Ni^{2+}-chelating chromatography were separated by SDS-PAGE on 10% polyacrylamide gel and stained with Coomassie blue. Lane 1, IPTG-induced cell lysate (4μL); lane 2, insoluble fraction of IPTG-induced cell lysate (2μL); lane 3, soluble fraction of IPTG-induced cell lysate (5μL); lane 4, first pass-through fraction (Pt1, 5μL); lane 5, second pass-through fraction (Pt2, 5μL); lane 6, proteins eluted with 50 mM imidazole (F50, 5μL); lanes 7–10, proteins eluted with 150 mM imidazole (F150A-D, 3μL); lane 11, first fraction of proteins eluted with 400 mM imidazole (F400A, 3μL); lane 12, desalted F400A (3μL); lanes 13–15, proteins eluted with 400 mM imidazole (F400B-D, 3μL); lane 16, proteins eluted with 500 mM imidazole (F500, 3μL); lane 17, BSA (3μg); lane 18, molecular mass markers (kDa)

3.6 Desalting of hFADS2 by Gel Filtration Chromatography

After IMAC purification, the purified recombinant proteins (normally fraction F400A) are desalted by gel filtration on a prepacked column. PD10 Desalting Columns contain Sephadex G25 Medium, which allows for rapid group separation of high molecular weight substances from low molecular weight substances (exclusion limit 5000 M_r). This procedure is useful to remove the buffer containing imidazole (400 mM) and the nickel ions which can promote protein oxidation and destabilization and lead to phenomena of precipitation. To prevent these events the column equilibration and separation Buffer B contains the reducing agent β-mercaptoethanol.

According to PD10-column manufacturer's protocol:

1. Equilibrate columns with 35 mL of Buffer B.
2. Add 2.5 mL aliquot of purified fractions on the top of the column (allow the sample to enter the packed bed completely before any addition of buffer for elution) (*see* **Note 10**).

3. Add 3.5 mL of Buffer B to elute proteins and collect the eluate.
4. Wash the column with 35 mL of 0.3% Sodium Azide (antibacterial solution) and keep the column in this solution.

These eluted protein samples are stable for at least 20 days at 4 °C, do not freeze. Now, determine the concentration, purity, and amount of FAD bound to 6His-hFADS2.

3.7 Protein Concentration Measurements and Analyses

3.7.1 Bradford Protein Assay

Determine protein concentration by using the Bradford assay [18]. This method consists in using Coomassie blue G-250. Without protein, the solution is red to brown in its acidic solution. When the dye binds proteins, its pKa shifts to blue. Measure the absorbance at 595 nm. Apply the following procedure, using Bradford Dye Reagent and preparing the blank sample (without proteins) and two duplicates of unknow samples:

1. Pipet 790–780μL of dH_2O in 1 mL plastic cuvette.
2. Add 200μL of Bradford Dye Reagent.
3. Add 10–20μL of protein sample and invert.
4. Incubate for 5 min.
5. Set the spectrophotometer at 595 nm. Zero the instrument with the blank sample (800μL of dH_2O + 200μL dye reagent) and measure the absorbance of unknown samples, averaging the values.
6. Use a calibration curve obtained with known concentrations of a BSA standard solution to trace the protein concentration in the sample.

Elution profiles and cell lysates can be validated by SDS-PAGE. Briefly, treat aliquots of eluted fractions or other protein samples according to [19] and apply proteins on SDS-PAGE (10% polyacrylamide gel) (Fig. 3). Opportune protocols for this analysis are described elsewhere in this series [20].

Coomassie Blue-stained protein bands can be analyzed and roughly quantitated by using a trans-UV imaging system and, possibly, a devoted software able to graphically manage gel images.

The purified protein appears as a single 56 kDa band as estimated by comparison with migration of molecular markers. The purity of purified protein is routinely checked, by measuring the percentage of purified protein band with respect to all the bands (if) visible in that lane. By comparing the colorimetric density of purified protein band with a known amount of standard protein (or bovine serum albumin, BSA, loaded for quantitative evaluation) band, the concentration of purified proteins could be approximately determined.

3.7.2 Ultraviolet-Visible Spectrophotometry

In an alternative rapid procedure, estimate protein concentration of the purified 6His-hFADS2 by its absorbance spectra. Generally, proteins in solution, thanks to their aromatic aminoacidic residues, absorb ultraviolet light with an absorbance maximum at 280 nm. Peptide bonds are, in turn, primarily responsible for a peak at 200–212 nm. The UV-Vis spectrum is graphically represented as absorbance as a function of wavelength and the height of the absorption peaks is directly proportional to the concentration of the species. The calculation of concentration is governed by the Lambert–Beer law:

$$A_\lambda = \varepsilon_\lambda \cdot c \cdot l$$

where c is the chromophore concentration, l is the optical path length (normally 1 cm) and ε_λ is the specific absorbance coefficient at each λ (nm).

To determine the absorption spectrum of the desalted purified fraction, apply the following procedure:

1. Calibrate to zero with Buffer B only, in UV quartz cuvette.
2. Transfer protein sample in UV quartz cuvette.
3. Measure the UV-Vis spectrum (260–600 nm) of protein sample and recover the sample (Fig. 4).

The purified recombinant 6His-hFADS2 shows a typical flavoprotein absorbance spectrum, with a main peak at 274 nm and two minor peaks at 350 and 450 nm (Fig. 4) due to both protein amino-acidic composition and bound FAD [6]. FAD UV-Vis spectrum, reported in the inset, shows three absorbance peaks at 263, 373, and 450 nm. FAD standard solution absorbance at 280 nm is higher than its absorbance at 450 nm with a ratio of A_{280}/A_{450} of 1.7.

To calculate protein concentrations, the contribution of the bound FAD at 280 nm has to be subtracted from the A_{280} readings. Because FAD is also responsible for the UV absorbance (inset of Fig. 4), the A_{280} actually due to the apoprotein may be calculated from the equation:

$$A_{280}\ \text{apoenzyme} = A_{280}\ \text{fraction} - (A_{450} \cdot 1.7)$$

The protein concentration is estimated by using ε_{280}, as calculated from the protein sequence by using the EXPASY PROT PARAM tool (i.e., extinction coefficients at 280 nm of 6His-hFADS2 is 56.435 $mM^{-1} \cdot cm^{-1}$, or 0.998 g/L).

Measurements made by either the spectrophotometric or the Bradford method differ by no more than 7%.

From the UV-Vis absorbance spectrum the [FAD]/[protein monomer] ratio can also be calculated and expressed as flavinylation (Fl%) which corresponds to % of FAD bound (holo) protein/total protein (*see* **Note 11**).

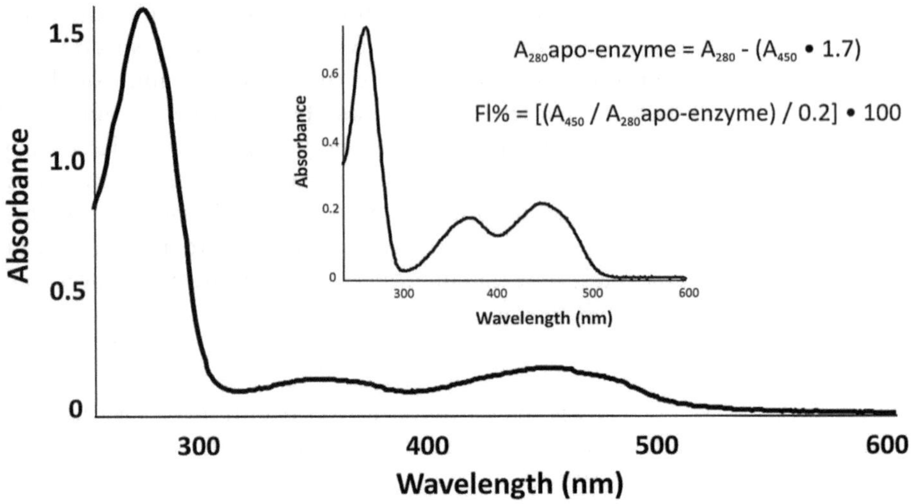

Fig. 4 The absorption spectrum of 6His-hFADS2 (19.7μM), purified to homogeneity, recorded in Buffer B. The protein concentration was measured as indicated in the Subheading 3.7.2. In the inset, the absorption spectrum of 20μM FAD standard solution (gray line) and the equation to determine spectrophotometrically the FAD/protein monomer ratio, indicated as Fl%

Considering the Lambert–Beer law (*see* above), the theoretic ε_{450} FAD of 11.3 $mM^{-1} \cdot cm^{-1}$ and the ε_{450} FAD/ε_{280} apoenzyme ratio equal to 0.2, calculate Fl%, using the following equation:

$$Fl\% = [(A_{450}/A_{280}\text{apoenzyme})/0.2] \cdot 100$$

Note that 0.2 is the value that arbitrarily indicates a degree of flavinylation of the protein equal to 100%, that is, one mole of FAD per mole of protein monomer.

The spectrum reported in Fig. 4 refers to a fraction performing a Fl% of 80%.

Even if very tightly, the cofactor is not-covalently bound to the apoprotein. Not-covalently bound FAD can be removed from the protein or by acidic treatment, which completely destroys the apoenzyme or by a conservative procedure based on the chaotropic effect of potassium bromide (KBr), which allows the apoenzyme to remain active. The latter treatment also suggests that protein–FAD interactions are mainly electrostatic. Both treatments were previously described in detail in [6].

HPLC analysis (*see* Chapter 7) of the supernatant obtained by acidic treatment of purified 6His-hFADS2 confirms that the bound cofactor is FAD. Measurements of [FAD]/[protein monomer] ratio made by the simple spectrophotometric approach described in this Chapter differ by no more than 10% from that performed by HPLC on acidic-treated supernatants.

4 Notes

1. Use Erlenmeyer flasks with a volume about four to five fold greater than that of the LB Broth to ensure Oxygen (O_2) availability to cells during growth.
2. Use about one half of the resin volume (2 mL) when starting with a half cell lysate.
3. OD_{600} lower than recommended value indicate that cells have some difficulties to grow: always start from a fresh (not more than 15 days) LB-Agar plate to avoid working with non-viable cells (only for preinoculum), check the incubator temperature or increase time of growth.
4. Avoid the addition of IPTG in a hot cell culture to ensure IPTG efficacy.
5. Further temperature reduction, even if increasing the amount of soluble protein, could reduce the total expression.
6. Prior to resuspend the cell pellet, weight 30 mL of Buffer A in a sterile Falcon tube, then the total final resuspended cell pellet. Calculate the cell wet weight by difference.
7. Do not resuspend the pellet in Buffer A if further purification from insoluble fraction is necessary.
8. Prior to pass to the next step, wait until the buffer drain from the column by gravity until the resin goes dry.
9. For 2 mL packed resin, apply the sample volumes, buffers, and imidazole solutions as indicated in the table:

	2 mL packed resin
Sample	20 mL
Buffer A/50 mM imidazole	18 mL
150/400 mM imidazole	10 mL
500 mM imidazole	2.5 mL

10. Since 6His-hFADS2 elutes mainly in the F400A fraction, which is a 5 mL fraction, use two different columns to desalt the whole fraction applying 2.5 mL each one.
11. The operational use of the term flavinylation made here, indeed, is not chemically correct; flavinylation literally means the post-translational process of covalent flavin incorporation into apoenzymes. For extensive reviews on flavinylation mechanism, *see* [21–23] and refs. therein.

Acknowledgments

This work was supported by "Progetto Competitivo" funds "Effetto di mutazioni di FLAD1 e di alterazioni dell'omeostasi delle flavine sullo stato redox e sulla biogenesi mitocondriale: uno studio integrato su fibroblasti umani", University of Bari "A. Moro" (to M.B.).

References

1. Brizio C, Galluccio M, Wait R, Torchetti EM, Bafunno V, Accardi R, Gianazza E, Indiveri C, Barile M (2006) Over-expression in Escherichia coli and characterization of two recombinant isoforms of human FAD synthetase. Biochem Biophys Res Commun 344 (3):1008–1016. https://doi.org/10.1016/j.bbrc.2006.04.003
2. Vergani L, Barile M, Angelini C, Burlina AB, Nijtmans L, Freda MP, Brizio C, Zerbetto E, Dabbeni-Sala F (1999) Riboflavin therapy. Biochemical heterogeneity in two adult lipid storage myopathies. Brain 122 (Pt 12):2401–2411
3. Gianazza E, Vergani L, Wait R, Brizio C, Brambilla D, Begum S, Giancaspero TA, Conserva F, Eberini I, Bufano D, Angelini C, Pegoraro E, Tramontano A, Barile M (2006) Coordinated and reversible reduction of enzymes involved in terminal oxidative metabolism in skeletal muscle mitochondria from a riboflavin-responsive, multiple acyl-CoA dehydrogenase deficiency patient. Electrophoresis 27(5–6):1182–1198. https://doi.org/10.1002/elps.200500687
4. Olsen RKJ, Konarikova E, Giancaspero TA, Mosegaard S, Boczonadi V, Matakovic L, Veauville-Merllie A, Terrile C, Schwarzmayr T, Haack TB, Auranen M, Leone P, Galluccio M, Imbard A, Gutierrez-Rios P, Palmfeldt J, Graf E, Vianey-Saban C, Oppenheim M, Schiff M, Pichard S, Rigal O, Pyle A, Chinnery PF, Konstantopoulou V, Moslinger D, Feichtinger RG, Talim B, Topaloglu H, Coskun T, Gucer S, Botta A, Pegoraro E, Malena A, Vergani L, Mazza D, Zollino M, Ghezzi D, Acquaviva C, Tyni T, Boneh A, Meitinger T, Strom TM, Gregersen N, Mayr JA, Horvath R, Barile M, Prokisch H (2016) Riboflavin-responsive and -non-responsive mutations in FAD synthase cause multiple Acyl-CoA dehydrogenase and combined respiratory-chain deficiency. Am J Hum Genet 98(6):1130–1145. https://doi.org/10.1016/j.ajhg.2016.04.006
5. Galluccio M, Brizio C, Torchetti EM, Ferranti P, Gianazza E, Indiveri C, Barile M (2007) Over-expression in Escherichia coli, purification and characterization of isoform 2 of human FAD synthetase. Protein Expr Purif 52(1):175–181. https://doi.org/10.1016/j.pep.2006.09.002
6. Torchetti EM, Bonomi F, Galluccio M, Gianazza E, Giancaspero TA, Iametti S, Indiveri C, Barile M (2011) Human FAD synthase (isoform 2): a component of the machinery that delivers FAD to apo-flavoproteins. FEBS J 278 (22):4434–4449. https://doi.org/10.1111/j.1742-4658.2011.08368.x
7. Leone P, Galluccio M, Barbiroli A, Eberini I, Tolomeo M, Vrenna F, Gianazza E, Iametti S, Bonomi F, Indiveri C, Barile M (2018) Bacterial production, characterization and protein modeling of a novel monofuctional isoform of FAD synthase in humans: an emergency protein? Molecules 23(1). https://doi.org/10.3390/molecules23010116
8. Torchetti EM, Brizio C, Colella M, Galluccio M, Giancaspero TA, Indiveri C, Roberti M, Barile M (2010) Mitochondrial localization of human FAD synthetase isoform 1. Mitochondrion 10(3):263–273. https://doi.org/10.1016/j.mito.2009.12.149
9. Leone P, Galluccio M, Brizio C, Barbiroli A, Iametti S, Indiveri C, Barile M (2019) The hidden side of the human FAD synthase 2. Int J Biol Macromol 138:986–995. https://doi.org/10.1016/j.ijbiomac.2019.07.138
10. Barile M, Giancaspero TA, Brizio C, Panebianco C, Indiveri C, Galluccio M, Vergani L, Eberini I, Gianazza E (2013) Biosynthesis of flavin cofactors in man: implications in health and disease. Curr Pharm Des 19(14):2649–2675
11. Giancaspero TA, Busco G, Panebianco C, Carmone C, Miccolis A, Liuzzi GM, Colella M, Barile M (2013) FAD synthesis and degradation in the nucleus create a local

flavin cofactor pool. J Biol Chem 288 (40):29069–29080. https://doi.org/10.1074/jbc.M113.500066

12. Miccolis A, Galluccio M, Giancaspero TA, Indiveri C, Barile M (2012) Bacterial overexpression and purification of the 3′phosphoadenosine 5′phosphosulfate (PAPS) reductase domain of human FAD synthase: functional characterization and homology modeling. Int J Mol Sci 13 (12):16880–16898. https://doi.org/10.3390/ijms131216880
13. Giancaspero TA, Colella M, Brizio C, Difonzo G, Fiorino GM, Leone P, Brandsch R, Bonomi F, Iametti S, Barile M (2015) Remaining challenges in cellular flavin cofactor homeostasis and flavoprotein biogenesis. Front Chem 3:30. https://doi.org/10.3389/fchem.2015.00030
14. Leone P, Galluccio M, Quarta S, Anoz-Carbonell E, Medina M, Indiveri C, Barile M (2019) Mutation of Aspartate 238 in FAD synthase isoform 6 increases the specific activity by weakening the FAD binding. Int J Mol Sci 20(24). https://doi.org/10.3390/ijms20246203
15. Miccolis A, Galluccio M, Nitride C, Giancaspero TA, Ferranti P, Iametti S, Indiveri C, Bonomi F, Barile M (2014) Significance of redox-active cysteines in human FAD synthase isoform 2. Biochim Biophys Acta 1844 (12):2086–2095. https://doi.org/10.1016/j.bbapap.2014.08.005
16. Balasubramaniam S, Christodoulou J, Rahman S (2019) Disorders of riboflavin metabolism. J Inherit Metab Dis. https://doi.org/10.1002/jimd.12058
17. Barile M, Brizio C, De Virgilio C, Delfine S, Quagliariello E, Passarella S (1997) Flavin adenine dinucleotide and flavin mononucleotide metabolism in rat liver--the occurrence of FAD pyrophosphatase and FMN phosphohydrolase in isolated mitochondria. Eur J Biochem 249(3):777–785
18. Bradford MM (1976) A rapid and sensitive method for the quantitation of microgram quantities of protein utilizing the principle of protein-dye binding. Anal Biochem 72:248–254. https://doi.org/10.1006/abio.1976.9999
19. Laemmli UK (1970) Cleavage of structural proteins during the assembly of the head of bacteriophage T4. Nature 227 (5259):680–685. https://doi.org/10.1038/227680a0
20. Kurien BT, Scofield RH (2015) Multiple immunoblots by passive diffusion of proteins from a single SDS-PAGE gel. Methods Mol Biol 1312:77–86. https://doi.org/10.1007/978-1-4939-2694-7_11
21. Heuts DP, Scrutton NS, McIntire WS, Fraaije MW (2009) What's in a covalent bond? On the role and formation of covalently bound flavin cofactors. FEBS J 276(13):3405–3427. https://doi.org/10.1111/j.1742-4658.2009.07053.x
22. Edmondson DE, Newton-Vinson P (2001) The covalent FAD of monoamine oxidase: structural and functional role and mechanism of the flavinylation reaction. Antioxid Redox Signal 3(5):789–806. https://doi.org/10.1089/15230860152664984
23. Decker K, Brandsch R (1997) Determining covalent flavinylation. Methods Enzymol 280:413–423. https://doi.org/10.1016/s0076-6879(97)80133-4

Chapter 7

Continuous and Discontinuous Approaches to Study FAD Synthesis and Degradation Catalyzed by Purified Recombinant FAD Synthase or Cellular Fractions

Piero Leone, Maria Tolomeo, and Maria Barile

Abstract

Riboflavin, or vitamin B2, is the precursor of flavin mononucleotide (FMN) and flavin adenine dinucleotide (FAD), essential redox (and sometimes non-redox) cofactors of a large number of flavoenzymes involved in energetic metabolism, protein folding, apoptosis, chromatin remodeling, and a number of other cell regulatory processes.

The cellular and subcellular steady-state concentrations of flavin cofactors, which are available for flavoprotein biogenesis and assembly, depend on carrier-mediated transport processes and on coordinated synthesizing/destroying enzymatic activities, catalyzed by enzymes whose catalytic and structural properties are still matter of investigation.

Alteration of flavin homeostasis has been recently correlated to human pathological conditions, such as neuromuscular disorders and cancer, and therefore we propose here protocols useful to detect metabolic processes involved in FAD forming and destroying.

Our protocols exploit the chemical-structural differences between riboflavin, FMN, and FAD, which are responsible for differences in the spectroscopic properties (mainly fluorescence) of the two cofactors (FMN and FAD); therefore, in our opinion, when applicable measurements of fluorescence changes in continuo represent the elective techniques to follow FAD synthesis and degradation. Thus, after procedures able to calibrate flavin concentrations (Subheading 3.1), we describe simple continuous and rapid procedures, based on the peculiar optical properties of free flavins, useful to determine the rate of cofactor metabolism catalyzed by either recombinant enzymes or natural enzymes present in cellular lysates/subfractions (Subheading 3.2).

Fluorescence properties of free flavins can also be useful in analytical determinations of the three molecular flavin forms, based on HPLC separation, with a quite high sensitivity. Assaying at different incubation times the molecular composition of the reaction mixture is a discontinuous experimental approach to measure the rate of FAD synthesis/degradation catalyzed by cell lysates or recombinant FAD synthase (Subheading 3.3). Continuous and discontinuous approaches can, when necessary, be performed in parallel.

Key words Synthesis, Degradation, Fluorescence changes, HPLC separation

Maria Barile (ed.), *Flavins and Flavoproteins: Methods and Protocols*, Methods in Molecular Biology, vol. 2280, https://doi.org/10.1007/978-1-0716-1286-6_7,

1 Introduction

As the precursor of the two enzymatic cofactors flavin mononucleotide (FMN) and flavin adenine dinucleotide (FAD), which participate in a wide variety of reactions, and in particular in mitochondrial oxidative metabolism and electron transport chain, the water-soluble vitamin B2 or riboflavin (Rf) is essential for energy generation in the aerobic cell. It also has a central role in ROS generation/defense and in several different cellular regulatory processes [1–6].

Cofactor availability in the cell and in subcellular compartments is expected to be strictly coordinated with flavoproteome maintenance. Consistently, derangements in this expected (but poorly understood) metabolic coordination has been recently correlated with severe neuromuscular disorders, namely LSMFLAD and BVVLs [7–11] and cancer [12–14].

Rf (7,8-dimethyl-10-ribitylisoalloxazine) is made up of a substituted isoalloxazine ring, whose N-10 atom is bound to a ribityl residue. In FMN the 5′ end of the ribityl moiety is esterified by a single phosphoryl group; adenylation of FMN gives rise to FAD. The chemical structures of Rf and of its derived cofactors are viewable in Fig. 1.

Rf is essential in the mammalian diet; three members of the SLC52A family of membrane transporters, recently characterized and named RFVT1–3, are responsible of cellular uptake of the vitamin, prior its conversion to cofactors (Fig. 1). Protocols to perform studies aimed at the molecular and kinetical characterization of human RFVT2 are described in Chapter 4. Further studies are urgent to unravel the mechanisms of subcellular distribution of Rf, which are still matter of debate [15–17].

The sequence of intracellular FAD forming reactions and the identity of the enzymes involved are summarized in the cartoon, reported as Fig. 1. Phosphorylation of Rf to FMN is catalyzed by Rf Kinase (RFK, ATP: riboflavin 5′ phosphotransferase, EC 2.7.1.26), which transfers a phosphoryl group from ATP to Rf to form FMN. A second step is catalyzed by FAD synthase (FADS or FMNAT, ATP: FMN adenylyltransferase, EC 2.7.7.2), which adenylates FMN to FAD.

In animals and lower eukaryotes, two physically distinct polypeptides, with either RFK or FADS activities, have been purified and characterized, as extensively reviewed [1, 11, 17, 18]. In bacteria both RFK and FADS activities are fused in a bifunctional enzyme, which is still named FAD synthetase (*see* refs. 19–22). Besides the bifunctional enzymes, some monofunctional RFKs have reported in prokaryotes [23].

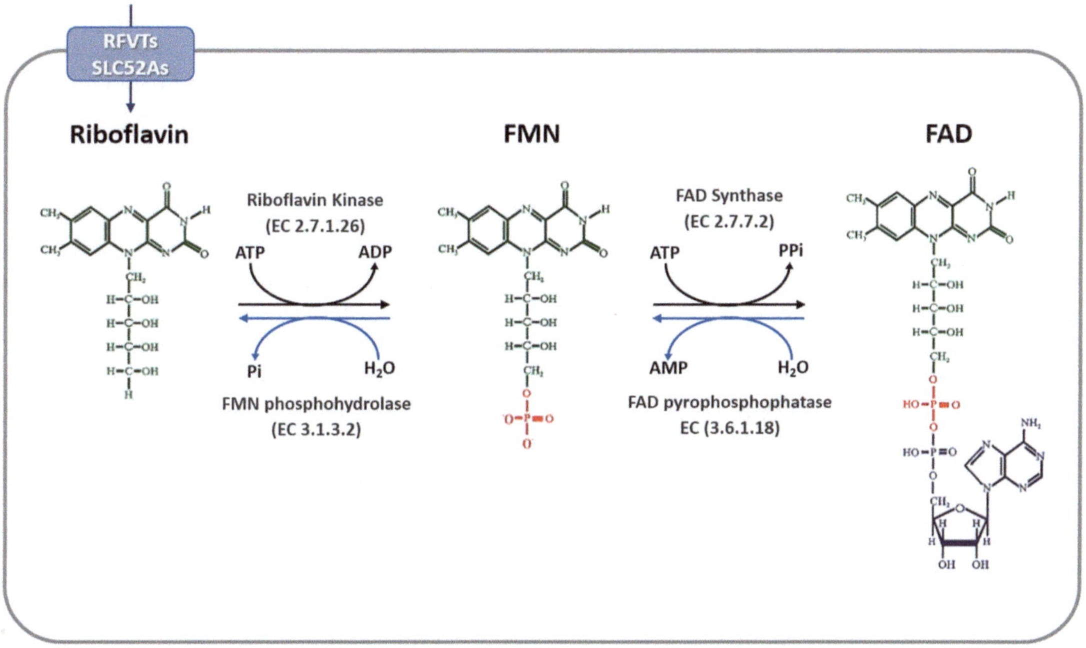

Fig. 1 Metabolic pathway and chemical structures. The chemical structures of riboflavin, FMN, and FAD are reported with the indication of enzymes involved in conversion of riboflavin to FAD and vice versa. In humans, riboflavin is taken up in a carrier-mediated process by three transporters named RFVT1–3 (or SLC52A1–3). Inside the cells, riboflavin conversion to flavin cofactors occurs in two steps catalyzed by riboflavin kinase forming FMN, and FAD Synthase forming FAD. Recycling of the vitamin can move from FAD in two steps, catalyzed by FAD pyrophosphatase, and FMN phosphohydrolase

Human FAD synthase exists in different isoforms with different subcellular localization; they are produced by the alternative splicing of *FLAD1* gene [6, 8, 11, 24].

The ATP dependent reaction catalyzed by FAD synthase is reversible in vitro; thus, in the presence of PPi, FMN can be generated by FAD, via the reverse reaction (pyrophosphorolysis). In the intracellular conditions, the rapid PPi hydrolysis by inorganic pyrophosphatase, presumably makes the reaction catalyzed by FADS unidirectional [25]. Two chapters in this book (Chapters 5 and 6) contain protocols devoted to produce and characterize different isoforms of hFADS.

Besides considering flavin cofactor forming events, which have been studied in some details in humans ([6, 11] and refs therein), also alternative and physiologically relevant FAD and FMN destroying events are indicated in Fig. 1. The state of knowledge on this topic is relatively poor, even if cofactor degradation, which allows both intestinal Rf absorption [1, 14, 26, 27] and intracellular Rf recycling [1, 17, 28–30], is relevant in regulating vitamin availability, both in the whole body and inside the cells/subcellular compartments.

FMN conversion to Rf is catalyzed by distinct hydrolases or monophosphoesterases (E.C. 3.1.3.2) even if FMN cleavage events in mammals are rather poor. Together with quite unspecific acid phosphatases abundant in lysosomes [31], specific FMN hydrolases located in mitochondria have been described [1, 28].

FAD conversion to FMN is catalyzed by pyrophosphatases or diphosphatases (E.C. 3.6.1.18. or 3.6.1.9), localized on the plasma membrane and in different subcellular compartments [28, 32–34]. FAD pyrophosphatase (FADPPase) identity, pH, and substrate specificities have not yet been completely elucidated; sometimes FAD degradation has been attributed to certain diphosphatases able to hydrolyze nucleotides, belonging to NUDIX (*NU*cleoside *DI*phosphate linked to some other moiety *X* hydrolase) family [35, 36].

Please note that, FADPPase hydrolytic reactions differ from the reverse reaction of FMN adenylation, catalyzed by FADS, that is, FAD pyrophosphorolysis, as diffusely described in [30, 37]. Only recently we discovered that isoform 2 of hFADS, can hydrolyze FAD at neutral pH, using the N-terminal domain of the protein which we named FADHy. FADHy diphosphatase activity is different from the hydrolytic reaction catalyzed by NUDIX hydrolases, because of its pH dependence, substrate affinity and inhibitor sensitivity [30]. However, we still not know if FAD hydrolysis catalyzed by the bifunctional FADSs is the sole pathway of cofactor degradation in cell/subcellular fraction or if other hydrolases, presumably belonging to the NUDIX family [35, 38] can contribute to FAD homeostasis in vivo. A rapid and sensitive method to follow FAD degradation catalyzed by crude biological samples, in the future will aid to solve this problem.

In relation to these considerations, experimental approaches described here respond to the need of studying mammalian FAD forming and destroying processes in health and diseases.

Our protocols can be applied to study FAD metabolism either in crude biological samples as cell/subcellular fractions, or by using recombinant over-expressed and purified hFADSs.

Both approaches useful to study flavin cofactors metabolism take a big advantage from the natural optical properties of these molecules, as addressed at Subheadings 3.1 and 3.2.1.

The absorption properties of the oxidized isoalloxazine ring of flavins confer to these molecules a typical bright yellow color. The UV/Vis spectrum of FAD, free in solution, is reported in Fig. 2a. Flavin cofactors can undergo one-electron and two-electron transfer processes, giving rise to the semireduced (only stabilized by the action of the protein environment) or fully reduced forms, which can be differentiated from one another and from the oxidized flavins based on their optical properties. While the oxidized flavins are yellow and the fully reduced molecules are colorless, the half-reduced forms can be red or blue depending on pH [6, 39, 40].

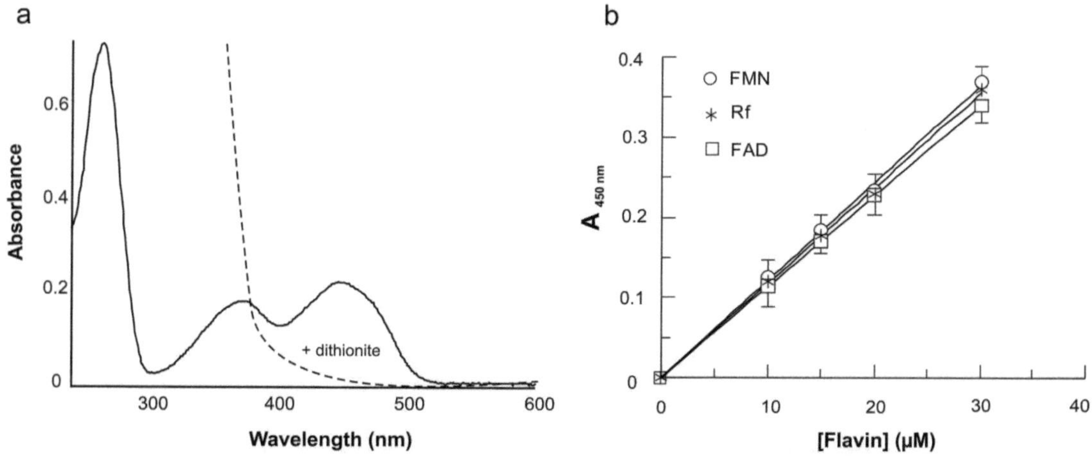

Fig. 2 Spectrophotometric calibration of flavin solutions. (**a**) 20 μM FAD UV/Vis absorption spectrum performed at 25 °C in 20 mM Tris–HCl pH 7.0 in the absence (straight line) or in the presence of dithionite (dotted line). (**b**) Calibration curves of Rf, FMN, and FAD obtained in a 0–30 μM *expected* concentration range at 25 °C in 20 mM Tris–HCl pH 7.0. Theoretical $\varepsilon_{450\ nm}$ values are 11.3 $mM^{-1} \bullet cm^{-1}$ for FAD and 12.2 $mM^{-1} \bullet cm^{-1}$ for FMN and Rf. Each point represents the mean value of three different measurements

Another optical property distinctive of the oxidized free flavins and absent in the reduced flavins is a strong fluorescence. The spectra of fluorescence of FAD are reported in Fig. 3a. The strong visible fluorescence of free Rf/FMN (Fig. 3b), can be very useful when studying the metabolic conversions of Rf derivatives [25, 28, 41] and biogenesis of flavoenzymes [42–45].

In the first experimental approach proposed here, FMN to FAD conversion (synthesis, forward direction) and vice versa, that is, FAD degradation (either via the reverse direction, or pyrophosphorolysis, and via hydrolysis) are continuously followed by monitoring fluorescence changes of the reaction assays.

In the second experimental approach proposed here, FMN to FAD conversion and vice versa are followed by a discontinuous HPLC (high-performance liquid chromatography) approach: it implies acidic treatments of the reaction mixture, both to stop the reaction at the appropriate time points and to obtain the not-covalently bound cofactor in its free form. Neutralized samples are analyzed and quantitated by HPLC analysis.

2 Materials

Prepare all solutions using ultrapure water (prepared by purifying deionized water, to attain a sensitivity of 18 MΩ-cm at 25 °C) unless otherwise indicated, and analytical grade reagents. Diligently follow all waste disposal regulations when disposing of waste materials. We do not add sodium azide to reagents.

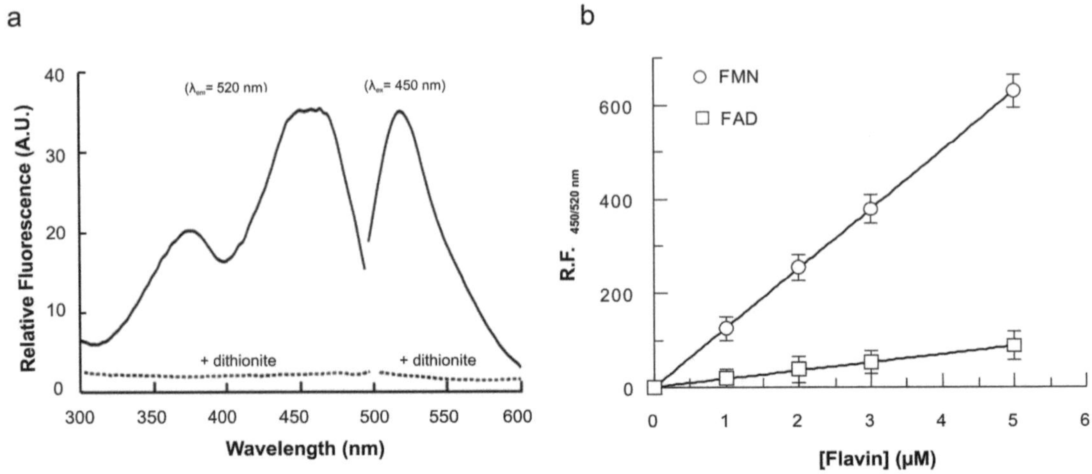

Fig. 3 Fluorescence properties of FAD and comparison of FAD and FMN specific fluorescence. (**a**) 2 μM FAD excitation spectrum (emission wavelength at 520 nm) and emission spectrum (excitation wavelength at 450 nm) performed at 25 °C in 50 mM Tris–HCl pH 7.5 in the absence (straight line) or in the presence of dithionite (dotted line). (**b**) Fluorescence specific calibration of FAD and FMN at 25 °C in 50 mM Tris–HCl pH 7.5. Fluorescence intensities of spectrophotometrically calibrated solutions were measured and expressed in Arbitrary Units (A.U.). Under our experimental conditions the slope, that is, FAD fluorescence constant (K_{FAD}) is about seven times lower than that of FMN (K_{FMN}). Each point represents the mean value of three different measurements

Applying the equation $g = MW(\text{g/mol}) \cdot M(\text{mol/L}) \cdot V(\text{L})$, prepare the following stock solutions.

Consider powder's purity (if it is not 100%) and adjust its weight properly.

Chemicals used for HPLC protocol are of the highest grade commercially available (HPLC grade).

2.1 Buffer and Reagents for Subheading 3.1

Before use of standard flavin stock solutions in Subheading 3.2 or Subheading 3.3, they must be carefully calibrated, by means of spectrophotometric calibration, as described in Subheading 3.1.

1. 500 mM Tris–HCl pH 7: dissolve 30.28 g of Trizma in 400 mL of ultrapure water, mix and adjust pH to 7 with HCl, then reach a final volume of 500 mL with ultrapure water. Store at −20 °C (*see* **Note 1**). For working buffer mix 2 mL of 500 mM Tris–HCl pH 7 in 48 mL of ultrapure water to obtained 20 mM Tris–HCl pH 7 solution.
2. Standard stock flavin solutions (*see* **Note 2**):
   (a) *1 mM FAD*: prepare a 10 mM (10×) starting solution by dissolving 0.0083 g of FAD in 1 mL of ultrapure water, mix and adjust pH to about 7, using a pH-indicator paper, with few crystals of Trizma. Then mix 0.1 mL of the 10 mM FAD and 0.9 mL of ultrapure water.

The final *expected* concentration (1 mM) will be corrected using a spectrophotometric calibration procedure. Store at −20 °C. The solution is stable for 1 month.

(b) *1 mM FMN*: prepare a 10 mM (10×) starting solution by dissolving 0.0048 g of FMN in 1 mL of ultrapure water, mix and adjust pH to about 7, using pH-indicator paper, with few crystals of Trizma. Then mix 0.1 mL of the 10 mM FMN and 0.9 mL of ultrapure water.

The final *expected* concentration (1 mM) will be corrected using a spectrophotometric calibration procedure. Store at −20 °C. The solution is stable for 1 month.

(c) *0.2 mM Riboflavin*: dissolve 0.0015 g of riboflavin in 20 mL of ultrapure water. Mix for about 4 h, under continuous magnetic stirring and store at 4 °C when the solution seems uniform. Adjust pH to about 7, using pH-indicator paper, with few crystals of Trizma.

The *expected* concentration will be corrected using a spectrophotometric calibration procedure. The solution is stable for 1 month.

2.2 Buffer and Reagents for Subheading 3.2

1. 500 mM Tris–HCl pH 7.5: dissolve 30.28 g of Trizma in 400 mL of ultrapure water, mix and adjust pH to 7.5 with HCl, then reach a final volume of 500 mL with ultrapure water. Store at −20 °C (*see* **Note 1**). For working buffer dilute ten-fold in ultrapure water.
2. Calibrated flavin working solutions: from spectrophotometric calibrated stock flavin solutions (Subheading 3.1) prepare 1 mL of fresh working solutions by diluting ten-fold 1 mM FAD or FMN and two-fold 0.2 mM riboflavin in ultrapure water to obtain 0.1 mM final concentration.
3. 100 mM NaATP: dissolve 0.1100 g of ATP in 2 mL of distilled water, mix and adjust pH to about 7, using pH-indicator paper, with few crystals of Trizma. Store at −20 °C. For working solution dilute ten-fold in distilled water to obtained 10 mM NaATP solution. The solution is stable for 6 months.
4. 100 mM $MgCl_2$: dissolve 0.4480 g of $MgCl_2$ in 20 mL of ultrapure water and mix. Store at 4 °C. The solution is stable almost for 2–3 months.
5. 100 mM NaPPi: dissolve 0.0222 g of NaPPi in 1 mL of distilled water and mix. Store at 0–4 °C during experimental procedure. The solution is not stable for more than 1 day, therefore, prepare fresh solution every time you need.

6. 100 mM $CoCl_2$: dissolve 0.0238 g of $CoCl_2$ in 1 mL of ultrapure water and mix. Store at 4 °C. The solution is stable for 1 month.
7. 2 M KCl: dissolve 1.49 g of KCl in 10 mL of ultrapure water and mix. Store at 4 °C. The solution is stable for 2–3 months.

2.3 Buffer and Reagents for Subheading 3.3

1. The following stock and working solutions prepared as for Subheading 3.2 (*see* Subheading 2.2):

   50 mM Tris–HCl pH 7.5, calibrated 0.1 mM flavin solutions, 100 mM NaATP, 100 mM $MgCl_2$.
2. Home-made lysis buffer (*see* **Note 3**):
   (a) Dissolve 0.0021 g of NaF in 0.5 mL of ultrapure water to obtain 100 mM NaF.
   (b) Dissolve 0.1742 g of PMSF in 10 mL of ethanol (100%) to obtain 100 mM PMSF.
   (c) Add in a Falcon tube:
   - 50 mM Tris–HCl pH 7.5 (to reach a final volume of 5 mL),
   - 1% Triton X-100 (50 μL),
   - 5 mM ß-mercaptoethanol (1.7 μL of 14.3 M ß-mercaptoethanol),
   - 1 mM NaF (50 μL of 100 mM NaF),
   - 0.1 mM PMSF (5 μL of 100 mM PMSF),
   - Protease inhibitor cocktail (PIC, added according to the producer's protocol).
3. 10% $HClO_4$: starting from concentrated, commercially available, perchloric acid (65%) add 2.3 mL of perchloric acid to 12.7 mL of ultrapure water up to 15 mL final volume.
4. 5 M KOH/2 M KH_2PO_4 (neutralizing solution): dissolve 28.05 g of KOH in 100 mL of ultrapure water and mix; dissolve 27.20 g of KH_2PO_4 in 100 mL of ultrapure water and mix; then mix both solutions in a 1:1 ratio to obtain 200 mL solution (*see* **Note 4**).
5. 10 mM KH_2PO_4/K_2HPO_4 pH 6 (phosphate buffer for HPLC): to prepare 800 mL of buffer dissolve 1.0887 g of KH_2PO_4 in 600 mL of ultrapure water by magnetic stirring (acid solution); dissolve 0.3484 g of K_2HPO_4 in 200 mL of ultrapure water by magnetic stirring (basic solution). Mix them together by adding the basic solution (about 100 mL will be enough) to the acidic solution until pH 6 is obtained, then add ultrapure water to reach the final volume of 800 mL.
6. Methanol for HPLC: purity ≥99.9%.

3 Methods

3.1 Spectrophotometric Calibration of Flavin Solution Concentrations

The spectrophotometric calibration of flavin solution concentrations described in this method should be always performed before running Subheading 3.2 or Subheading 3.3.

3.1.1 UV/Vis Spectra of Flavins

UV/Vis spectra of oxidized flavins, performed at 25 °C, pH 7 show three absorbance peaks, at 266, 373, and 445 nm for Rf/FMN and 263, 373, and 450 for FAD. The latter is not present in the reduced form of flavins (Fig. 2a).

Instrumental Settings

- Spectrophotometer.
  - For spectra: λ interval = 240–600 nm.
  - Scan speed = 1800 nm/min.

Procedure for Spectra Registration

In each assay:

- Add 2 mL of 20 mM Tris–HCl pH 7 in a quartz cuvette;
- Measure background;
- Add 0.04 mL of 1 mM FMN/FAD or 0.2 mL of 0.2 mM Rf;
- Mix and start absorbance spectra measurements;
- Add few crystals of dithionite in each cuvette and register the zero of absorbance.

Procedure for Calibration

1. Set spectrophotometer at λ = 450 nm.
2. Prepare flavin solutions in a glass (or plastic) cuvette as described below.

   For FAD and FMN calibration:

   Starting from 1 mM *expected* flavin stock solutions prepare 10, 15, 20, and 30 μM solutions in a final volume of 2 mL, as follows:

   (a) 10 μM (1.980 mL of 20 mM Tris–HCl pH 7 + 0.020 mL of 1 mM FAD or FMN),

   (b) 15 μM (1.970 mL of 20 mM Tris–HCl pH 7 + 0.030 mL of 1 mM FAD or FMN),

   (c) 20 μM (1.960 mL of 20 mM Tris–HCl pH 7 + 0.040 mL of 1 mM FAD or FMN),

   (d) 30 μM (1.940 mL of 20 mM Tris–HCl pH 7 + 0.060 mL of 1 mM FAD or FMN).

For riboflavin calibration:

Starting from 0.2 mM *expected* riboflavin stock solution prepare 10, 15, 20, and 30 μM solutions in a final volume of 2 mL, as follows:

(a) 10 μM (1.900 mL of 20 mM Tris–HCl pH 7 + 0.100 mL of 0.2 mM Riboflavin),

(b) 15 μM (1.850 mL of 20 mM Tris–HCl pH 7 + 0.150 mL of 0.2 mM Riboflavin),

(c) 20 μM (1.800 mL of 20 mM Tris–HCl pH 7 + 0.200 mL of 0.2 mM Riboflavin),

(d) 30 μM (1.700 mL of 20 mM Tris–HCl pH 7 + 0.300 mL of 0.2 mM Riboflavin).

3. In each assay:

   (a) Add 2 mL of 20 mM Tris–HCl pH 7 in a glass (or plastic) cuvette;

   (b) Measure background;

   (c) Register $A_{450\ nm}$ of each flavin solution prepared;

   (d) Add few crystals of dithionite in each cuvette and register the zero of absorbance.

   Repeat each assay in triplicate.

4. Calculations: Plot $A_{450\ nm}$ vs *expected* concentration (0, 10, 15, 20, and 30 μM) for each flavin and in the linear regression line, calculate the intercept and the slope, which corresponds to the *experimental* $\varepsilon_{450\ nm}$ (Fig. 2b).

   The ratio "*experimental* $\varepsilon_{450\ nm}$ (slope)/*theoretical* $\varepsilon_{450\ nm}$" gives a correction index (Ci) for each flavin species.

   Theoretical $\varepsilon_{450\ nm}$ values, at pH 7 and 25 °C, are 11.3 $mM^{-1}\ cm^{-1}$ for FAD and 12.2 $mM^{-1}\ cm^{-1}$ for FMN and Rf.

   The correct final concentrations of stock flavin solutions (1 mM FAD/FMN or 0.2 mM Rf) are determined by multiplying the *expected* flavin concentration by the correction index.

   Starting from calibrated stock flavin solutions prepare operative or working solutions each day for Subheading 3.2 or Subheading 3.3.

3.2 Assaying FAD Metabolic Conversion In Continuo by Spectrofluorimetric Methods

3.2.1 Fluorescence Spectra Registration and Fluorescence Constant Assessment

Oxidized flavins perform natural peculiar fluorescence properties, useful to high sensitivity quantitative determinations. Exciting free FAD solutions in a variable λ interval (300–500 nm), and registering its fluorescence emission at 520 nm (in the excitation spectrum) two peaks at 370 and 450 nm can be observed. When excitation λ is fixed at 450 nm, free flavins show one emission peak at 515–520 nm. Reduced flavins, instead, do not show any fluorescence peak. The spectra of fluorescence excitation (at $\lambda_{em} = 520$ nm) and emission (at $\lambda_{ex} = 450$ nm) of FAD are reported in Fig. 3a.

Instrumental Settings

- Spectrofluorometer.
- Excitation spectra: set λ emission at 520 nm; λ excitation interval = 300–500 nm.
- Emission spectra: set λ excitation at 450 nm; λ emission interval = 480–600 nm.
- Em bandwidth = 5 nm; Ex bandwidth = 5 nm.
- Scan speed = 1000 nm/min.
- Response = 0.05 s.

Procedure for Spectra Registration

In each assay:

- Add 1.980 mL of 50 mM Tris–HCl pH 7.5 in a quartz cuvette (four optical faces);
- Measure fluorescence background;
- Add 0.02 mL of 0.1 mM flavin working solution;
- Mix and start fluorescence (excitation and emission) spectra measurements;
- Add few crystals of dithionite in each cuvette and register fluorescence (excitation and emission) spectra.

Procedure for Calibration

1. Set spectrofluorometer at λ emission = 520 nm, λ excitation = 450 nm.
2. Starting from 0.1 mM FAD or FMN working solutions prepare for each flavin in a glass (or plastic cuvette, four optical faces) 1, 2, 3, and 5 μM solutions in a final volume of 2 mL, as follows:
   (a) 1 μM (1.980 mL of 50 mM Tris–HCl pH 7.5 + 0.02 mL of 0.1 mM working solution),
   (b) 2 μM (1.960 mL of 50 mM Tris–HCl pH 7.5 + 0.04 mL of 0.1 mM working solution),
   (c) 3 μM (1.940 mL of 50 mM Tris–HCl pH 7.5 + 0.06 mL of 0.1 mM working solution),
   (d) 5 μM (1.900 mL of 50 mM Tris–HCl pH 7.5 + 0.10 mL of 0.1 mM working solution).

3. In each assay:
   (a) Add 2 mL of 50 mM Tris–HCl pH 7.5 in a glass (or plastic, four optical faces) cuvette;
   (b) Measure fluorescence background;
   (c) Measure $F_{450/520\ \mathrm{nm}}$ of each flavin solution prepared.

Repeat each assay in triplicate.

F value is normally expressed in Arbitrary Units (A.U.), since it is not an absolute value and can be influenced by several parameters, among which lamp energy efficiency. In our experimental conditions, Rf 2 μM is used as an internal standard and, before each experiment, its fluorescence is set at 250 A.U., using the specific instrument adjusting procedure. Now, *F* is expressed as relative fluorescence (R.F.).

Plot $\mathrm{R.F.}_{450/520\ \mathrm{nm}}$ vs concentration and calculate the intercept and the slope of the fitted linear regression line. Slope represents the fluorescence constant $K_{450/520}$ which is expressed in μM^{-1}. FMN and Rf do not differ significantly in their fluorimetric properties; conversely their $K_{450/520}$ is about 7–8 times higher than that of FAD [41, 46] (Fig. 3b).

For each single solution, the ratio $\mathrm{R.F.}_{450/520\ \mathrm{nm}}/K_{450/520}$ multiplicated by the final volume of the cuvette (in mL) gives the total amount of flavin in nmol.

3.2.2 Enzymatic Fluorimetric Continuous Assays

Instrumental Settings

- Spectrofluorometer.
- Time Drive measurements: set λ excitation at 450 nm and λ emission at 520 nm.
- Em bandwidth = 5 nm; Ex bandwidth = 5 nm.
- Response = 0.5 s.

FAD Synthesis Assay

When using purified recombinant FAD synthases (purified as described in Chapter 6) FMN conversion to FAD can be quite easily and rapidly fluorimetrically followed in continuo.

Time drive measurements (set the excitation and emission wavelengths at 450 nm and 520 nm, respectively) are usually performed at 37 °C in 50 mM Tris-HCl pH 7.5 in the presence of the substrate pair ATP and FMN. The reaction requires $MgCl_2$ and starts by the addition of the recombinant enzyme. Optimum substrate pair and $MgCl_2$ concentrations, normally used to measure enzyme activity, are described in the procedure below.

Before each experiment, FAD and FMN fluorescence values are individually calibrated as described above at Subheading 3.2.1.

Procedure

1. Add in a glass (or plastic cuvette, four optical faces) with the following order:

   (a) 50 mM Tris–HCl pH 7.5 (to reach a final volume of 2 mL),
   (b) 5 mM $MgCl_2$ (100 μL of 100 mM $MgCl_2$),
   (c) 100 μM ATP (20 μL of 10 mM ATP).
2. Measure fluorescence background, then add:
   (a) 2 μM FMN (40 μL of 0.1 mM FMN working solution).
3. Start fluorescence time course measurement (follow for about 30 s to ascertain that initial fluorescence is constant, under our calibration conditions it corresponds to 250 A.U.) then add:
   (a) the appropriate volume (based on protein concentration/activity) of the purified recombinant protein (*see* **Note 5**).
4. Follow the fluorescence decrease for almost 3–5 min and then stop recording.

In this interval time we expect about 35% fluorescence decrease. Please take into account that if the reaction is complete (at the end-point, if all FMN is converted into FAD), you may expect at maximum about a 75% fluorescence decrease.

To calculate the rate of fluorescence decrease, choose the initial part of the experimental curve (Fig. 4a).

Depending on the instruments used, you can manage appropriately the ordinate axis by devoted analysis tools. Thus, set your axis opportunely at the beginning of the observation. Sometimes analysis tool will open a window showing experimental curve, allowing scale adjusting at the end.

Please mind that the stoichiometry of the reaction is 1:1 and that at the initial point the whole fluorescence (Fi) of the reaction mixture is due to FMN, thus:

$$\text{Fi} = K_{\text{FMN}} \cdot [\text{FMNi}]$$

At any time (x) during the reaction progress:

$$Fx = K_{\text{FMN}} \cdot [\text{FMN residual}] + K_{\text{FAD}} \cdot [\text{FAD formed}]$$

where [FMN residual] is equal to [FMNi] – [FAD formed].

If we define: $\Delta F = \text{Fi} - Fx$

$$\Delta F = K_{\text{FMN}} \cdot [\text{FMNi}] - K_{\text{FMN}} \cdot [\text{FMNi}] + K_{\text{FMN}} \cdot [\text{FAD formed}] - K_{\text{FAD}} \cdot [\text{FAD formed}]$$

and yet:

$$\Delta F = (K_{\text{FMN}} - K_{\text{FAD}}) \cdot [\text{FAD formed}]$$

Thus, the rate of FAD synthesis, v_0, that is, the rate of FAD formation, is calculated from the rate of fluorescence decrease ΔF, measured as the tangent to the initial part of the experimental curve by applying the following equation:

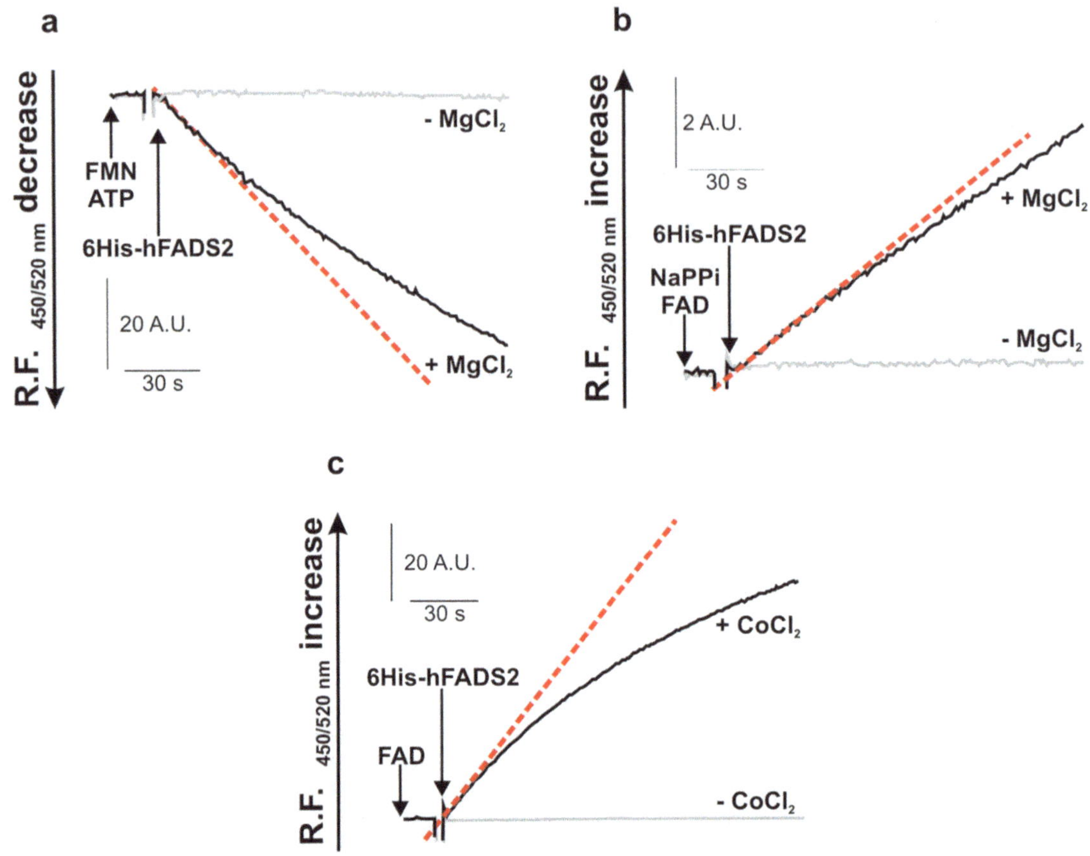

Fig. 4 Typical progress curves of the continuous fluorimetric assays of FAD synthesis or degradation, as catalyzed by purified recombinant hFADS2. (**a**) FAD synthesis (forward reaction) is followed at 37 °C in 2 mL of 50 mM Tris–HCl, pH 7.5, containing 5 mM $MgCl_2$ (black line) and the substrate pair 2 μM FMN and 100 μM ATP. The reaction starts by the addition of 6His-hFADS2 (10 μg, 0.17 nmol). As a control, $MgCl_2$ is omitted (light gray line). (**b**) The reverse reaction (i.e., pyrophosphorolysis) is followed under the same conditions, by using, as substrates, 0.5 μM FAD and 1 mM NaPPi, (black line) and started by the addition of 6His-hFADS2 (10 μg, 0.17 nmol). As control, $MgCl_2$ is omitted (light gray line). (**c**) FAD hydrolysis reaction, which strictly depends on $CoCl_2$, is followed at 37 °C in 2 mL of 50 mM Tris–HCl, pH 7.5, in the presence of 50 mM KCl, using as substrate the sole 1 μM FAD (black line). The reaction starts by the addition of 6His-hFADS2 (10 μg, 0.17 nmol). As control, $CoCl_2$ is omitted (light gray line). In each panel, v_0 is calculated from the rate of fluorescence ($\lambda_{ex} = 450$ nm, $\lambda_{em} = 520$ nm) decrease (**a**) or increase (**b** and **c**), respectively, measured as tangent to the initial part of the experimental curve (dotted red line). Following calibration procedures described in Fig. 3 at Subheading 3.2.1, results are finally expressed as nmol FAD formed or degraded per minute and, then, referred to the amount of protein present in each assay

$$v_0 = (\Delta F/\Delta K \cdot V)/\Delta t \cdot m$$

where ΔF is fluorescence expressed in Arbitrary Units, $\Delta K = K_{FMN} - K_{FAD}$ is expressed as μM^{-1}, V is the volume expressed in mL, t is time expressed in min, and m is the mass of protein in mg.

Thus, v_0 is expressed as $nmol \cdot min^{-1} \cdot mg\ protein^{-1}$.

FAD Pyrophosphorolysis

The reverse reaction of FAD synthesis, that is, the FAD pyrophosphorolysis, catalyzed by purified recombinant hFADS can be also fluorimetrically followed, essentially under the same experimental conditions and instrument settings, described above. The reaction is followed at 37 °C in the presence of $MgCl_2$ using as substrates the products of the forward reaction, that is, FAD and NaPPi. The optimal concentrations usually used to measure V_{max} of the reaction are described in the procedure.

Procedure

1. Add in a glass (or plastic cuvette, four optical faces) with the following order:
   (a) 50 mM Tris–HCl pH 7.5 (to reach a final volume of 2 mL),
   (b) 5 mM $MgCl_2$ (100 μL of 100 mM $MgCl_2$),
   (c) 1 mM NaPPi (20 μL of 100 mM NaPPi).
2. Measure fluorescence background, then add:
   (a) 0.5 μM FAD (10 μL of 0.1 mM FAD working solution).
3. Start fluorescence time course measurement (follow for about 30 s to ascertain that fluorescence is constant, under our calibration conditions it corresponds to 9 A.U.), then add:
   (a) the appropriate volume (based on protein concentration/ activity) of purified recombinant protein (*see* **Note 6**).
4. Follow the fluorescence increase almost for 3–5 min and then stop recording.

FAD to FMN conversion implies a significant flavin fluorescence increase, according to a higher value of the product with respect to the substrate.

To calculate the rate of fluorescence increase, choose the initial part of the experimental curve (Fig. 4b).

In this case the initial fluorescence is due only to FAD concentration according its constant of fluorescence K_{FAD}, while at any time (x) during the progress curve, Fx is equal to $K_{FAD} \cdot$ [FAD residual] + $K_{FMN} \cdot$ [FMN formed] which corresponds to [FAD cleaved]. Thus, ΔF is fluorescence increase which can be put in relationship as above to ΔK to calculate the amount of flavin converted measured by:

$$\Delta F = (K_{FMN} - K_{FAD}) \bullet [\text{FMN formed}].$$

Thus, the rate of FAD pyrophosphorolysis, v_0, that is, the rate of FAD cleavage, expressed as nmol $\cdot$ min^{-1} $\cdot$ mg protein^{-1}, is calculated from the rate of fluorescence increase ΔF, measured as the tangent to the initial part of the experimental curve by applying the following equation:

$$v_0 = (\Delta F/\Delta K \cdot V)/\Delta t \cdot m$$

where ΔF is fluorescence expressed in Arbitrary Units, $\Delta K = K_{FMN} - K_{FAD}$ is expressed as μM^{-1}, V is the volume expressed in mL, t is time expressed in min, and m is the mass of protein in mg.

FAD Hydrolysis by Purified hFADS

FAD can be converted to FMN also by hydrolases or diphosphatases (E.C. 3.6.1.18) which differ from adenylyl transferases (E.C. 2.7.7.2), which, in vitro, in the reverse direction, are responsible for FAD pyrophosphorolysis (*see* Subheading 1). Purified recombinant isoform 2 of hFADS, containing at its N-terminus an FADHy domain (i.e., a molybdopterin-binding resembling domain) catalyzes FAD hydrolysis to FMN, which can be followed in continuo fluorimetrically ([30, 37] *see* also Chapter 6).

FAD hydrolysis, as catalyzed by FADHy domain, occurs only in the presence of $CoCl_2$, and it is strongly stimulated in the presence of KCl. It can be followed in continuo by this fluorimetric approach.

Procedure

1. Add in a glass (or plastic cuvette, four optical faces) with the following order:
   (a) 50 mM Tris–HCl pH 7.5 (to reach a final volume of 2 mL),
   (b) 1 mM $CoCl_2$ (20 μL of 100 mM $CoCl_2$),
   (c) 50 mM KCl (50 μL of 2 M KCl).
2. Measure fluorescence background, then add:
   (a) 5 μM FAD (100 μL of 0.1 mM FAD working solution).
3. Start fluorescence time course measurement (follow for about 30 s to ascertain that fluorescence is constant, under our calibration conditions it corresponds to 90 A.U.), then add:
   (a) the appropriate volume (based on protein concentration/activity) of purified recombinant protein (*see* **Note 7**).
4. Follow the fluorescence increase almost for 3–5 min and stop recording.

To calculate the rate of fluorescence increase, choose the initial part of the experimental curve (Fig. 4c).

Thus, the rate of FAD hydrolysis, v_0, that is, the rate of FAD cleavage, expressed as nmol · min^{-1} · mg protein^{-1}, is calculated from the rate of fluorescence increase ΔF, measured as the tangent to the initial part of the experimental curve by applying the following equation:

$$v_0 = (\Delta F/\Delta K \cdot V)/\Delta t \cdot m$$

where ΔF is fluorescence expressed in Arbitrary Units, $\Delta K = K_{FMN} - K_{FAD}$ is expressed as μM^{-1}, V is the volume expressed in mL, t is time expressed in min, and m is the mass of protein in mg.

The continuous fluorimetric methods described above are useful for steady-state kinetic studies since they are good approaches to determine the initial rate (v_0) of flavin conversion; therefore, they were used to describe kinetical features (k_{cat}, K_m, inhibitor sensitivities, and others) of different isoforms of human FADSs [25, 30, 47–49].

FAD Hydrolysis by Cell, Subcellular Fractions

The rapid, continuous, fluorimetric approach described above is also applicable to crude cellular/subcellular fractions, which have a significant FAD hydrolytic activity [1, 28, 29]. We successfully used this strategy in the past, to identify FAD hydrolase activity in both the outer membrane of mitochondria and nucleus from rat liver and human muscle/cells [17, 29, 32, 46]. FAD metabolism in *Saccharomyces cerevisiae* and worms were also detected in this way [50, 51].

To measure the rate of FAD hydrolysis in crude cellular/subcellular fractions the same procedure used to measure the rate of FAD hydrolysis with recombinant proteins could be used, although pH optimum of the buffer can range from 7.5 to 8.5 and the temperature from 25 °C to 37 °C and incubation times are normally longer.

Procedure

1. Add in a glass (or plastic cuvette, four optical faces) with the following order:
   (a) 50 mM Tris–HCl pH 7.5 (to reach a final volume of 2 mL),
   (b) 1 mM $MgCl_2$ (20 μL of 100 mM $MgCl_2$).

   (*see* **Note 8**)
2. Measure fluorescence background, then add:
   (a) 5 μM FAD (100 μL of 0.1 mM FAD working solution).
3. Start fluorescence time course measurement (follow for about 30 s to ascertain that fluorescence is constant, under our calibration conditions it corresponds to 90 A.U.), then add:
   (a) the appropriate volume (based on protein concentration) of cell lysate (0.01–0.10 mg).
4. Follow the fluorescence increase almost for 10–15 min and stop recording.

To calculate the rate of fluorescence increase, choose the initial part of the experimental curve.

Thus, the rate of FAD hydrolysis, v_0, that is, the rate of FAD cleavage, expressed as nmol • min^{-1} • mg $protein^{-1}$, was calculated from the rate of fluorescence increase ΔF (as in hydrolysis performed by recombinant protein), measured as the tangent to the initial part of the experimental curve by applying the following equation:

$$v_0 = (\Delta F/\Delta K \cdot V)/\Delta t \cdot m$$

where ΔF is fluorescence expressed in Arbitrary Units, $\Delta K = K_{FMN} - K_{FAD}$ is expressed as μM^{-1}, V is the volume expressed in mL, t is time expressed in min, and m is the mass of protein in mg in cell lysate.

Anyway, this protocol alone is not able to discriminate the nature of the flavin formed, that is, between FMN or Rf, since they have superimposable fluorescence properties. In this case, a control by analytical methods should be necessary, preferably by HPLC, as described below in Subheading 3.3 (*see* Subheading 3.3.3).

3.2.3 Limitations to Continuous Fluorimetric Approach and Alternative Strategies

The major limitation of the fluorimetric protocol is that it cannot be simply used with crude extracts, when measuring the FAD synthase activity, since this implies a small decrease in fluorescence over a high background. In addition, the presence of a great variety of contaminants in the extracts from biological samples, could contribute to a big noise and affect the background or blank.

Vice versa, as outlined at the end of Subheading 3.2.2 the rapid fluorimetric approach fits quite well to follow the rate of FAD hydrolysis in crude samples, which implies magnification of fluorescence over time.

Alternative strategy to continuously follow FAD formation in crude biological samples have been applied successfully in the past. This high sensitivity (amplificative) enzymatic strategy is based on spectrophotometric or polarographic measurements; it makes use of an enzymatic FAD detection system (FADS D.S.), which is based on the rate of reconstitution of newly synthesized FAD with a home-prepared apo-D-amino acid oxidase [17, 24, 25, 46, 47]. Commercial kits are now available, where the activity of the oxidase reconstituted with FAD is followed by a probe generating color and fluorescence.

Another useful strategy to measure the rate of FAD synthesis, as performed either by natural FADSs in cellular/subcellular fractions or by purified recombinant proteins, is described below, in Subheading 3.3.

It is a discontinuous assay that makes use of high-performance liquid chromatography (HPLC) followed by fluorimetric detection of separated flavins.

3.3 Assaying FAD Metabolic Conversion by Discontinuous HPLC Methods

In mammalian, worm, plant, and yeast cells (or subcellular fractions) the rate of flavin conversion, that is, FAD synthesis or degradation (as well as the amount of total endogenous flavin content) has been measured by HPLC [8, 28, 29, 32, 50–53]. HPLC analysis can also be useful (in parallel with fluorimetric Method described above) to analyze and validate the progress of FAD formation reaction by recombinant purified FADS or to quantitate the amount of FAD bound to the monomer of the apo-protein [25], as reported in Chapter 6.

Up to now, this can be considered the election procedure chosen in many laboratories, among which ours, to quantitatively analyze metabolic conversion of flavins in crude biological samples, even though we expect in the future the advent of metabolomics to perform this analysis in a wider perspective [54].

In between, we propose, in particular because of the high natural fluorescence of flavins, the combination of HPLC with fluorimetric detection, that allows in our protocol a sensitivity in the order of pmol.

3.3.1 Cell Lysis

Prior to perform chromatographic analysis in cell contest an opportune procedure of cell lysis, which both maintains the proteins in their enzymatically active form and prevents the alteration of the steady-state level of endogenous flavins, is required.

Here is described the procedure to obtain mammalian cellular extracts by a home-made lysis buffer.

Procedure

- Resuspend fresh or frozen cell pellets (4×10^6 cells) in 150 μL of home-made lysis buffer.
- Pass through a 22G needle (ten passages).
- Pass through a 26G needle (ten passages).
- Incubate for 30 min on ice.
- Pass through a 26G needle (ten passages).
- Centrifuge the cell suspension at 13,000 × *g* for 10 min at 4 °C and recover the supernatant as cell lysate.

3.3.2 FAD Synthesis Assay

Appropriate amounts of freshly obtained cell lysate (or recombinant proteins) are incubated at 37 °C in a reaction mixture consisting of 50 mM Tris–HCl pH 7.5, $MgCl_2$, ATP and FMN to allow for FAD formation. At appropriate time points reaction is stopped by treatment with perchloric acid of the sample [8, 46]. Omit substrate pair addition, if you want to measure total cellular endogenous flavin content (i.e., 0 time of reaction).

Acidic treatment causing protein precipitation, allows the removal of not-covalently bound cofactor from protein components, and avoids their injection in the column.

Procedure

For each assay, prepare the reaction mixture by adding in the eppendorf tube with the following order:

- 50 mM Tris–HCl pH 7.5 (to reach a final volume of 600 μL),

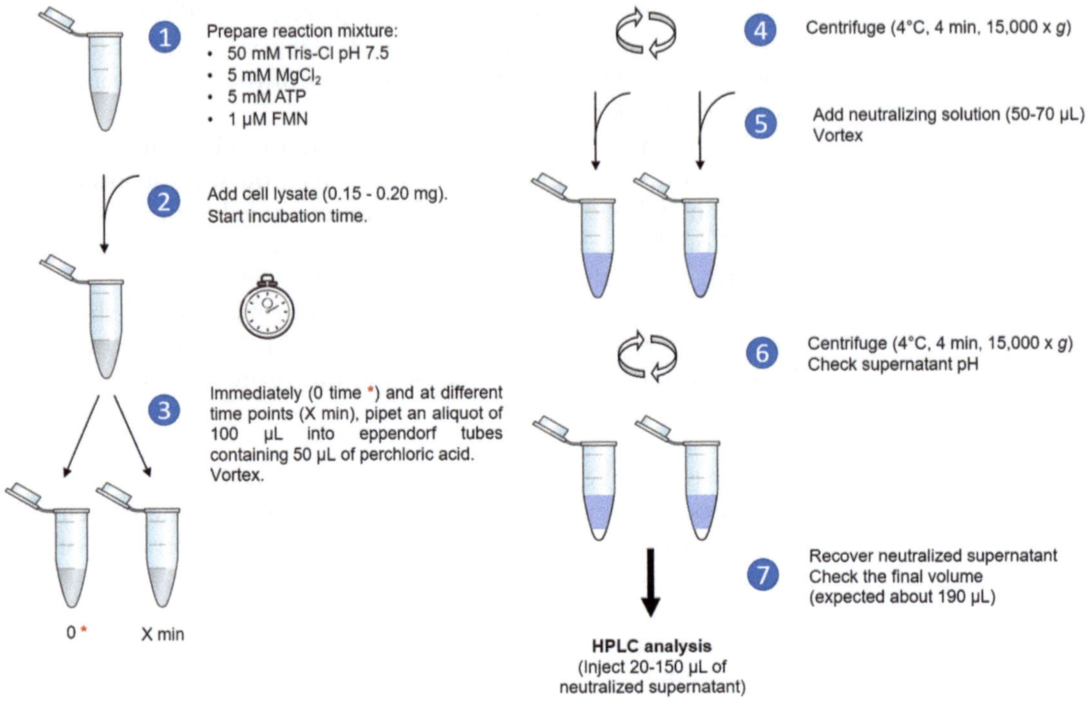

Fig. 5 Scheme of the discontinuous assay of FAD synthesis and flavin extraction. *Note that at 0 time the amount of each flavin corresponds to its endogenous content in the cell lysate

- 5 mM $MgCl_2$ (30 μL of 100 mM $MgCl_2$),
- 5 mM ATP (30 μL of 100 mM ATP),
- 1 μM FMN (6 μL of 0.1 mM FMN working solution),
- the appropriate volume (based on protein concentration) of cell lysates (0.15–0.20 mg) and start incubation time.

At different time points (from 0 to 60 min), treat five aliquots (100 μL) taken from reaction mixture with perchloric acid to extract flavins for HPLC measurements as described at Subheading 3.3.4 and in Fig. 5.

If necessary, this procedure can be adopted to investigate reactions catalyzed by recombinant FAD synthases from different sources as described in [22, 25, 55, 56].

3.3.3 FAD Hydrolysis Assay

As discussed at Subheading 3.2.3, FAD hydrolysis can be rapidly quantitated by the fluorimetric continuous methods (*see* Subheading 3.2); a parallel HPLC analysis (as in Fig. 6) allows to discriminate between the two sequential possible products of FAD hydrolysis in crude extract, that is, FMN and Rf, which are isofluorescent.

In this case:

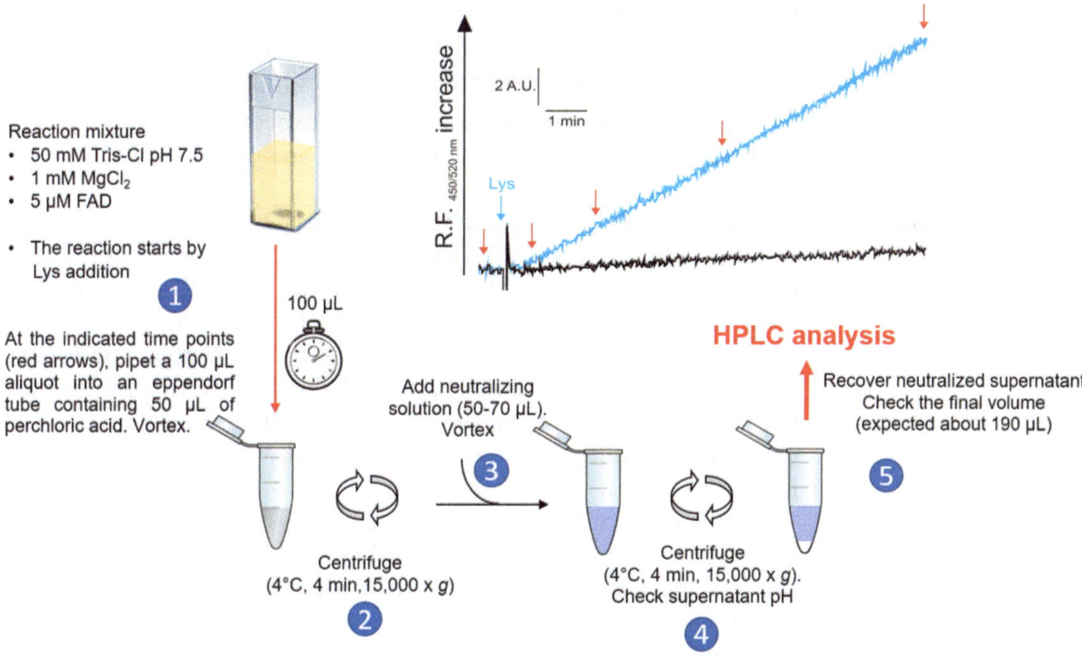

Fig. 6 Scheme of the continuous assay of FAD hydrolysis catalyzed by cell lysates coupled with HPLC analysis of the reaction products. In the top right an experimental curve of FAD hydrolysis, followed at 25 °C in 2 mL of reaction mixture (blue line), catalyzed by human fibroblast cell lysate (0.04 mg) [57] is reported. As control, the reaction was followed omitting cell lysate (black line). Numbers 1–5 indicate each step of the procedure of flavin extraction and neutralization prior to HPLC analysis carried out as in Fig. 5

At different time points (e.g., 0.5, 2, 5, and 10 min), treat aliquots (100 µL), taken from the cuvette used in fluorimetric continuous method, with perchloric acid to extract flavins. Longer times allow to observe the reaction at completion (not shown). On the neutralized extracts, HPLC measurements will be carried out as described at Subheading 3.3.4 and in Fig. 6.

3.3.4 Reaction Stopping, Perchloric Flavin Extraction, and Neutralization

Reactions of FAD synthesis or degradation are blocked and flavin extractions from cell/protein samples are obtained by means of perchloric acid treatment, as described in [17, 46, 58] and in Figs. 5 and 6.

Procedure

- Add 100 µL of sample to eppendorf tubes containing 50 µL of 10% $HClO_4$ and rapidly vortex.
- Centrifuge at 15,000 × *g* for 4 min at 4 °C.
- Add an adequate volume (50–70 µL) of neutralizing solution (5 M KOH/2 M KH_2PO_4 1:1), to obtain a pH of about 7 (*see* **Note 9**) and rapidly vortex.

- Centrifuge again at 15,000 × *g* for 4 min at 4 °C and check that supernatant pH is about 7, using pH-indicator paper. If it is not, add 5 μL of neutralizing solution and centrifuge again.
- Recover the neutralized supernatant (expected about 190 μL) in dark glass cones for chromatographic analysis in HPLC.

3.3.5 HPLC Separation and Analysis

Sample Injection

Opportune volumes of neutralized supernatant (V_{inj}) (20–150 μL) suitably treated as described in Subheading 3.3.4, are injected into the column, by means of the autosampler system, and eluted at room temperature. The different polarity of the flavins gives them a different partition coefficient between the non-polar stationary phase and the polar liquid phase, and this feature is exploited for the purpose of separation: the FAD, being the most polar (adenine ring), elutes first, followed by FMN (only one phosphate) and finally riboflavin.

Separation of flavins we propose here is based on reverse phase chromatography, using a 5-μm ODS3 column whose stationary phase is octadecylsilane in the form of microporous particles of 5 μm diameter (25 cm · 4.6 mm) endowed with a precolumn (1.5 cm · 4.6 mm).

Elution

Elution of the flavin species is carried out using an elution gradient in which the composition of the mobile phase changes during the chromatographic run. We use a linear gradient consisting of a mixture of 10 mM pH 6 potassium phosphate buffer and methanol. At the beginning, the percentages are 35% methanol and 65% phosphate buffer, after 1 min methanol increases linearly and pass to 45% in 7 min. The composition of the mobile phase (45% methanol–55% phosphate buffer) remains stationary for 2 min, then in 1 min, the organic phase returns to the initial value and remains stationary for 4 minutes till the end of the run, which lasts 15 min in total.

The flow rate of the mobile phase is 1 mL/min.

Before sample injection the column is equilibrated with 35% methanol–65% phosphate buffer for 30 min at 0.5 mL/min and then for 30 min at 1 mL/min and the background of fluorescence is measured (Autozero).

Fluorimetric Detection

Eluted flavins are detected fluorimetrically by a spectrofluorometer devoted to HPLC.

Instrumental settings

- Set λ excitation at 450 nm and λ emission at 520 nm.
- Set sensitivity (GAIN = 100).
- Response: 3 s.

After the chromatographic run, fluorimetrically revealed FAD, FMN, and Rf are recognized by comparison of retention times of

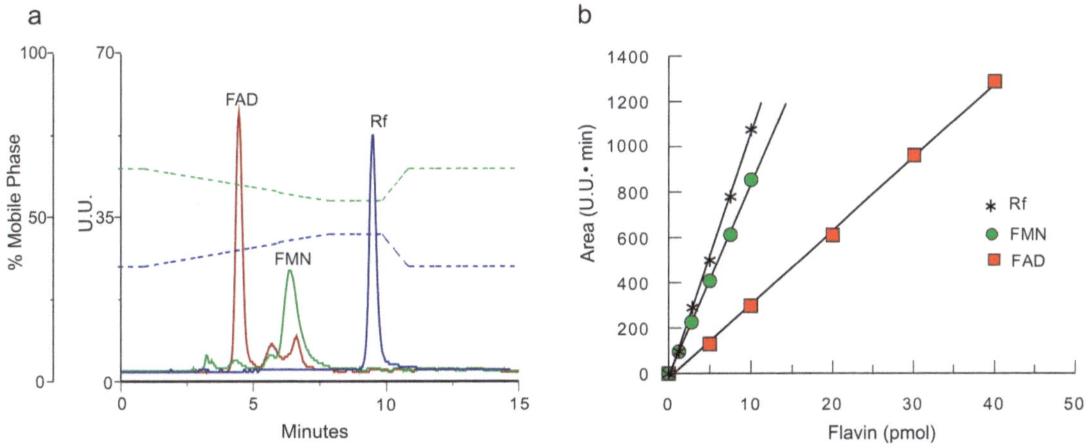

Fig. 7 Flavin analysis by HPLC. (**a**) A typical chromatogram of standard flavins (10 pmol for FMN/Rf and 40 pmol for FAD) revealed by fluorimetric detector. Fluorescence intensity (mV) is here expressed in User Units (U.U.). Dotted lines represent the elution gradient (methanol % in purple, phosphate buffer % in green). (**b**) Calibration curves obtained by plotting chromatographic peak area vs the amount (in pmol) of standard FAD, FMN, and Rf injected into the column. The slope of the straight lines, namely K_A (expressed in U.U. · min/pmol) of FAD (32.4) is about three times lower of that of FMN (84.6) and Rf (108.5)

standard flavins. Under our conditions they correspond to about 4.5, 6.5, and 9 min, respectively (*see* below Subheading "Calibration" and Fig. 7a).

Integration of standard peak areas in sample chromatograms allows for measurements of each flavin species in the extracts.

Calibration

Calibration curves are obtained daily by carrying out the chromatographic separation of standard flavin solutions, under the same instrumental conditions used for the samples to be analyzed.

Flavin standard solution concentrations used for the calibration of the chromatographic analysis are spectrophotometrically determined as previously described in Subheading 3.1.

Procedure

1. Starting from 20 μM spectrophotometrically calibrated flavin solutions, prepare in eppendorf tubes 2 μM FAD, 0.5 μM FMN, and 0.5 μM Rf in a final volume of 1 mL of ultrapure water, as described below:
   (a) 2 μM FAD (0.900 mL of ultrapure water + 0.100 mL of 20 μM FAD).
   (b) 0.5 μM FMN or Rf (0.975 mL of ultrapure water + 0.025 mL of 20 μM FMN or Rf).
2. Perform perchloric flavin extraction of 100 μL aliquots as described at Subheading 3.3.4.
3. Before recovering the neutralized supernatant in dark glass cones for chromatographic analysis in HPLC add ultrapure water to reach the volume of 200 μL and a final concentration of 1 μM FAD and 0.25 μM FMN/Rf.

Table 1
Calibration of standard flavin solutions

Flavin	Injected volume (V_{inj})	Injected pmol ($pmol_{inj}$)	Calibration level
1 μM FAD	5	5	1
	10	10	2
	20	20	3
	30	30	4
	40	40	5
0.25 μM FMN/Rf	5	1.25	1
	10	2.5	2
	20	5	3
	30	7.5	4
	40	10	5

4. Calculations: we establish five calibration levels setting the software for the injection of the opportune volume of supernatant in HPLC as described in the Table 1.

Dedicated softwares allow for integration of each peak area (Area), which is proportional to the amount of the species according to the equation:

$$\text{Area} = k_A \cdot \text{pmol}$$

Fluorescence intensity (mV) is here expressed in User Units (U.U.). In our experimental conditions, the peak area (U.U. · min) corresponding to 10 pmol of Rf is used as an internal standard and it is set at 1000 U.U. · min.

Plot Area vs pmol (for each flavin) as in Fig. 7b.

Calibration curve is a line passing from the origin of the axes, whose slope, or constant, (indicated as k_A) is expressed as U. U. · min/pmol of flavin in the injected volume.

3.3.6 Quantitation of Flavin Levels

To calculate the amount of flavin ($pmol_{inj}$) in the injection volume (V_{inj}) and, then, in the original assay sample (the reaction mixture), please use the calibration curves. Thus, for each flavin species:

$$\text{pmol}_{inj} = \text{Area}/k_A$$

Please take into account that V_{inj} (20–150 μL) is a fraction of the volume of neutralized supernatant (190 μL) obtained after perchloric flavin extraction, which derives from the treatment of 100 μL aliquots taken from reaction mixture (600 μL) (Fig. 5). Thus, in the original assay sample:

$$\text{pmol} = \left(\text{pmol}_{inj}/V_{inj}\right) \cdot 1.9 \cdot 600\,\mu\text{L}$$

where 1.9 corresponds to the dilution factor.

The total amount of flavin is usually referred to protein amount and expressed as pmol • mg protein^{-1}, where mg protein corresponds to the amount of cell lysate proteins in the reaction mixture.

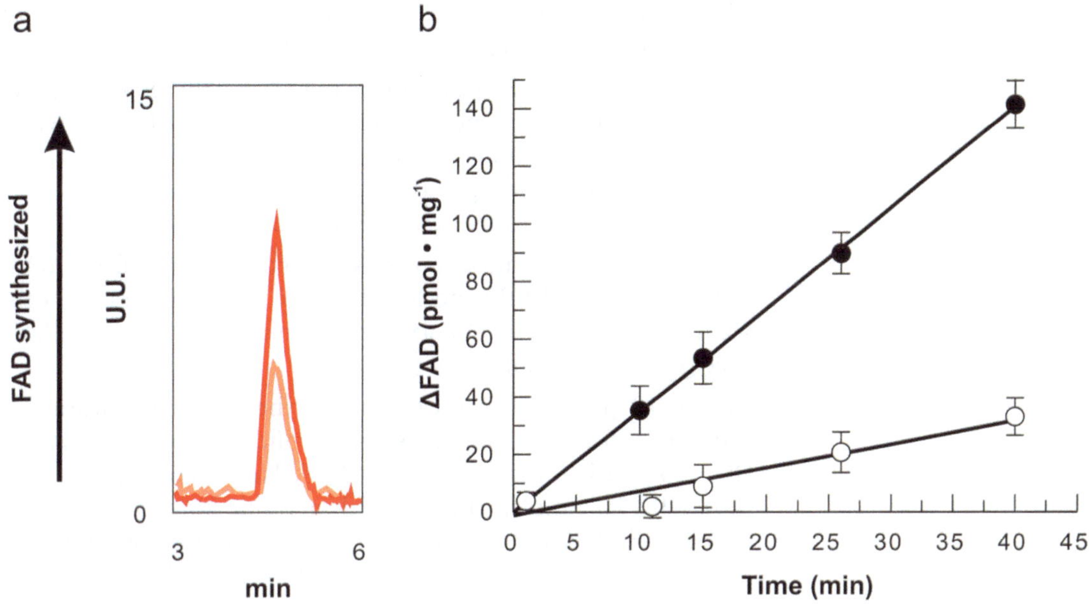

Fig. 8 Measuring FAD synthesis rate in human fibroblasts. (**a**) Overlapping chromatographic peaks of FAD at 0 time (endogenous content, in light red) and total FAD present in the reaction mixture following 10 min incubation (in dark red), under the conditions described in Fig. 5. From the difference between the areas the newly synthesized FAD at each time can be calculated (pmol = Area/K_A) and indicated as ΔFAD and referred to the amount (mg) of the cell lysate present in the mixture. In (**b**) typical time courses of FAD synthesis catalyzed by human fibroblasts obtained from either a control (filled circle) or a patient suffering for LMSFLAD (open circle) are reported [57]. The amount of FAD synthesized increases linearly with incubation time; thus, the rate can be calculated by the slope of the experimental line

FAD synthesis rate:

FAD formed (ΔFAD) is calculated by subtracting the endogenous FAD amount (reaction time 0 min) from the amount of FAD calculated at each time.

Plot ΔFAD (pmol · mg protein^{-1}) vs time to calculate the rate of FAD formation as in the typical discontinuous plot curve (Fig. 8). The initial rate v_0 corresponds to the slope of the experimental line and it is expressed as pmol · mg protein^{-1} · min^{-1}.

If the reaction rate is high (the curve reaches a plateau) as in the case of FADS overproduced in fibroblasts [8] the values of ΔFAD (pmol · mg protein^{-1}) as a function of time can be fitted with Grafit program (or equivalent) to 1st order rate equation. The initial rate v_0 is calculated by multiplying the Limit value by the Rate constant value.

4 Notes

1. 500 mM Tris–HCl pH 7 (Subheading 3.1) or 7.5 (Subheading 3.2) solution is stored at −20 °C in 5–10 mL aliquots in 15 mL sterile Falcon tubes.

2. The solutions are wrapped in aluminum foil to protect them from light during preparation, use, and storage.
3. The solution is not stable for more than 1 day therefore prepare fresh solution every time you need. PMSF and PIC (protease inhibitor cocktail) must be added at the last moment before use. Please avoid adding NaF (diphosphatase inhibitor) to lysis buffer if you want to measure FAD degradation.
4. The final neutralizing solution is not completely solved until both solution of KOH and KH_2PO_4 are mixed, then mix both solutions to obtain it. Do not use the solution when crystalline precipitates are visible.
5. Add 5–10 μg if you use one out of the different natural or mutant isoforms of hFADSs we purified in our laboratory [8, 24, 25, 48, 49, 59–61]. They correspond to 0.085–0.170 nmol protein, which are able to form at maximum 0.35–0.70 nmol FAD/min, in the case of 6His-hFADS2 (MW 56,537.32 g/mol, k_{cat} 4.1 min^{-1}) [25]. In the case of 6His-hFADS6, MW and k_{cat} value correspond to 38,266.46 g/mol and 2.9 min^{-1}, respectively [48].
6. Add 5–10 μg if you use one out of the different natural or mutant isoforms of hFADSs we purified in our laboratory [8, 24, 25, 48, 49, 59–61]. They correspond to 0.085–0.170 nmol protein, which are able to form at maximum 0.026–0.053 nmol FMN/min, in the case of 6His-hFADS2 (MW 56,537.32 g/mol, k_{cat} 0.3 min^{-1}) [25].
7. Add 5–10 μg of 6His-hFADS2 (MW 56,537.32 g/mol, k_{cat} 6.9 min^{-1}) [30, 37]. They correspond to 0.085–0.170 nmol protein, which are able to form at maximum 0.59–1.18 nmol FMN/min [30].
8. If you work with isolated mitochondria add sucrose to the buffer to obtain isosmotic medium.
9. To determine the adequate volume of neutralizing solution, add 50 μL of 10% $HClO_4$ to 100 μL of 50 mM Tris–HCl pH 7.5 (the buffer of reaction mixture), vortex and centrifuge, then add 50 μL of neutralizing solution, vortex, centrifuge, and measure pH; if pH value is lower than 7, repeat the procedure increasing the volume of neutralizing solution till pH is 7. Generally, no more than 70 μL of neutralizing solution is needed. The pH measurement can be done using pH-indicator paper.

Acknowledgments

These protocols were used in studies supported by "Progetto Competitivo" funds "Effetto di mutazioni di FLAD1 e di alterazioni dell'omeostasi delle flavine sullo stato redox e sulla biogenesi mitocondriale: uno studio integrato su fibroblasti umani"

University of Bari "A. Moro" to (M.B.) and by a grant from Cure RTD (Year 2019) http://curertd.org/news/new/ "Alterations of Rf transport and metabolism in Brown-Vialetto-Van-Laere Syndrome (BVVLS)" (to M.B.).
The technical assistance of Vito Giannoccaro (University of Bari "A. Moro") is gratefully acknowledged.

References

1. McCormick DB (1989) Two interconnected B vitamins: riboflavin and pyridoxine. Physiol Rev 69(4):1170–1198. https://doi.org/10.1152/physrev.1989.69.4.1170
2. Massey V (2000) The chemical and biological versatility of riboflavin. Biochem Soc Trans 28 (4):283–296
3. Massey V (1995) Introduction: flavoprotein structure and mechanism. FASEB J 9 (7):473–475. https://doi.org/10.1096/fasebj.9.7.7737454
4. Merrill AH Jr, Lambeth JD, Edmondson DE, McCormick DB (1981) Formation and mode of action of flavoproteins. Annu Rev Nutr 1:281–317. https://doi.org/10.1146/annurev.nu.01.070181.001433
5. Depeint F, Bruce WR, Shangari N, Mehta R, O'Brien PJ (2006) Mitochondrial function and toxicity: role of the B vitamin family on mitochondrial energy metabolism. Chem Biol Interact 163(1–2):94–112. https://doi.org/10.1016/j.cbi.2006.04.014
6. Barile M, Giancaspero TA, Brizio C, Panebianco C, Indiveri C, Galluccio M, Vergani L, Eberini I, Gianazza E (2013) Biosynthesis of flavin cofactors in man: implications in health and disease. Curr Pharm Des 19(14):2649–2675
7. O'Callaghan B, Bosch AM, Houlden H (2019) An update on the genetics, clinical presentation, and pathomechanisms of human riboflavin transporter deficiency. J Inherit Metab Dis. https://doi.org/10.1002/jimd.12053
8. Olsen RKJ, Konarikova E, Giancaspero TA, Mosegaard S, Boczonadi V, Matakovic L, Veauville-Merllie A, Terrile C, Schwarzmayr T, Haack TB, Auranen M, Leone P, Galluccio M, Imbard A, Gutierrez-Rios P, Palmfeldt J, Graf E, Vianey-Saban C, Oppenheim M, Schiff M, Pichard S, Rigal O, Pyle A, Chinnery PF, Konstantopoulou V, Moslinger D, Feichtinger RG, Talim B, Topaloglu H, Coskun T, Gucer S, Botta A, Pegoraro E, Malena A, Vergani L, Mazza D, Zollino M, Ghezzi D, Acquaviva C, Tyni T, Boneh A, Meitinger T, Strom TM, Gregersen N, Mayr JA, Horvath R, Barile M, Prokisch H (2016) Riboflavin-responsive and -non-responsive mutations in FAD synthase cause multiple Acyl-CoA dehydrogenase and combined respiratory-chain deficiency. Am J Hum Genet 98(6):1130–1145. https://doi.org/10.1016/j.ajhg.2016.04.006
9. Balasubramaniam S, Christodoulou J, Rahman S (2019) Disorders of riboflavin metabolism. J Inherit Metab Dis 42(4):608–619. https://doi.org/10.1002/jimd.12058
10. Jaeger B, Bosch AM (2016) Clinical presentation and outcome of riboflavin transporter deficiency: mini review after five years of experience. J Inherit Metab Dis 39 (4):559–564. https://doi.org/10.1007/s10545-016-9924-2
11. Barile M, Giancaspero TA, Leone P, Galluccio M, Indiveri C (2016) Riboflavin transport and metabolism in humans. J Inherit Metab Dis 39(4):545–557. https://doi.org/10.1007/s10545-016-9950-0
12. Powers HJ (2003) Riboflavin (vitamin B-2) and health. Am J Clin Nutr 77 (6):1352–1360. https://doi.org/10.1093/ajcn/77.6.1352
13. Suwannasom N, Kao I, Pruss A, Georgieva R, Baumler H (2020) Riboflavin: the health benefits of a forgotten natural vitamin. Int J Mol Sci 21(3). https://doi.org/10.3390/ijms21030950
14. Tutino V, Defrancesco ML, Tolomeo M, V DEN, Lorusso D, Paleni D, Caruso MG, Notarnicola M, Barile M (2018) The expression of riboflavin transporters in human colorectal cancer. Anticancer Res 38 (5):2659–2667. https://doi.org/10.21873/anticanres.12508
15. Yonezawa A, Inui K (2013) Novel riboflavin transporter family RFVT/SLC52: identification, nomenclature, functional characterization and genetic diseases of RFVT/SLC52. Mol Asp Med 34(2–3):693–701. https://doi.org/10.1016/j.mam.2012.07.014
16. Console L, Tolomeo M, Colella M, Barile M, Indiveri C (2019) Reconstitution in proteoliposomes of the recombinant human riboflavin

transporter 2 (SLC52A2) overexpressed in E. coli. Int J Mol Sci 20(18). https://doi.org/10.3390/ijms20184416

17. Barile M, Brizio C, Valenti D, De Virgilio C, Passarella S (2000) The riboflavin/FAD cycle in rat liver mitochondria. Eur J Biochem 267 (15):4888–4900
18. McCormick DB (2000) A trail of research on cofactors: an odyssey with friends. J Nutr 130 (2S Suppl):323S–330S. https://doi.org/10.1093/jn/130.2.323S
19. Yatsyshyn VY, Ishchuk OP, Voronovsky AY, Fedorovych DV, Sibirny AA (2009) Production of flavin mononucleotide by metabolically engineered yeast Candida famata. Metab Eng 11(3):163–167. https://doi.org/10.1016/j.ymben.2009.01.004
20. Frago S, Martinez-Julvez M, Serrano A, Medina M (2008) Structural analysis of FAD synthetase from Corynebacterium ammoniagenes. BMC Microbiol 8:160. https://doi.org/10.1186/1471-2180-8-160
21. Herguedas B, Martinez-Julvez M, Frago S, Medina M, Hermoso JA (2010) Oligomeric state in the crystal structure of modular FAD synthetase provides insights into its sequential catalysis in prokaryotes. J Mol Biol 400 (2):218–230. https://doi.org/10.1016/j.jmb.2010.05.018
22. Bacher A, Eberhardt S, Fischer M, Kis K, Richter G (2000) Biosynthesis of vitamin b2 (riboflavin). Annu Rev Nutr 20:153–167. https://doi.org/10.1146/annurev.nutr.20.1.153
23. Herguedas B, Lans I, Sebastian M, Hermoso JA, Martinez-Julvez M, Medina M (2015) Structural insights into the synthesis of FMN in prokaryotic organisms. Acta Crystallogr D Biol Crystallogr 71(Pt 12):2526–2542. https://doi.org/10.1107/S1399004715019641
24. Brizio C, Galluccio M, Wait R, Torchetti EM, Bafunno V, Accardi R, Gianazza E, Indiveri C, Barile M (2006) Over-expression in Escherichia coli and characterization of two recombinant isoforms of human FAD synthetase. Biochem Biophys Res Commun 344 (3):1008–1016. https://doi.org/10.1016/j.bbrc.2006.04.003
25. Torchetti EM, Bonomi F, Galluccio M, Gianazza E, Giancaspero TA, Iametti S, Indiveri C, Barile M (2011) Human FAD synthase (isoform 2): a component of the machinery that delivers FAD to apo-flavoproteins. FEBS J 278 (22):4434–4449. https://doi.org/10.1111/j.1742-4658.2011.08368.x
26. Said HM (2011) Intestinal absorption of water-soluble vitamins in health and disease. Biochem J 437(3):357–372. https://doi.org/10.1042/BJ20110326
27. Subramanian VS, Ghosal A, Kapadia R, Nabokina SM, Said HM (2015) Molecular mechanisms mediating the adaptive regulation of intestinal riboflavin uptake process. PLoS One 10(6):e0131698. https://doi.org/10.1371/journal.pone.0131698
28. Barile M, Brizio C, De Virgilio C, Delfine S, Quagliariello E, Passarella S (1997) Flavin adenine dinucleotide and flavin mononucleotide metabolism in rat liver—the occurrence of FAD pyrophosphatase and FMN phosphohydrolase in isolated mitochondria. Eur J Biochem 249(3):777–785
29. Vergani L, Barile M, Angelini C, Burlina AB, Nijtmans L, Freda MP, Brizio C, Zerbetto E, Dabbeni-Sala F (1999) Riboflavin therapy. Biochemical heterogeneity in two adult lipid storage myopathies. Brain 122(Pt 12):2401–2411
30. Leone P, Galluccio M, Brizio C, Barbiroli A, Iametti S, Indiveri C, Barile M (2019) The hidden side of the human FAD synthase 2. Int J Biol Macromol 138:986–995. https://doi.org/10.1016/j.ijbiomac.2019.07.138
31. Chen CH, Chen SC (1988) Evidence of acid phosphatase in the cytoplasm as a distinct entity. Arch Biochem Biophys 262 (2):427–438. https://doi.org/10.1016/0003-9861(88)90394-3
32. Giancaspero TA, Busco G, Panebianco C, Carmone C, Miccolis A, Liuzzi GM, Colella M, Barile M (2013) FAD synthesis and degradation in the nucleus create a local flavin cofactor pool. J Biol Chem 288 (40):29069–29080. https://doi.org/10.1074/jbc.M113.500066
33. Shin HJ, Mego JL (1988) A rat liver lysosomal membrane flavin-adenine dinucleotide phosphohydrolase: purification and characterization. Arch Biochem Biophys 267(1):95–103. https://doi.org/10.1016/0003-9861(88)90012-4
34. Kim JK, Ezaki J, Himeno M, Kato K, Kim S (1993) Purification and characterization of flavine-adenine dinucleotide phosphohydrolase from rat liver lysosomal membranes. J Biochem 114(1):126–131. https://doi.org/10.1093/oxfordjournals.jbchem.a124127
35. McLennan AG (2006) The Nudix hydrolase superfamily. Cell Mol Life Sci 63(2):123–143. https://doi.org/10.1007/s00018-005-5386-7

36. Abdelraheim SR, Spiller DG, McLennan AG (2017) Mouse Nudt13 is a mitochondrial nudix hydrolase with NAD(P)H pyrophosphohydrolase activity. Protein J 36(5):425–432. https://doi.org/10.1007/s10930-017-9734-x
37. Giancaspero TA, Galluccio M, Miccolis A, Leone P, Eberini I, Iametti S, Indiveri C, Barile M (2015) Human FAD synthase is a bi-functional enzyme with a FAD hydrolase activity in the molybdopterin binding domain. Biochem Biophys Res Commun 465 (3):443–449. https://doi.org/10.1016/j.bbrc.2015.08.035
38. Carreras-Puigvert J, Zitnik M, Jemth AS, Carter M, Unterlass JE, Hallstrom B, Loseva O, Karem Z, Calderon-Montano JM, Lindskog C, Edqvist PH, Matuszewski DJ, Ait Blal H, Berntsson RPA, Haggblad M, Martens U, Studham M, Lundgren B, Wahlby C, Sonnhammer ELL, Lundberg E, Stenmark P, Zupan B, Helleday T (2017) A comprehensive structural, biochemical and biological profiling of the human NUDIX hydrolase family. Nat Commun 8(1):1541. https://doi.org/10.1038/s41467-017-01642-w
39. Gibson QH, Massey V, Atherton NM (1962) The nature of compounds present in mixtures of oxidized and reduced flavin mononucleotides. Biochem J 85:369–383. https://doi.org/10.1042/bj0850369
40. Murataliev MB (1999) Application of electron spin resonance (ESR) for detection and characterization of flavoprotein semiquinones. Methods Mol Biol 131:97–110. https://doi.org/10.1385/1-59259-266-X:97
41. King TE, Howard RL, Wilson DF, Li JC (1962) The partition of flavins in the heart muscle preparation and heart mitochondria. J Biol Chem 237:2941–2946
42. Ghisla S, Massey V, Lhoste JM, Mayhew SG (1974) Fluorescence and optical characteristics of reduced flavines and flavoproteins. Biochemistry 13(3):589–597. https://doi.org/10.1021/bi00700a029
43. Brizio C, Brandsch R, Douka M, Wait R, Barile M (2008) The purified recombinant precursor of rat mitochondrial dimethylglycine dehydrogenase binds FAD via an autocatalytic reaction. Int J Biol Macromol 42(5):455–462. https://doi.org/10.1016/j.ijbiomac.2008.03.001
44. Singer TP, Edmondson DE (1978) Flavoproteins (overview). Methods Enzymol 53:397–418. https://doi.org/10.1016/s0076-6879(78)53045-0
45. Caldinelli L, Iametti S, Barbiroli A, Bonomi F, Piubelli L, Ferranti P, Picariello G, Pilone MS, Pollegioni L (2004) Unfolding intermediate in the peroxisomal flavoprotein D-amino acid oxidase. J Biol Chem 279(27):28426–28434. https://doi.org/10.1074/jbc.M403489200
46. Barile M, Passarella S, Bertoldi A, Quagliariello E (1993) Flavin adenine dinucleotide synthesis in isolated rat liver mitochondria caused by imported flavin mononucleotide. Arch Biochem Biophys 305(2):442–447. https://doi.org/10.1006/abbi.1993.1444
47. Giancaspero TA, Colella M, Brizio C, Difonzo G, Fiorino GM, Leone P, Brandsch R, Bonomi F, Iametti S, Barile M (2015) Remaining challenges in cellular flavin cofactor homeostasis and flavoprotein biogenesis. Front Chem 3:30. https://doi.org/10.3389/fchem.2015.00030
48. Leone P, Galluccio M, Barbiroli A, Eberini I, Tolomeo M, Vrenna F, Gianazza E, Iametti S, Bonomi F, Indiveri C, Barile M (2018) Bacterial production, characterization and protein modeling of a novel monofuctional isoform of FAD synthase in humans: an emergency protein? Molecules 23(1). https://doi.org/10.3390/molecules23010116
49. Leone P, Galluccio M, Quarta S, Anoz-Carbonell E, Medina M, Indiveri C, Barile M (2019) Mutation of aspartate 238 in FAD synthase isoform 6 increases the specific activity by weakening the FAD binding. Int J Mol Sci 20(24). https://doi.org/10.3390/ijms20246203
50. Giancaspero TA, Dipalo E, Miccolis A, Boles E, Caselle M, Barile M (2014) Alteration of ROS homeostasis and decreased lifespan in S. cerevisiae elicited by deletion of the mitochondrial translocator FLX1. Biomed Res Int 101286:2014. https://doi.org/10.1155/2014/101286
51. Liuzzi VC, Giancaspero TA, Gianazza E, Banfi C, Barile M, De Giorgi C (2012) Silencing of FAD synthase gene in Caenorhabditis elegans upsets protein homeostasis and impacts on complex behavioral patterns. Biochim Biophys Acta 1820(4):521–531. https://doi.org/10.1016/j.bbagen.2012.01.012
52. Giancaspero TA, Locato V, de Pinto MC, De Gara L, Barile M (2009) The occurrence of riboflavin kinase and FAD synthetase ensures FAD synthesis in tobacco mitochondria and maintenance of cellular redox status. FEBS J 276(1):219–231. https://doi.org/10.1111/j.1742-4658.2008.06775.x
53. Giancaspero TA, Locato V, Barile M (2013) A regulatory role of NAD redox status on flavin cofactor homeostasis in S. cerevisiae mitochondria. Oxidative Med Cell Longev

2013:612784. https://doi.org/10.1155/2013/612784
54. Monteverde DR, Sylvan JB, Suffridge C, Baronas JJ, Fichot E, Fuhrman J, Berelson W, Sanudo-Wilhelmy SA (2018) Distribution of extracellular flavins in a coastal marine basin and their relationship to redox gradients and microbial community members. Environ Sci Technol 52(21):12265–12274. https://doi.org/10.1021/acs.est.8b02822
55. Serrano A, Ferreira P, Martinez-Julvez M, Medina M (2013) The prokaryotic FAD synthetase family: a potential drug target. Curr Pharm Des 19(14):2637–2648. https://doi.org/10.2174/1381612811319140013
56. Pedrolli DB, Nakanishi S, Barile M, Mansurova M, Carmona EC, Lux A, Gartner W, Mack M (2011) The antibiotics roseoflavin and 8-demethyl-8-amino-riboflavin from Streptomyces davawensis are metabolized by human flavokinase and human FAD synthetase. Biochem Pharmacol 82(12):1853–1859. https://doi.org/10.1016/j.bcp.2011.08.029
57. Ryder B, Tolomeo M, Nochi Z, Colella M, Barile M, Olsen RK, Inbar-Feigenberg M (2019) A novel truncating FLAD1 variant, causing multiple acyl-CoA dehydrogenase deficiency (MADD) in an 8-year-old boy. JIMD Rep. 45:37–44. https://doi.org/10.1007/8904_2018_139
58. Pallotta ML, Brizio C, Fratianni A, De Virgilio C, Barile M, Passarella S (1998) Saccharomyces cerevisiae mitochondria can synthesise FMN and FAD from externally added riboflavin and export them to the extramitochondrial phase. FEBS Lett 428 (3):245–249. https://doi.org/10.1016/s0014-5793(98)00544-4
59. Galluccio M, Brizio C, Torchetti EM, Ferranti P, Gianazza E, Indiveri C, Barile M (2007) Over-expression in Escherichia coli, purification and characterization of isoform 2 of human FAD synthetase. Protein Expr Purif 52(1):175–181. https://doi.org/10.1016/j.pep.2006.09.002
60. Miccolis A, Galluccio M, Giancaspero TA, Indiveri C, Barile M (2012) Bacterial over-expression and purification of the 3′phosphoadenosine 5′phosphosulfate (PAPS) reductase domain of human FAD synthase: functional characterization and homology modeling. Int J Mol Sci 13 (12):16880–16898. https://doi.org/10.3390/ijms131216880
61. Miccolis A, Galluccio M, Nitride C, Giancaspero TA, Ferranti P, Iametti S, Indiveri C, Bonomi F, Barile M (2014) Significance of redox-active cysteines in human FAD synthase isoform 2. Biochim Biophys Acta 1844 (12):2086–2095. https://doi.org/10.1016/j.bbapap.2014.08.005

Part III

Looking at Flavin Cofactors in Flavoenzymes

Chapter 8

Redox Titration of Flavoproteins: An Overview

Francesco Bonomi and Stefania Iametti

Abstract

Redox titration of flavoproteins allows to detect and analyze (1) the determinants of the stabilization of individual redox forms of the flavin by the protein; (2) the binding of the redox-active cofactor to the protein; (3) the effects of other components of the systems (such as micro- or macromolecular interactors) on parameters 1 and 2; (4) the pattern of electron flow to and from the flavin cofactor to other redox-active chemical species, including those present in the protein itself or in its physiological partners. This overview presents and discusses the fundamentals of the methodological approaches most commonly used for these purposes, and illustrates how data may be obtained in a reliable way, and how they can be read and interpreted.

Key words Flavin cofactors, Flavin semiquinone, Redox potential, Potentiometric titration, Electron paramagnetic resonance, Redox indicators

1 Introduction

1.1 Significance of Redox Chemistry in Flavoproteins

There are several elements of complexity that need to be taken into account when considering the redox behavior of flavin cofactors in proteins. In addition to the well-known ability of flavin cofactors to undergo both mono- and bielectronic transitions [1, 2], one has to consider the effects of the protein itself (and of the surrounding chemical environment) as for stabilizing a given redox status of the cofactor [3–6]. Also, protein-related effects are often complicated by the sensitivity of the protein conformation to the presence of effectors and to the effects ensuing from binding of substrate(s) and of inhibitors [7–12]. All these events are of fundamental relevance, and are significant in all cases where the so-called electron bifurcation may occur, in electron transfer chains where electrons may be transferred from the reduced species to vastly different (and often physically distant) redox partners [13], or when electron flow to and from flavin cofactors is associated with other functions, such as proton translocation [14].

Maria Barile (ed.), *Flavins and Flavoproteins: Methods and Protocols*, Methods in Molecular Biology, vol. 2280,
https://doi.org/10.1007/978-1-0716-1286-6_8,

In turn, redox changes in the flavin component may affect the protein structure itself. Besides modulating the affinity for substrates and products [15, 16] and those with redox partners [17], redox events in flavoproteins are emerging as a way of sensing and amplifying redox changes of remarkable physiological relevance through the ensuing conformational changes [18, 19]. Indeed, the redox-dependent (or light-excited, or again energy-dependent) changes of flavin and neighboring residues in proteins act as molecular "switches" that "turn on" various conformational changes in other proteins [20, 21]. In some cases, this requires some collaboration by proteins that sense intracellular concentrations of other bioactive species, as exemplified by the control of Nitric Oxide Synthase by Ca^{2+}/calmodulin [22].

Yet another level of complexity is evident when the dissociation constant of the flavin from the protein is high enough to result in a significant presence of free flavin species in equilibrium with protein-bound flavin species [5, 23]. In this frame, it may be relevant to note that different redox forms of the flavin may have different affinity for binding to the protein. Variable affinity issues, along the very noticeable differences in the redox behavior of free and protein-bound flavin cofactors, introduce further levels of complexity in analysis of these systems [5, 23, 24].

Finally, several flavoproteins contain additional redox cofactors, that may be bound (sometimes covalently, but—most frequently—noncovalently) to the same polypeptide chain that harbors the flavin itself, or be present in the active species as separate protein entities that form complexes of variable stability with the flavoprotein portion of the system, or just derive from association of the protein into oligomeric forms [25]. These additional components may be various flavin derivatives [14, 22, 26, 27], other organic cofactors such as dihydrobiopterin [28], along with heme iron [8, 29] or with inorganic cofactors such as molybdenum [28, 30], iron–sulfur clusters [10, 16, 31–35] or hemerythrin-like oxo-bridged diiron centers [36]. In all these cases, electrons entering the system will partition among components of the system according to the redox properties of each of the individual components as determined by the appropriate interactions with the host protein(s) and by the conformational considerations outlined above. Also in this case, the binding of substrate(s), the presence of effectors, or interaction with other protein components may impact on the redox properties of individual electron-transferring species, resulting in an additional level of complexity, but also in the possibility of addressing the possible physiological relevance of these measurements.

1.2 Fundamentals of Redox Chemistry

Redox chemistry is governed by the well-known Ernst equation [37]. The most common form of the equation used in biological systems (where the activity of water is considered unvariable and the pH is assumed to remain constant at 7.0) is.

$$E' = E'_0 + (RT/nF)\ \ln\ ([ox]/[red])$$

Thus, the ratio between oxidized and reduced species at a given value of potential may be calculated from the simple exponential,

$$([ox]/[red]) = e^{(E'-E'_0)(nF/RT)}$$

In the case of a flavin system, the possible redox events can be simplified into a series of monoelectronic redox transitions, each with its own E'_0,

$$\overset{E'_{0Fox/Fsq}}{F \leftrightarrows} F_{sq} \overset{E'_{0Fsq/FH2}}{\leftrightarrows} FH2$$

leading to the following relationships among the redox potential of the system and the ratio among involved species:

$$([F]/[Fsq]) = e^{\left(E'-E'_{0\ F/FSq}\right)\ (F/RT)}$$

$$([Fsq]/[FH2]) = e^{\left(E'-E'_{0\ Fsq/FH2}\right)\ (F/RT)}$$

It has to be noted that, in the equilibria reported above, E'_0 for the bielectronic transition between the fully oxidized and fully reduced form of the flavin ($E'_{0\ F/FH2}$) is numerically given by the average between those of the two monoelectronic half cells ($E'_{0\ F/FH2} = 1/2\ (E'_{0\ Fsq/FH2} + E'_{0\ F/FSq})$). Whenever formation or stabilization of the semiquinone form of the flavin is not allowed by the protein, and therefore escapes detection, the redox behavior of the system can be further simplified into.

$$([F]/[FH2]) = e^{\left(E'-E'_{0\ F/FH2}\right)(2F/RT)}$$

Based on these simple equations, the fraction of molecules being in a given redox state with respect to the total concentration of the cofactor (assuming $[F_{tot}] = [F_{ox}] + [F_{sq}] + [Fr_{ed}]$) may be easily inferred and analyzed by means of standard algorithms typically embedded in most of the graphical and statistical routines of programs commonly used for data analysis.

As shown by the simulated curves in the two panels of Fig. 1, formation of the one-electron semiquinone intermediate relates to the distance between the redox potentials of individual half-cells. The higher the difference between E'_0 (F_{sq}/Fr_{ed}) and E'_0 (F_{ox}/F_{sq}) the higher the amount of semiquinone that may be formed at

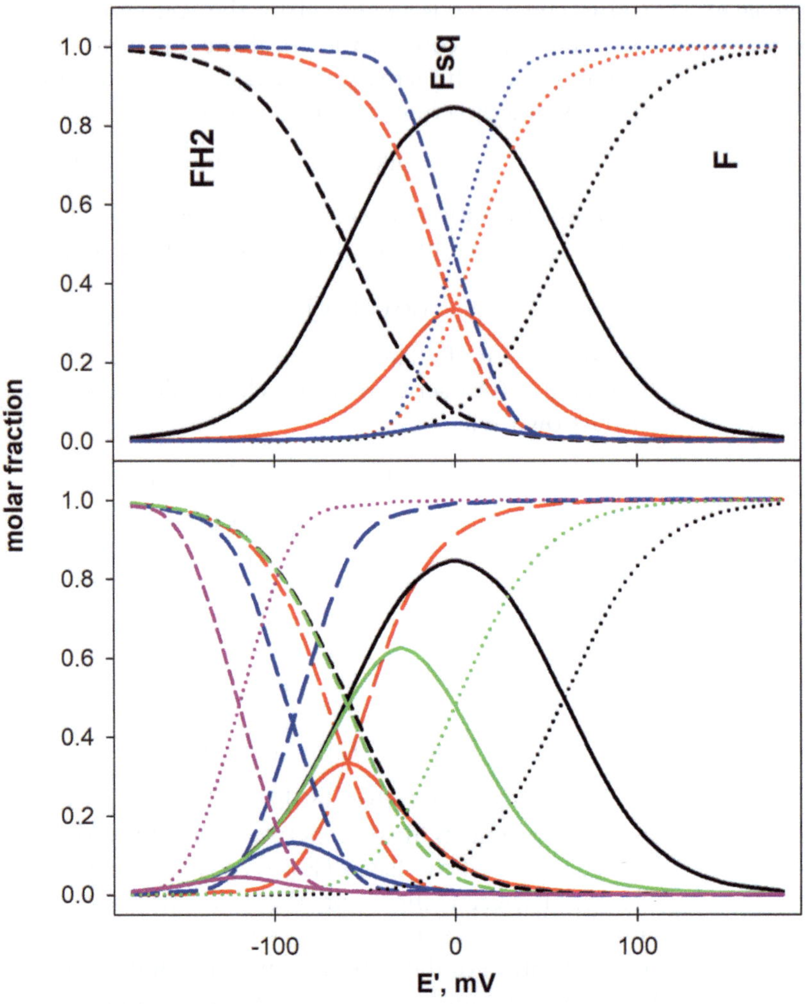

Fig. 1 A simulation of the distribution of redox forms of flavin as dependent on the applied redox potential and on the E'_0 of individual redox events. *Top*. The E'_0 value for the bielectronic F/FH2 transition was assumed as constant at 0 mV in all cases. Values for the Fsq/Fred and Fox/Fsq monoelectronic half reactions were modified in 60 mV increments/decrements as follows: black, $E'_{0\ F/Fsq} = +60$ mV, $E'_{0\ Fsq/FH2} = -60$ mV; red, $E'_{0\ F/Fsq} = E'_{0\ Fsq/FH2} = 0$ mV; blue, $E'_{0\ F/Fsq} = -60$ mV, $E'_{0\ Fsq/FH2} = +60$ mV. *Bottom*. The E'_0 value for the Fsq/FH2 monoelectronic half reaction was kept constant at -60 mV, whereas the E'_0 value for the F/Fsq monoelectronic half reaction was decreased in 60 mV increments as follows: black, +60 mV ($E'_{0\ F/FH2} = 0$ mV); green, 0 mV ($E'_{0\ F/FH2} = -30$ mV); red, -60 mV ($E'_{0\ F/FH2} = -60$ mV); blue, -120 mV ($E'_{0\ F/FH2} = -90$ mV); magenta, -180 mV ($E'_{0\ F/FH2} = -120$ mV)

intermediate potential values. However, in all cases, the actual formation of the semiquinone form(s) of the flavin is dictated by the extent to which the products of monoelectronic transitions are stabilized by suitable interactions with the protein.

2 Methodologies

2.1 Titration of Flavoproteins with Reductants

Under this subheading are comprised all the procedures in which a strong reductant (most commonly, sodium dithionite, $Na_2S_2O_4$) is used to supply electrons to redox components in a flavoprotein under conditions where no side reactions may occur (i.e., under strict anaerobiosis). In the absence of oxygen, the dithionite anion splits into two bisulfite anions, with simultaneous release of two protons and two electrons, whereas the presence of oxygen makes dithionite ineffective as a source of "reducing power," as exemplified by the reactions reported below:

$$S_2O_4{}^{2-} + 2\ H_2O \rightarrow 2\ HSO_3{}^- + 2\ e^- + 2\ H^+$$

$$S_2O_4{}^{2-} + H_2O + O_2 \rightarrow HSO_3{}^- + HSO_4{}^-$$

At physiological pH values the redox potential of dithionite is low enough (−660 mV vs NHE at pH 7.0) to ensure that virtually all reducible species even in the most complex protein will be reduced. Also advantageous from a practical standpoint is that reduction by dithionite of the most common organic and inorganic cofactors in proteins occurs in most cases on a very short (sec to msec) time scale. There are exceptions, of course, to these general rules, but addressing them is beyond the scope of this presentation. It has to be noted that sulfite (but not bisulfite, from which sulfite originates at relatively high pH values) may form—at high concentrations—adducts at the N5 position of the flavin isoalloxazine ring in a number of flavoproteins of different classes (*see* for instance [32]). Thus, the use of excessive concentrations of this reductant should be avoided.

Also of practical interest is that this methodology does not imply the use of additional components in the system (such as the redox mediators discussed in a later section). The requirement for strict anaerobiosis is usually met by thorough degassing of the sample (and of the titrant!) through repeated cycles of evacuation and through replacement of air with oxygen-free Ar. Ar has a higher density than O_2 or N_2, and thus "blankets" samples to be kept anaerobic in a very convenient way. Commercial Ar is much more oxygen-free than commercial N_2 (as a consequence of the very small difference in boiling points between O_2 and N_2). Incidentally, this latest feature also helps in prolonging the operating life of the catalysts used for removal of any traces of residual oxygen in standard vacuum lines and in anaerobic chambers.

Once samples are made anaerobic, the effects of the addition of reductants may then be assessed by suitable spectroscopic techniques, such as spectrophotometry or electron paramagnetic resonance (EPR) also known as electron spin resonance (ESR) [8, 10, 16, 30–36]. Other types of spectrometry (circular dichroism [38],

Raman [26], and NMR) have also been exploited to analyze some specific cases, also in combination with one or both the techniques mentioned above [30].

Spectrophotometry is possibly the easiest and most popular choice for this type of approach, in view of the large difference in the absorption coefficients of oxidized and reduced flavin at convenient wavelengths (typically, 445 nm for oxidized flavin, with an extinction coefficient around 12,500 M^{-1} cm^{-1}), and of the neutral and anionic forms of the flavin semiquinone, that have absorption bands centered at 540 nm for the "red" anionic form and at 620 nm for the "blue" neutral form. Please note that there are no reported extinction coefficients for either form of the semiquinone, also because of the possibility of pH-dependent conversion of the two forms (although each class of flavoproteins is reportedly able to stabilize only one of them). However, the extinction coefficient for the "red" anionic form is much higher than that of the "blue" neutral form. A convenient way of addressing the simultaneous presence of multiple species is to consider ratios of absorbance values measured at convenient wavelength pairs (e.g., 380/474 nm, or 400/460 nm). For the sake of simplicity, pH dependence of all the redox events involving flavins (reviewed in [2]) will not be addressed here.

There are several excellent reviews on these topics [2, 18, 19], and a number of milestone papers—most of them appearing in the accompanying list of references—have addressed these issues in great detail. Some of these reports also addressed changes ensuing from "capture" of free flavins by the appropriate apoprotein (either native or conveniently mutated), providing very detailed analyses of the thermodynamic and structural features of these systems [4, 5, 11, 23–25].

The typical outcome of this type of titration is presented in Fig. 2. Data used to construct these curves are the same used for the simulations already presented in Fig. 1.

In principle, similar approaches may be used to test what happens when electrons are supplied through the sequential addition of a redox-active substrate (or cosubstrate, such as NADH or NADPH as appropriate). NADH and NADPH are often used to supply electrons either alone or in the presence of catalytic amounts of accessory redox-active protein. Ferredoxin and ferredoxin: NADPH oxidoreductase are a common example of proteins being used—in trace amounts—for these purposes [5].

However, one should be aware that—when substrates are used as the redox-active species—the otherwise simple spectrophotometric analysis of the results may be quite often complicated by the possible formation of charge-transfer complexes. Unfortunately, in several cases, the spectrophotometric signals from charge

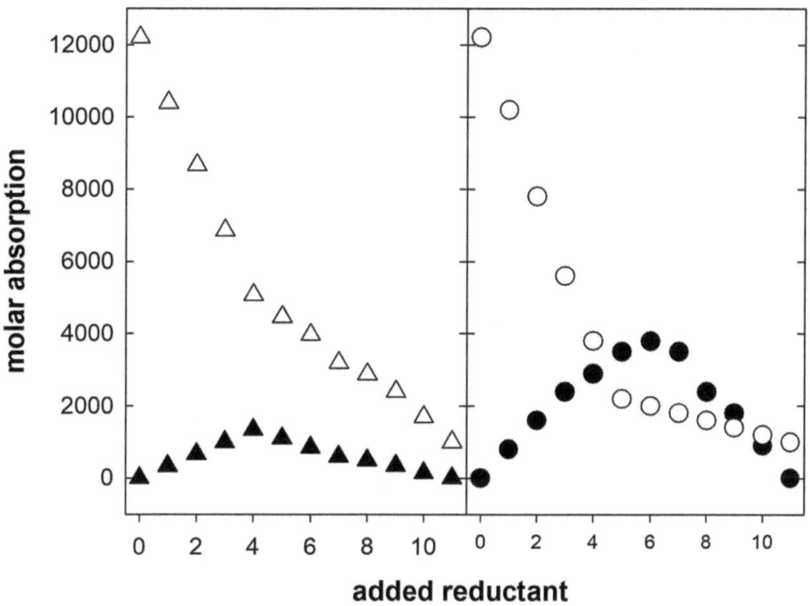

Fig. 2 A simulation of the absorbance changes ensuing from the progressive addition of a strong reductant to flavoproteins. Open symbols, absorbance at 445 nm, indicative of the concentration of fully oxidized flavin (assuming $A_{445} = 12{,}200$); full symbols, absorbance at 540 nm, indicative of concentration of the anionic (red) form of the flavin semiquinone (assuming $A_{540} \cong 4000$). *Left panel*: E'_0 values for the Fsq/FH_2 and F/Fsq monoelectronic reactions were assumed to be equal at +60 mV, resulting in a maximum semiquinone formation corresponding to 33% of the total flavin. *Right panel*: E'_0 values for the Fsq/FH_2 and F/Fsq monoelectronic reactions were assumed to have values of −60 ($E'_{0\ F/FH2}$) and +60 mV ($E'_{0\ F/Fsq}$), resulting in a maximum semiquinone formation corresponding to 92% of the total flavin

transfer complexes involving the flavin cofactor are overlapping those ensuing from flavin-related redox events [4, 9, 11, 33].

2.2 Assessing the Redox Potential of Individual Transitions

Redox potentials of individual transitions in simple and complex flavoproteins (as well as changes in their values ensuing from flavin–protein interactions and by modulation exerted by other interactors, as amply discussed above) may be assessed by simultaneous monitoring of spectroscopic changes and of the actual redox potential of the solution [39].

Methods available for poising the redox potential at a given (and controlled) are listed here below, and will be considered separately in subsequent subheadings:

(a) using appropriate concentrations of redox-active substrate and product,

(b) monitoring the redox status of the system by using appropriate working and reference electrodes in the presence of redox mediators,

(c) monitoring the redox status of the system through the use of redox indicators.

2.2.1 Using Redox-Active Substrate/Product Couples

In theory, setting the redox potential by providing appropriate concentration of components in the substrate/product half-cell does not suffer from the technical complexities related to maintenance of an anaerobic environment to ensure stability of the potential. However, as stated in the previous subheading, the use of substrate/product couple to set a given redox potential is somewhat limited, in view of the concurrent formation of charge transfer complex that pose practical limitation to interpretation of the data, in particular when using spectrophotometric techniques [9]. Indeed, such an approach has been used when other types of spectroscopies may be used to assess the redox state of the flavin. Both circular dichroism and EPR have been used with some success to analyze factors affecting redox parameters in enzymes that contain covalently bound flavin along with other nonflavin redox cofactors, such as purified mitochondrial succinate dehydrogenase [7, 10].

Please note that the uptake/release of protons is often associated with the substrate/product half-cell in these systems, so that attention must be paid to pH effects on the estimated value of E'_0 and/or when comparing results obtained at slightly different pH values. A shift in one pH unit results in a 30 mV change in redox potential for redox couples involving uptake/release of a single proton.

2.2.2 Potentiometric Titration Using Electrodes

Anaerobiosis is strictly required when using the other two methods listed above for assessing or poising the redox potential. Again, use of specialized glassware and thorough removal of oxygen is recommended, and preference should be given to oxygen-free Ar as the inert gas in place of the most commonly used N_2. In some cases, continuous removal of oxygen may be necessary, in particular when long-term stability of the potential is a requisite. In this frame, catalytic amounts of oxidases and a suitable excess of their substrates may help to scavenge any trace of oxygen that could permeate the system.

Independently of whether the potential is measured by potentiometric techniques or by monitoring the redox state of indicator dyes, several sources of electrons can be used. The most common procedures involve either the controlled addition of dithionite from a fairly concentrated aqueous solution (0.2–0.5 M) or the addition of deazaflavin (0.002–0.005 mM) and EDTA (1–5 mM) followed by exposure of the resulting mixture to a suitable source of visible light for photoreduction [40]. In the times when they were a common commodity, slide projectors were often moved from classrooms or seminar rooms to the lab for this very purpose.

In most cases, it has been common practice to promote full reduction of the system, and then to raise the potential by addition of small amounts of an oxidant. Aqueous solutions of potassium ferricyanide in the low mM range have been the most common oxidizing agent used as a "back titrant." Nonspecific oxidants such as ferricyanide have often been used in the presence of compounds (such as phenazine methosulfate) that reportedly facilitate electron transfer between the flavin and "generic" electron donors/acceptors. Other "redox mediators" could be used, including the redox indicator dyes discussed in what follows.

Please note that all these compounds should be used as sparingly as possible (typically, not exceeding the protein concentration—i.e., in the 0.01–0.02 mM range) in consideration of the possibility that they may interact nonspecifically with the protein under scrutiny, and affect its behavior in a nonspecific way. It should also be noted that some of the species are highly toxic, and proper caution should be exerted in their handling. This applies in particular to the whole family of viologen dyes, that are close relatives of the well-known herbicide Paraquat®.

Redox mediators are also useful when using electrodes in assessing the actual redox potential of the system, as they allow fast and complete equilibration between the components of the system and the measuring electrode (most typically, a Pt electrode). Common reference electrodes are based either on the Hg^0/Hg^+ couple (saturated calomel electrode, SCE, +244 mV vs the standard hydrogen electrode (SHE) at 25 °C) or the almost equivalent—but cheaper—Ag^0/Ag^+ electrode (+197 mV vs SHE at 25 °C, when using saturated KCl as the salt solution).

All these electrodes are available in microformat, making them suitable for use in custom-designed anaerobic cells, that may include a cuvette-shaped end suitable for direct spectrophotometric or CD measurements as well as ports suitable for either addition of titrants or for withdrawal of aliquots of the solution. Simpler systems based on standard nonanaerobic components may be used when a reliable anaerobic chamber is available for hosting all the required equipment: from the titration cell itself to the spectrophotometer used to monitor flavin-related changes.

When EPR spectroscopy is used to monitor formation of the flavin semiquinone and/or redox events that affect other components of the system (such as the various types of metal clusters present in complex flavoproteins), samples may be withdrawn from the titration cell by using a gastight syringe with a needle of length appropriate to transfer of the sample to the bottom of an EPR tube. The sample can then be frozen in a suitable cryogenic mixture (e.g., isopentane and dry ice). Once frozen, the samples can be stored until measurement in the frozen state with no special requirement for anaerobiosis.

If an anaerobic chamber is not available for carrying out the required manipulations, a second needle may be inserted into the EPR tube to be filled, thus providing a positive flow of Ar through the tube itself. This makes it possible to carry out transfer of the sample and the subsequent freezing steps under reasonably strict anaerobic conditions. Needless to say, proper handling of these (tricky) steps is most critical when working at extremely low potentials. Literature reports also describe a number of custom set-ups for EPR titrations (*see*, for instance, [30]), that may be used to circumvent the technical difficulties associated to these measurements and to improve data reproducibility.

In the frame of flavin chemistry at large, it is also worth noting that EPR spectroscopy may be useful in assessing the formation of either the ionic or the nonionic form of the flavin semiquinone even in the presence of signals (such as those originating from metal clusters or from charge-transfer complexes) that impair the detection of the spectral features of the semiquinone by other spectrometric techniques. The half-signal linewidth of the EPR signal of the flavin semiquinone in its anionic (red) form is reportedly sensibly narrower (1.4–1.6 mT, equivalent to 14–15 G when referring to units used in the past) than that of the neutral (blue) form of the same species (1.9–2.1 mT, or 19–22 G).

Figure 3 (redrawn from a "vintage" report from one of the Authors [10]) shows how these approaches have been used to assess the redox potential of transitions involving the covalently bound flavin and of one of the multiple inorganic Fe-S cofactors in mitochondrial succinate dehydrogenase [31]. These same data also show how the redox behavior of individual components of the system (and the overall activity of the enzyme, as a consequence [7]) may be modulated through conformational changes ensuing from the binding of physiological effectors, and highlight the significance of the redox measurements and of the approaches reported here as for understanding their nature and their amplitude, thus providing a clue as for their molecular determinants. Notably, very similar approaches have been used much more recently to address the issue of substrate uptake and product release in a closely related enzyme, and have led to a sub-set of investigations on the relationship between the redox state of the flavin and the ability of flavoproteins to interact with substrates and products [16].

2.2.3 Potentiometric Titration with Redox Indicators

As anticipated, most of the practical complications related to the use of electrodes for assessing the redox potential of the system could be avoided by using redox indicators. This approach offers the fundamental advantage of being suitable for relatively simple spectrophotometric measurements, that can be miniaturized whenever the amount of sample is a matter of concern [41].

A necessarily partial list of the most commonly used redox indicators is given in Table 1, along with their values of E'_0 at pH 7.0. Please note that all these values are increasing as pH is decreased (+30 mV or +60 mV for each pH unit, depending of the

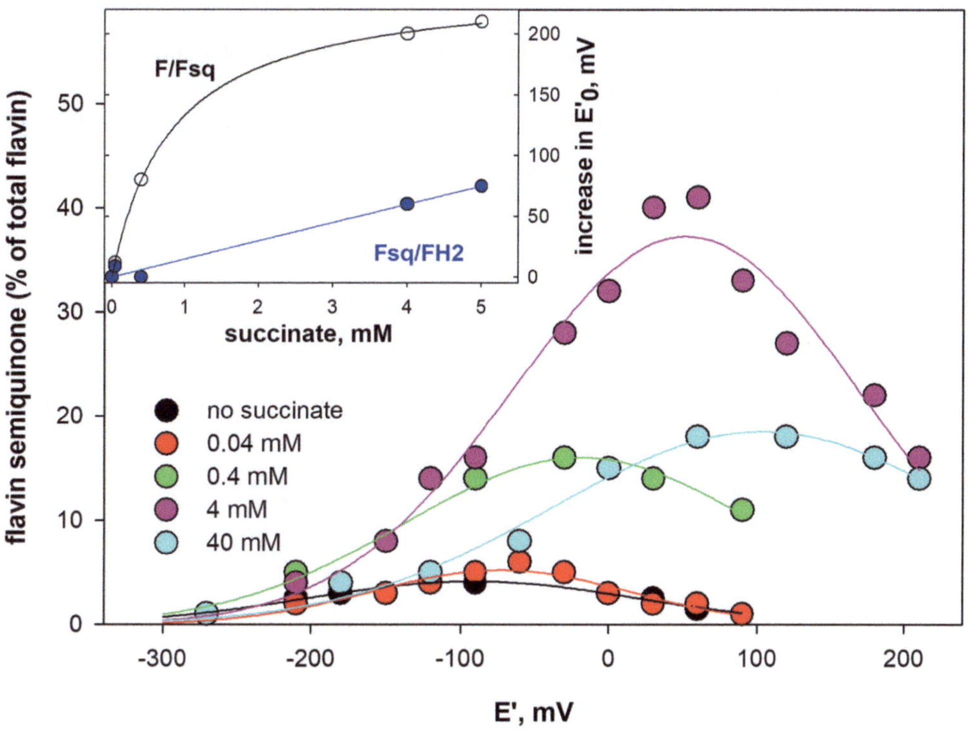

Fig. 3 Effect of added succinate on the redox parameters of the covalently bound flavin in mitochondrial succinate dehydrogenase. Potentiometric titration of the intensity of the $g = 2.00$ EPR signal of the flavin semiquinone in the presence of the given concentrations of succinate. The amount of semiquinone formed at each given potential was assessed by double integration of the signal and comparison with a $CuSO_4$ standard. Inset. Changes in E'_0 for individual Flavin monoelectronic half-cells as a function of the effector concentration. Original data in either panel are from [5]

number of involved protons). The concentration of each redox form of the indicator may be easily assessed spectrophotometrically (all these dyes have extinction coefficients in excess of 20,000 in the visible range), and are related to the actual potential of the system (E′) through the simple equation:

$$E' = E'_0 + RT/(nF) * \ln\ ([ox]/[red])$$

Redox indicators belonging to the indigo sulfonates and the anthraquinone sulfonates families are most commonly used in spectrophotometric measurements on flavoproteins, due to the fact that: (1) their redox potentials are in a range relevant to the flavin redox transitions; (2) maxima and isosbestic points in their absorption spectra minimize the chance of interference with flavin-related spectral features; (3) their highly anionic nature minimizes changes of nonspecific binding of the dye to the protein under scrutiny at physiological pH values. Please note that all the indicators listed in Table 1 undergo bielectronic processes, with the only

Table 1
Main features of common redox indicators and mediators used in potentiometric titrations

Indicator	*n*	E'_0, mV	Color, oxidized	Color, reduced
2,6-Dichlorophenol-indophenol	2	+220	Blue	Colorless
o-Cresol indophenol	2	+190	Blue	Colorless
Thionine	2	+60	Violet	Colorless
Indigo tetrasulfonate	2	−50	Blue	Colorless
Indigo trisulfonate	2	−80	Blue	Colorless
2-Hydroxy-1,4-naphthoquinone	2	−145	Blue	Colorless
Indigo monosulfonate	2	−160	Blue	Colorless
Anthraquinone 2,6-disulfonate	2	−184	Blue	Colorless
Anthraquinone 2-sulfonate	2	−225	Blue	Colorless
Phenosafranin	2	−255	Red	Colorless
Safranin T	2	−276	Red-violet	Colorless
Neutral red	2	−330	Red	Colorless
Benzyl viologen	1	−359	Colorless	Purple-blue
2-Hydroxyethyl viologen	1	−408	Colorless	Purple-blue
Methyl viologen	1	−446	Colorless	Purple-blue

exception of those in the viologen family. In the latter case, the colored product formed upon monoelectronic reduction of the colorless indicator is a radical species.

Common practice in spectrophotometric measurements involves the preparation of a mixture of the protein and of the chosen indicator prior to any further manipulation. Ideally, each component of the mixture should be at concentrations in the 0.02–0.05 mM range. The resulting mixture is made anaerobic, and a spectrum of the fully oxidized system is taken in the visible range. Components of the mixture are then fully reduced, either chemically (by dithionite) or photochemically (by deazaflavin/EDTA). The absence of any color in the fully reduced mixture (if not using dyes in the viologens family) provides an immediate visual clue as for completion of electron uptake by all the components of the system. Spectra are then taken at various stages of the reoxidation process, and changes at wavelengths suitable for assessing the [ox]/[red] ratio of the indicator and of individual redox species are monitored and analyzed.

Figure 4 reports the simulated results of this type of titration for a flavin-containing system where indigo trisulfonate and phenosafranine were added as indicators, thus offering the possibility of

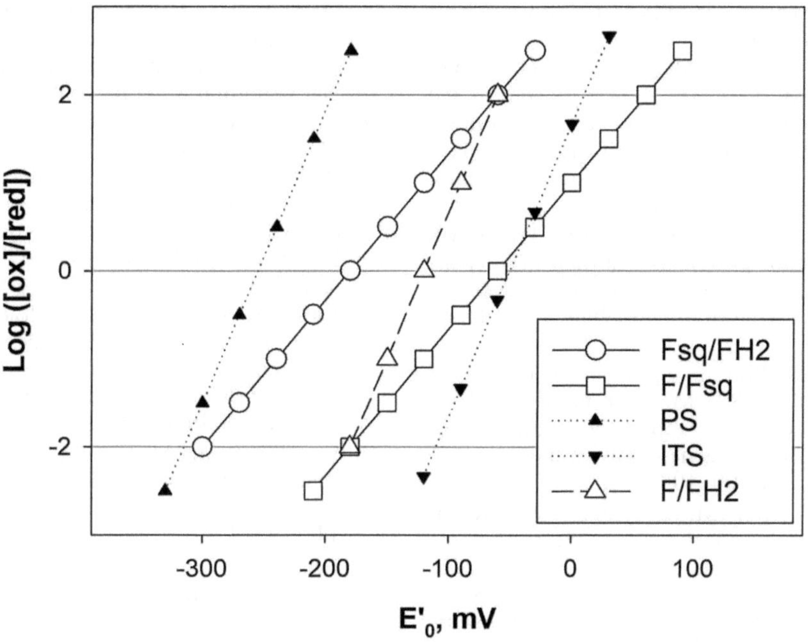

Fig. 4 Representative Nernst plot for the simulated potentiometric titration of flavoproteins as monitored spectrophotometrically in the presence of redox indicators. The semilog plots report the concentration of the redox species present in a simulated potentiometric titration of a flavoprotein ($E'_{0\ F/Fsq} = -60$ mV, $E'_{0\ Fsq/FH2} = -180$ mV; $E'_{0\ F/FH2} = -120$ mV) as monitored spectrophotometrically at neutral pH in the presence of the redox indicators indigo tetrasulfonate (ITS; blue/colorless, with $A_{590} = 22{,}800$ (ox); $n = 2$; $E'_0 = -50$ mV) and phenosafranin (PS; red/colorless, with $A_{520} = 36{,}500$ (ox); $n = 2$; $E'_0 = -255$ mV)

monitoring their redox state independently, in view of the position and intensity of their absorption maxima. The resulting semilog plot allows very easy detection of the mid-point potential (Em, a term often used to distinguish phenomenological observations by "true" thermodynamic parameters) for all the involved redox pairs.

The different slope of individual tracings in Fig. 4 relates to the number of involved electrons in individual half cells, as evident from the Nernst equation, allowing a quick visual inspection of the quality of the data and of possible environmental effects, as related—for instance—to proton dissociation from some of the involved species [2].

References

1. Bruice TC (1980) Mechanisms of flavin catalysis. Acc Chem Res 13:256–262
2. Mayhew SG (1999) The effects of pH and semiquinone formation on the oxidation-reduction potentials of flavin mononucleotide—a reappraisal. Eur J Bioch 265:698–702
3. Edmondson DE, Tollin G (1983) Semiquinone formation in flavo- and metalloflavoproteins. In: Radicals in biochemistry. topics in current chemistry, vol 108. Springer, Berlin, pp 109–138
4. Tedeschi G, Chen S, Massey V (1995) DT-diaphorase—redox potential, steady-state, and

rapid reaction studies. J Biol Chem 270:1198–1204

5. Ludwig ML, Pattridge KA, Metzger AL, Dixon MM, Eren M, Feng Y, Swenson RP (1997) Control of oxidation-reduction potentials in flavodoxin from Clostridium beijerinckii: the role of conformation changes. Biochemistry 36:1259–1280
6. Pollegioni L, Porrini D, Molla G, Pilone MS (2000) Redox potentials and their pH dependence of D-amino-acid oxidase of Rhodotorula gracilis and Trigonopsis variabilis. Eur J Bioch 267:6624–6632
7. Gutman M, Bonomi F, Pagani S, Cerletti P, Kroneck P (1980) Modulation of the flavin redox potential as mode of regulation of succinate-dehydrogenase activity. Biochim Biophys Acta 591:400–408
8. Tegoni M, Janot JM, Labeyrie F (1986) Regulation of dehydrogenases one-electron transferases by modification of flavin redox potentials—effect of product binding on semiquinone stabilization in yeast flavocytochrome-b2. Eur J Biochem 155:491–501
9. Ramsay RR, Hunter DJB (2002) Inhibitors alter the spectrum and redox properties of monoamine oxidase A. Biochim Biophys Acta 1601:178–184
10. Bonomi F, Pagani S, Cerletti P, Giori C (1983) Modification of the thermodynamic properties of the electron-transferring groups in mitochondrial succinate-dehydrogenase upon binding of succinate. Eur J Biochem 134:439–445
11. Stewart RC, Massey V (1985) Potentiometric studies of native and flavin-substituted Old Yellow Enzyme. J Biol Chem 260:13639–13647
12. Ishikita H (2013) Contributions of protein environment to the reduction potentials of flavin-containing proteins. Handbook of flavoproteins: complex flavoproteins, dehydrogenases and physical methods, vol 2. De Gryter, Berlin, pp 321–334
13. Zhang P, Yuly JL, Lubner CE, Mulder DW, King PW, Peters JW, Beratan DN (2017) Electron bifurcation: thermodynamics and kinetics of two-electron brokering in biological redox chemistry. Acc Chem Res 50:2410–2417
14. Buey RM, Arellano JB, Lopez-Maury L, Galindo-Trigo S, Velàzquez-Campoy A, Revuelta JL, de Pereda JM, Florencio FJ, Schuermann P, Buchanan BB, Balsera M (2017) Unprecedented pathway of reducing equivalents in a diflavin-linked disulfide oxidoreductase. Proc Natl Acad Sci U S A 114:12725–12730
15. Werther T, Wahlefeld S, Salewski J, Kuhlmann U, Zebger I, Hildebrandt P, Dobbek H (2017) Redox-dependent substrate-cofactor interactions in the Michaelis-complex of a flavin-dependent oxidoreductase. Nat Commun 8:Art nr 16084
16. Cheng VWT, Piragasam RS, Rothery RA, Maklashina E, Cecchini G, Weiner JH (2015) Redox state of flavin adenine dinucleotide drives substrate binding and product release in Escherichia coli succinate dehydrogenase. Biochemistry 54:1043–1052
17. Senda M, Kishigami S, Kimura S, Fukuda M, Ishida T, Senda T (2007) Molecular mechanism of the redox-dependent interaction between NADH-dependent ferredoxin reductase and Rieske-type [2Fe-2S] ferredoxin. J Mol Biol 373:382–400
18. Senda T, Senda M, Kimura S, Ishida T (2009) Redox control of protein conformation in flavoproteins. Antiox Redox Signal 11:1741–1766
19. Becker DF, Zhu W, Moxley MA (2011) Flavin redox switching of protein functions. Antiox Redox Signal 14:1079–1091
20. Vaidya AT, Top D, Manahan CC, Tokuda JM, Zhang S, Pollack L, Young MW, Crane BR (2013) Flavin reduction activates Drosophila cryptochrome. Proc Natl Acad Sci U S A 110:20455–20460
21. Samanta D, Widom J, Borbat PP, Freed JH, Crane BR (2016) Bacterial energy sensor Aer modulates the activity of the chemotaxis kinase CheA based on the redox state of the flavin cofactor. J Biol Chem 291:25809–25814
22. Guan ZW, Kamatani D, Kimura S, Iyanagi T (2003) Mechanistic studies on the intramolecular one-electron transfer between the two flavins in the human neuronal nitric-oxide synthase and inducible nitric-oxide synthase flavin domains. J Biol Chem 278:30859–30868
23. Curley GP, Carr MC, Mayhew SG, Voordouw G (1991) Redox and flavin-binding properties of recombinant flavodoxin from *Desulfovibrio vulgaris* (Hildenborough). Eur J Biochem 202:1091–1100
24. Zhou ZM, Swenson RP (1995) Electrostatic effects of surface acidic amino acid-residues on the oxidation-reduction potentials of the flavodoxin from Desulfovibrio vulgaris (Hildenborough). Biochemistry 34:3183–3192
25. Fantuzzi A, Artali R, Bombieri G, Marchini N, Meneghetti F, Gilardi G, Sadeghi SJ, Cavazzini D, Rossi GL (2009) Redox properties and crystal structures of a Desulfovibrio vulgaris flavodoxin mutant in the monomeric and homodimeric forms. Biochim Biophys Acta 1794:496–505

26. Finn RD, Basran J, Roitel O, Wolf CR, Munro AW, Paine MJ, Scrutton NS (2003) Determination of the redox potentials and electron transfer properties of the FAD- and FMN-binding domains of the human oxidoreductase NR1. Eur J Bioch 270:1164–1175
27. Brenner S, Hay S, Munro AW, Scrutton NS (2008) Inter-flavin electron transfer in cytochrome P450 reductase—effects of solvent and pH identify hidden complexity in mechanism. FEBS J 275:4540–4557
28. Kay CJ, Barber MJ, Notton BA, Solomonson LP (1989) Oxidation reduction midpoint potentials of the flavin, heme and Mo-pterin centers in spinach (*Spinacia oleracea* L) nitrate reductase. Biochem J 263:285–287
29. Tegoni M, Gervais M, Desbois A (1997) Resonance Raman study on the oxidized and anionic semiquinone forms of flavocytochrome b2 and L-lactate monooxygenase. Influence of the structure and environment of the isoalloxazine ring on the flavin function. Biochemistry 36:8932–8946
30. Porras AG, Palmer G (1982) The room-temperature potentiometry of xanthine oxidase—pH-dependent redox behavior of the flavin, molybdenum, and iron-sulfur centers. J Biol Chem 257:1617–1626
31. Ohnishi T, King TE, Salerno JC, Blum H, Bowyer JR, Maida T (1981) Thermodynamic and electron paramagnetic resonance characterization of flavin in succinate dehydrogenase. J Biol Chem 256:5577–5582
32. Ravasio S, Curti B, Vanoni MA (2001) Determination of the midpoint potential of the FAD and FMN flavin cofactors and of the 3Fe-4S cluster of glutamate synthase. Biochemistry 40:5533–5541
33. Pellett JD, Becker DF, Saenger AK, Fuchs JA, Stankovich MT (2001) Role of substrate/product in *Megasphaera elsdenii* short-chain acyl-coenzyme A dehydrogenase. Biochemistry 40:7720–7728
34. Hirasawa M, Robertson DE, Ameyibor E, Johnson MK, Knaff DB (1992) Oxidation-reduction properties of the ferredoxin-linked glutamate synthase from spinach leaf. Biochim Biophys Acta 1100:105–108
35. Paulsen KE, Orville AM, Frerman FE, Lipscomb JD, Stankovich MT (1992) Redox properties of electron-transfer flavoprotein ubiquinone oxidoreductase as determined by EPR-spectroelectrochemistry. Biochemistry 31:11755–11761
36. Silaghi-Dumitrescu R, Ng KY, Viswanathan R, Kurtz DM (2005) A flavo-diiron protein from Desulfovibrio vulgaris with oxidase and nitric oxide reductase activities. Evidence for an in vivo nitric oxide scavenging function. Biochemistry 44:3572–3579
37. Clark WM (1960) Oxidation-reduction potentials of organic systems. The Williams & Wilkins, Baltimore, MD
38. Edmondson DE, Tollin G (1971) Flavoprotein chemistry. 1. Circular dichroism studies of flavin chromophore and of relation between redox properties and flavin environment in oxidases and dehydrogenases. Biochemistry 10:113–116
39. Dutton P (1978) Redox potentiometry: determination of midpoint potentials of oxidation/reduction components of biological electron-transfer systems. Methods Enzymol 54:411–435
40. Massey V, Hemmerich P (1978) Photoreduction of flavoproteins and other biological compounds catalyzed by de-aza flavins. Biochemistry 17:9–16
41. Vogt S, Schneider M, Schaefer-Eberwein H, Noell G (2014) Determination of the pH dependent redox potential of glucose oxidase by spectroelectrochemistry. Anal Chem 86:7530–7535

Chapter 9

Anaerobic Stopped-Flow Spectrophotometry with Photodiode Array Detection in the Presteady State: An Application to Elucidate Oxidoreduction Mechanisms in Flavoproteins

Patricia Ferreira and Milagros Medina

Abstract

Anaerobic stopped-flow (SF) spectrophotometry is a powerful biophysical tool that allows a complete kinetic characterization of protein interactions with other molecules when they are in different redox states, as well as of the redox processes consequence of such interactions. Differences in the absorption spectroscopic properties of oxidized, semiquinone and hydroquinone states of flavoproteins, as well as the appearance of transient spectroscopic features produced by the flavin cofactor during substrate binding and electron transfer processes, have made SF a suitable technique for kinetically dissecting their mechanisms of reaction. In addition, SF coupled to photodiode array detection, enables kinetic data collection in a wavelength range. When such type of data are available for a flavoprotein reaction, they allow for obtaining detailed information of individual reaction steps, including intermolecular dissociation constants as well as electron transfer rate constants. Methodologies for the mechanistic characterization of flavoproteins involved in redox processes by SF spectrophotometry are described in this chapter.

Key words Stopped-flow, Photodiode array detection, Fast kinetics, Flavoprotein reduction, Flavoprotein oxidation, Electron transfer, Binding

1 Introduction

Stopped-flow (SF) is a spectroscopic technique used in the study of kinetics and mechanisms of fast chemical reactions in solution over timescales of milliseconds and widely used to evaluate biochemical processes [1–3]. The SF instrument is a rapid mixing device where two solutions are quickly forced together into a mixing chamber from two independent drive syringes. The mixed solution rapidly fills the optical observation cell displacing the previous contents with freshly mixed reactants. The exit of the observation cell is connected to a third syringe, the stop syringe, that collects the old mixed solution limiting the volume spent with each

Maria Barile (ed.), *Flavins and Flavoproteins: Methods and Protocols*, Methods in Molecular Biology, vol. 2280, https://doi.org/10.1007/978-1-0716-1286-6_9,

experiment. The main goal of this syringe is to abruptly stop the solution flow when its plunger moves back to its limit position while triggering data collection. Thus, the overall process starts by compressed air coupling the displacement of the stop syringe piston and the movement of the stop valve to allow emptying of a fixed volume from this syringe. Then, a sudden compressed air impulse of the two drive syringe pistons makes them to liberate the same volume (when both syringes have the same diameter) into the mixing chamber, flushing the old mixture towards the stop syringe. As this syringe only accepts the initially emptied volume, the flow is stopped at the measuring chamber and drive syringes while the measuring device is triggered. Then, fresh reactants in the observation cell are illuminated by a light source and the change, as a function of time, in the selected optical property can be measured by the adequate detector (absorbance, fluorescence, light scattering, turbidity, fluorescence anisotropy, etc...), which can be mounted either perpendicular or parallel to the path of incoming light depending on the optical property. Just prior to the flow stopping, a steady state flow is achieved at the solution entering the measuring chamber, which is only a few milliseconds old. The age of this reaction volume is known as the dead time of the SF system, usually it is 1–2 ms, but some new devices have it as short as 0.3–0.6 ms.

Most of the metabolic transformations that use flavoenzymes are mainly related to reduction–oxidation (redox) processes, because their FMN and FAD cofactors contain an isoalloxazine ring that can reversible accept and donate one or two electrons [4–9]. This later property confers many of them the ability to mediate obligatory processes of two-electron transfers with those involving a single-electron transfer. This is due to the isoalloxazine ring being a redox agent that can exist in three different states within the protein environment, fully oxidized or quinone (ox), one-electron reduced or semiquinone (sq), and two-electron reduced or hydroquinone (hq). Interestingly, these redox states are in addition easily distinguished by their different absorption and fluorescence properties in the visible region of the electromagnetic spectrum [10]. Moreover, when mixed with electron donors/acceptors the isoalloxazine ring of some flavoproteins can stabilize transient and spectroscopically detectable long-wavelength-absorption bands that relate to flavin-ligand reaction intermediates, such as excimers or charge transfer complexes (CTC)) [11, 12]. Redox reactions of flavoproteins involve in general a reductive half reaction where the enzyme-bound oxidized flavin accepts electrons from a donor substrate (as for example, $\text{P-FAD}_{ox} + AH_2 \Leftrightarrow \text{P-FADH}_{hq} + H^+ + A^+$ for a two electrons donor substrate or $\text{P-FAD}_{ox} + AH \Leftrightarrow \text{P-FAD(H/}^-)_{sq} + H^+ + A^+$ for a single electron donor substrate), and an oxidative half reaction where the reduced flavin is reoxidized by the electron acceptor substrate (e.g.,

P-FADH$_{hq}$ + B$^+$ + H$^+$ ⇔ P-FAD$_{ox}$ + BH$_2$ or P- FAD (H/$^-$)$_{sq}$ + (H$^+$) + B$^+$ ⇔ P-FAD$_{ox}$ + BH) [6]. In these processes A and B can be either different metabolites, substrates, NAD(P)$^+$/H coenzymes, or different electron transfer proteins, with the reaction equilibria establishment following the basic kinetics and thermodynamic principles [12, 13]. Because of these observations, changes in absorption and fluorescence properties have been used since mid-twentieth century to investigate reductive and oxidative processes involving flavoenzyme reactions, having the SF spectrophotometry a tremendous impact on kinetically dissecting individual steps in the mechanisms of an important number of flavoproteins and flavoenzymes [14–21]. Nonetheless, due to the propensity of most semiquinone and reduced flavoprotein states to oxidation by molecular oxygen and to the, in general, lower affinity of apoflavoproteins for the reduced form of flavins, most of these experiments must be carried out under strict anaerobic conditions. Single-wavelength absorption detection methods were used at early days, but sometimes they did not allow to un-ambiguously identify intermediate and final reaction species [13]. Later on, photodiode array (PDA) detection appeared as a better choice to follow full spectral time evolution [12, 13, 22–34]. Simultaneous data collection in wavelength and time ranges, allow using global and numerical integration methods to depict the composition of intermediate and final species of the reactions, as well as the kinetics of these processes. In this way, analysis of the resulting kinetic transients at multiple wavelengths allow determining rates for complex formation, electron and hydride transfer, as well as obtaining spectral information on initial and final species and on short-lived intermediates along the course of the reaction.

2 Materials

2.1 Specialized Materials and Instruments to Produce and to Ensure Anaerobic Conditions

Materials and instruments must allow replacing the air atmosphere of samples and buffers by an oxygen free Ar/N$_2$ atmosphere. Similarly, they must provide the way to make oxygen-free the drive syringes and the pipeline of the SF instrument.

1. Home-made all-glass tonometers similar to that of Fig. 1. Tonometers will contain the liquid samples that need being made anaerobic. In addition, after making solutions anaerobic inside them, tonometers will be used to introduce anaerobic samples into the SF system syringes. Tonometers must have a vacuum stopcock valve at each one of their two ends (Fig. 1). The top valve ends in a glass tube of fixed diameter that is used to connect the tonometer through a butyl rubber tube to the anaerobic manifold system (Fig. 2). The bottom one ends in a glass tube whose diameter is decreased to fit into the female Luer-lock fitting at the entrance valves of the SF drive syringes.

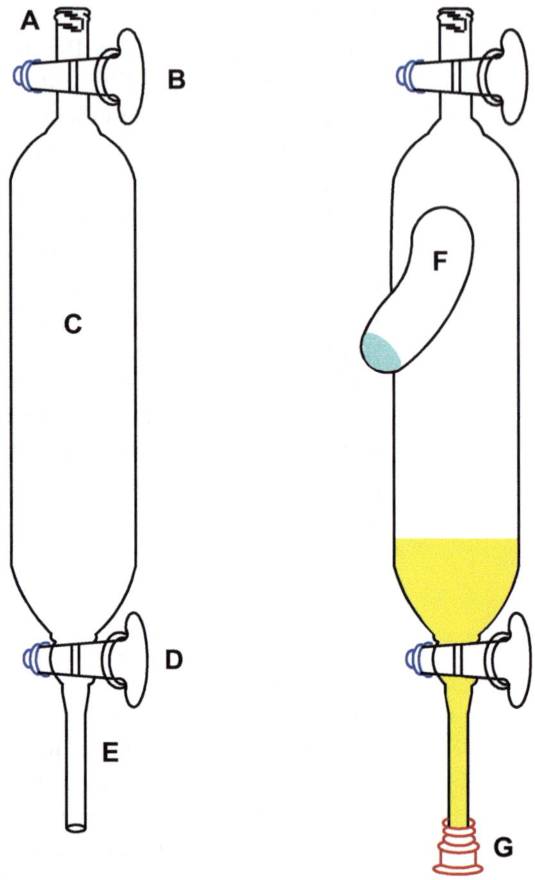

Fig. 1 Anaerobic tonometers. (**a**) Glass end to connect through butyl rubber tubing to the anaerobic train system. (**b**) Vacuum stopcock valve to open/close the connection of the tonometer content to the anaerobic system. (**c**) Main tonometer chamber to contain sample (2–4 ml). (**d**) Vacuum stopcock valve to flush sample from the tonometer chamber to the SF system. (**e**) Tonometer end cone with the exact dimensions to fit in the female Luer-lock fitting at the entrance valves of the SF drive syringes and to which the Suba-Seal rubber septum will be attached. (**f**) Side-arm to place a reagent solution (shown in blue) separated from the main sample solution (in yellow), in order to mix them after both have been made anaerobic. (**g**) Suba-Seal septum

2. Home-made manifold system (anaerobic train) (*see* **Note 1**). It will allow gases to be alternately evacuated from liquid samples using a vacuum pump, and then filling the samples with oxygen-free Ar (or N_2) coming from a gas cylinder (Fig. 2).
   (a) The system consists of a glass manifold column ($\approx 40 \times 2.5$ cm) (A) ending at one side in a glass tube fussed to a three-way vacuum stopcock (1). This stopcock is key to either fill in the cylinder with an anaerobic gas overpressure or connect it to the vacuum pump that allows evacuating gases from the system. In addition, the

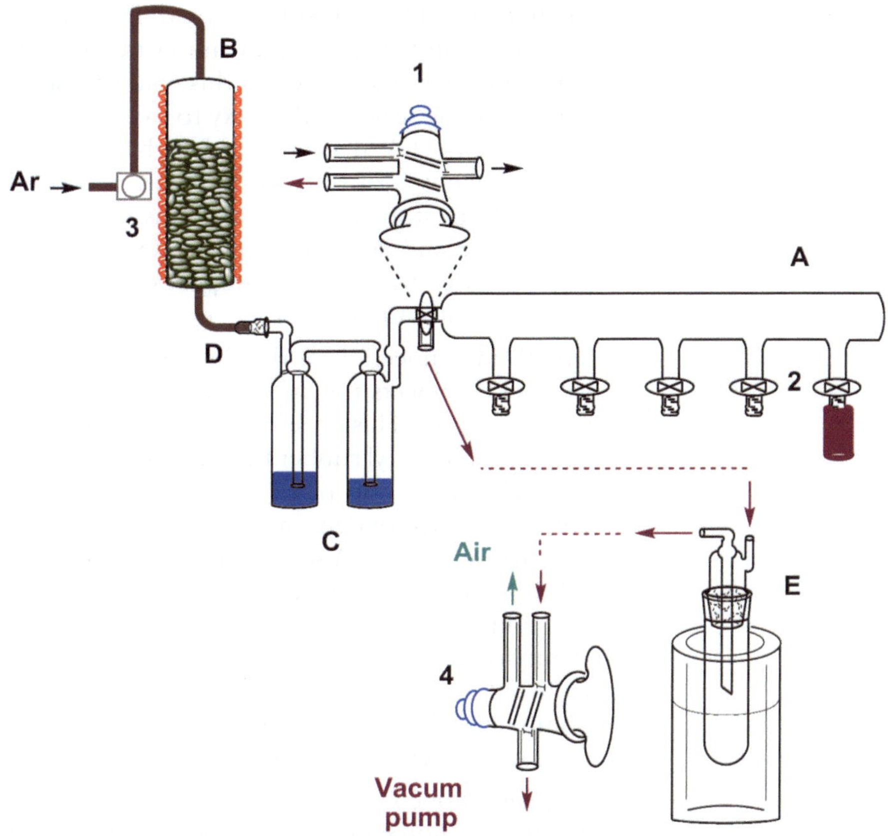

Fig. 2 Schematic representation of the gas flow and the vacuum line in the anaerobic manifold system. Main components: (**a**) Central glass column of the anaerobic manifold system. (**b**) Glass column two/thirds filled with BASF PuriStar® catalyst (R3–11 or R3–11G) and heated to 110–130 °C with a heating tape (in red). (**c**) In tandem 250 ml wash-bottles containing a methyl viologen solution. (**d**) Coils of 1/8 in. copper tubing sealed to glass tubing with epoxy cement. (**e**) Cold trap formed by a thick round tub with a ground-glass joint and a cap with the corresponding ground-glass connections, and placed into an open Dewar flask. Valves: (1) three-way vacuum stopcock valve connecting either Ar or vacuum to the manifold system, (2) stopcock vacuum valves connecting the anaerobic manifold system to the tonometers through butyl rubber (in red), (3) pressure regulator outlet valve, and (4) three-way vacuum stopcock valve connecting vacuum either to the anaerobic system or to the atmospheric air

glass manifold must have four to five exit glass tubes, each one containing a vacuum stopcock valve (2), after which a butyl rubber tubing of the adequate size is attached.

(b) Before reaching the three-way vacuum stopcock, the anaerobic gas from the gas cylinder enters into the system through a pressure regulator outlet (3) that provides a pressure slightly greater than 0 bar but well below 0.5 bar (*see* **Note 2**). The copper tubing coming out from the pressure regulator outlet then fills in the terminal tube (sealed with epoxy resin) of a glass column (B) ($\approx$34 × 5 cm) two/thirds filled with BASF PuriStar®

catalyst (R3–11 or R3–11G) used for the removal of oxygen impurities from gases and heated to 110–130 °C with a heating tape. Ar leaves this column using a second copper tubing sealed with epoxy to two in tandem 250 ml wash-bottles (C), that are filled with a methyl viologen mixture, whose exit is fussed by a glass tube to the three-way vacuum stopcock (1) that forms the gateway to the anaerobic manifold system. In this way, the Ar gas reaching the three-way vacuum stopcock is oxygen free and saturated with water vapor.

(c) The third connection of the three-way vacuum stopcock valve (1) connects the anaerobic manifold system to the vacuum pump. This system is formed by a vacuum pump connected by butyl rubber tubing to the vacuum stopcock through a cold trap placed into a liquid nitrogen open Dewar flask (to condense all liquated vapors from reaching the vacuum pump). A second three-way vacuum stopcock valve (4) is recommended between the vacuum pump and the cold trap. It allows connecting directly the vacuum pump to the air atmosphere when switching it on and off, to avoid liquid from the trapping flask getting to the pump or oil from the pump moving toward the anaerobic train when vacuum is released.

3. A high-quality stopcock grease, such as Apiezon Type N (*Apiezon Products, Manchester, UK*), for greasing all taps and ground-glass joints.
4. The methyl viologen mixture used in the wash-bottles supplying Ar to the anaerobic train contains 1 mM methyl viologen, 10 mM EDTA, and 5 μM 5-deazariboflavin, in a 200 mM Tris/HCl pH 8.0 buffer. After flowing Ar exiting from the catalyst column, this solution is exposed to visible light until reduced and turned blue. This mixture maintains the blue color as oxygen is trapped and is used as an anaerobic indicator. The bubbling wash-bottles are also used for humidifying the gas.
5. Sodium dithionite to make the SF pipeline anaerobic.
6. Suba-Seal rubber septa with small-inner diameters that tightly fit at the end cone of the tonometers (Fig. 1).
7. An Ar/N_2 gas cylinder containing O_2 impurities below 2 ppm.
8. A heating tape to maintain the desired temperature of the glass column and a thermocouple to control its temperature.

2.2 A Stopped-Flow Equipment Coupled to a Photodiode Array Detector

Basic SF instruments have two identical drive syringes where the reactant solutions are placed within a thermostatic chamber connected to a recirculating water bath that allows controlling temperature, and a stop syringe (Fig. 3). A pipeline connects drive syringes with the stop syringe, containing in between the mixing

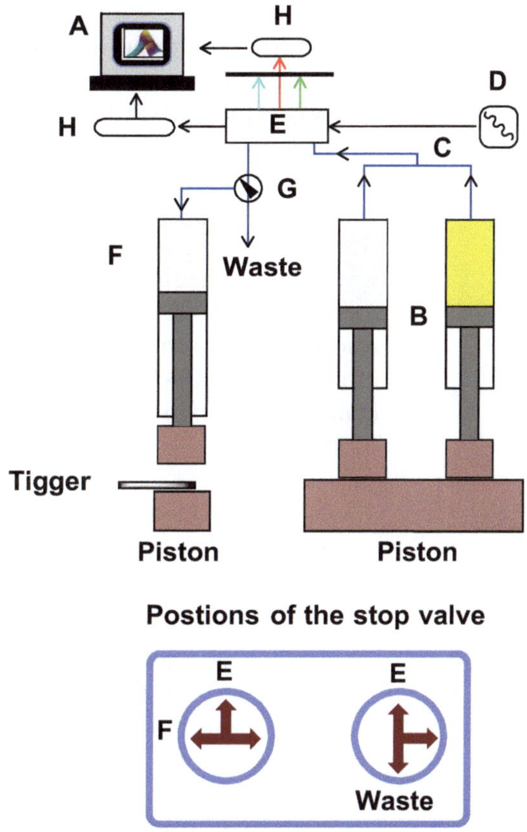

Fig. 3 Main components and working scheme of the stopped-flow equipment. Top panel: (**a**) Computer for the control of the equipment, data acquisition and analysis; (**b**) Drive syringes; (**c**) Mixing chamber; (**d**) UV/Visible lamp, power lamp supply and monochromator (not active when using PDA detection). (**e**) Observation chamber; (**f**) Stop syringe; (**g**) Stop valve; (**h**) Detectors: PDA or single-wavelength absorption detectors are situated in line with excitation lamp, while fluorescence detector is placed perpendicular to the excitation lamp. The bottom panel shows a detail of the two positions of the stop-valve, connecting the mixing chamber either to the stop syringe or to the waste. Measurement starts when an impulse of the stop-syringe piston by compressed air empties a fixed volume from this syringe to the waste. Quickly the stop-valve moves to connect the observation chamber with the stop-syringe while the drive syringes piston is pushed by compressed air. This makes the two drive syringes to liberate the same volume that was emptied to the waste in the stop-syringe. Thus, flow from the drive syringes mixes in the mixing chamber, filling up the pipeline to the observation chamber, and displacing the older solution toward the stop-syringe. When this syringe fills in, its plunger moves back and triggers data acquisition in the observation cell. As both solutions mix in the mixing chamber, reaction starts and it is followed in the observation cell using the adequate detector coupled to the computer

chamber and the observation cell. Drive syringes typically are Hamilton syringes of identical volume (around 2 ml) (*see* **Note 3**). Valves and pistons of drive syringes, as well as the pipeline inside the system, must be of a low O_2 permeable material, such as polyaryletheretherketone (PEEK), to maintain anaerobic conditions after purging the SF flow circuit with oxygen free solutions. The system is typically controlled with a computer and the corresponding software, and it can be coupled to either single wavelength absorption/fluorescence detectors or to a PDA detector. The PDA detector enables to record multiple-wavelength time-resolved measurements from a single stopped-flow drive. Usually, a minimum of 120–180 μl is required to fully replace the old solution from the two exits of the drive syringes to the entrance of the stop syringe.

2.3 Buffers and Substrates

1. Suitable donor and acceptor substrates of the flavoprotein whose redox process is going to be studied.
2. Suitable buffers (such as 50 mM Tris/HCl pH 8.0, 50 mM sodium phosphate pH 7.0, or 50 mM HEPES pH 7.0).
3. If the working protein either binds O_2 or has high affinity for it, you will need a system trapping it to ensure its total removal from samples. You must add glucose (10 mM) to your solutions (buffers, substrates and protein), make some vacuum-Ar cycles and then add glucose oxidase (10 U/ml). For oxidative half reaction assays, it is recommended to place the flavoprotein containing the glucose (10 mM) in the main chamber of the tonometer, while the reductant substrate must be mixed with glucose oxidase (10 U/ml) and placed in the side arm (Fig. 1). After making the tonometer anaerobic, mix the side arm content with the protein solution of the main compartment of tonometer.

3 Methods

3.1 Preparation of Anaerobic Solutions

1. With the BASF catalyst activated (*see* **Note 4**) and the heating tape set at 110–130 °C, flow Ar into the manifold system (Fig. 2). Set the three-way vacuum stopcock valve (1) to the position that allows entrance of Ar in the manifold system and open one of the valves at the exit of the anaerobic manifold (2). Check the gas bubbles in the tandem wash-bottles as indicative of Ar flowing through the whole system. Use the pressure regulator valve (and the flowmeter if you have it) to regulate the gas flow, making it minimal to ensure maximal contact time with the oxygen trap.
2. Illuminate the methyl viologen mixture contained in the tandem wash-bottles until it gets dark blue.

3. Close valve (2) (Fig. 2) and maintain the system under Ar pressure. Bubbling at the methyl viologen solution in the wash-bottles must stop.
4. Fill the open Dewar flask in with a small volume of liquid nitrogen. Introduce the cold trap in the Dewar and seal its top with laboratory absorbent paper to avoid evaporation of liquid nitrogen.
5. With the three-way stopcock valve (4) open to the air switch on the vacuum pump. Then, turn the valve to connect the vacuum pump to the end of the vacuum pipeline.
6. To prepare the tonometers, grease their glass-ground joints lightly avoiding the holes of the stopcocks. Twist each stopcock back and forth to make sure that it turns easily.
7. Take a tonometer (Fig. 1), close its stopcock valve D and fit a Suba-Seal rubber septum (G) at the end of the cone (E).
8. Open the stopcock valve B and use a Pasteur pipette to fill in the tonometer with 4–5 ml of the sample you want to make anaerobic via the glass tubing A connected to the stopcock valve B.
9. Connect the tonometer via its glass end tube A in the stopcock valve B to the butyl rubber tube of one of the stopcock valves (Fig. 2, (2)) attached to the manifold system (*see* **Note 5**).
10. Open the manifold valve (2) and then sequentially the valves B and D of the tonometer. The tonometer will be then filled with Ar while the blue solution will bubble for a short time.
11. When bubbling stops turn the three-way vacuum stopcock (4) 180° to connect the vacuum pump to the cold trap system, and then turn carefully the three-way vacuum stopcock (1) 180° to make vacuum in the manifold system and, as a consequence to the connected tonometers.
12. As the gas gets evacuated from the sample solution, bubbles will form in it, try to minimize them by sacking the tonometer and closing valve B several times.
13. Turn the three-way vacuum stopcock (1) slowly 180° to allow now the Ar from the pipeline to enter in the manifold system and in your tonometer.
14. Repeat alternate cycles of Ar/vacuum for at least one hour to ensure oxygen removal from the tonometer solution.
15. In the last cycle, increase slightly the Ar pressure to end up with a slight overpressure in the whole system and, particularly, in the tonometer.
16. Close the tonometer valve D, then the tonometer valve B and finally valve (2) of the manifold system.
17. The tonometer, filled with your buffer or sample in anaerobic gas, is ready to load the drive syringes of the SF equipment.

3.2 Making the SF Equipment Anaerobic

To ensure anaerobic conditions during kinetic measurements all the components of the SF instrument that will enter in contact with samples must be in advance settle under anaerobic conditions.

1. Fill in the titrant-Erlenmeyer flask with 50 ml of your working buffer and connect it to the anaerobic manifold system through its vacuum tap join. Follow procedures **7–14** explained in Subheading 3.1 to make this buffer anaerobic.
2. Open the titrant-Erlenmeyer flask and add a spatula of solid sodium dithionite to saturate the buffer in this compound. Quickly close the flask and follow again procedures in Subheading 3.1, **steps** 7–**14** to evacuate oxygen that entered when pouring the sodium dithionite.
3. With a Pasteur pipette, and as quick as possible, transfer 4–5 ml of the anaerobic sodium dithionite solution to a tonometers as explained in **step 6** of Subheading 3.1. Prepare a second identical tonometer.
4. Prepare two additional tonometers filled with the working buffer.
5. Follow again procedures in Subheading 3.1, **steps 7–15** to evacuate oxygen from the sodium dithionite and buffer solutions contained in these four tonometers.
6. Take one of the tonometers containing the anaerobic sodium dithionite with both vacuum valves closed to the SF system. Remove the Suba-Seal rubber septum from the end cone and quickly fit this tonometer end cone into the female Luer-lock fitting at the entrance of one of the valves of the SF drive syringes.
7. Open valve D of your tonometer and turn the SF valve connecting your tonometer to the drive syringe to the load position. Slowly pull down the drive syringe piston and allow it to be filled with the solution contained in the tonometer. When the drive syringe is filled, turn its valve to the closed position, and then close the tonometer valve B.
8. Take the second tonometer containing anaerobic sodium dithionite and follow the same procedure to load the second drive syringe of the SF system.
9. Set the computer software in the acquisition mode and select the data collection time (*see* **Note 6**). Activate the shooting mode that triggers the compressed air flow to the stop and drive syringes pistons (Fig. 3), and as consequence the start of a SF cycle. Continue shooting repeatedly until the drive syringes get empty (*see* **Note** 7).
10. Reload the drive syringes with the sodium dithionite solutions remaining in the tonometers still connected to the valves of the drive syringes following procedure in Subheading 3.2, **step** 7.

Then follow procedures described in Subheading 3.2, **step 9** to pass these new solutions through the SF system to fully eliminate oxygen traces from the pipeline.

11. Remove now the two tonometers containing sodium dithionite from the SF system. Take two new tonometers filled with anaerobic buffer and follow procedures in Subheading 3.2, **steps 6–10** to wash out the sodium dithionite solution from the pipeline system by replacing it with the anaerobic buffer solution.
12. Use the second buffer run of the drive syringes to collect the baseline in the whole-absorption range allowed by your PDA detector. Save it and check for its stability along the selected measuring times. This will be the baseline you will be using for the rest of your experiment.
13. The system is ready to start your experiment, but do not remove buffer tonometers from the SF system until the moment you are ready to load the sample tonometers.

3.3 Measurement of a Flavoprotein Reductive Half Reaction

1. Follow Subheading 3.1 to prepare a tonometer containing 4–5 ml of an anaerobic solution of the flavoenzyme at a known concentration (a solution absorbing 0.2–0.4 at the flavin Band-I is recommended). Similarly prepare a tonometer containing 4–5 ml of an anaerobic solution of the reductant substrate at a known given concentration. Start with at least a molar 1:1 ratio regarding the flavoenzyme.
2. Set the valve of one of the drive syringes to the load position, and then open the buffer tonometer valve D. Empty any buffer content of the syringe by pushing its piston upward to transfer the buffer solution back into the tonometer. Close the valve D of the tonometer, and then set the drive valve to the close position. Remove this buffer tonometer from the SF valve and replace it with the tonometer containing the anaerobic flavoenzyme solution. Load the content of this drive syringe with the protein solution as explained before.
3. Move to the other drive syringe and reload it with the anaerobic buffer solution still contained in the corresponding tonometer.
4. Using the software, trigger the shooting system to start a cycle by mixing both solutions and use the PDA detector to collect full-range absorption data, at least in the 340–1100 nm range, for a short time (collection of 100–400 spectra in 0.1–1 s or spectra every 2.5 ms will be appropriated). This will provide you with a set of spectra along the time, which must be identical (since no reaction must be produced under these conditions) and represent the initial spectrum of your flavoenzyme in the oxidized state, before the reaction starts and under your particular SF experimental conditions.

5. Switch the valve of the SF drive syringe filled with buffer to connect it to the buffer tonometer, and push up the piston of this syringe to transfer its content back to the tonometer (procedure as described in Subheading 3.3.2). Close the valves.
6. Replace the buffer tonometer by the tonometer containing the reducing agent, and load this drive syringe with the reducing agent sample as before explained.
7. Once both tonometers are filled with the reaction reactants (flavoprotein and reductive agent, respectively), you can start a data acquisition run to evaluate changes in absorption spectra upon protein reduction. Evaluate shoot by shoot data to choose the acquisition time that allows you to follow the whole reaction.
8. Using the shots that the equipment will allow you (around 20), explore several acquisition times to collect multiple wavelength absorption kinetic data containing the different features of the process in appropriate temporal scales for accurate determination of the observed rate constants (k_{obs}) of the potential different processes (*see* **Notes 8** and **9**) (Fig. 4). Usually it is adequate to get at least five replicas for each timescale and experimental condition that you want to evaluate.
9. If you wish to determine the dependence of k_{obs} values on the reductant concentration, you need to perform a set of assays by repeating the procedures above described. Use flavoenzyme tonometers of the same concentration to load the corresponding drive syringe, and load the reductant in the other drive syringe using tonometers with increasing reducing agent concentrations (*see* **Notes 10** and **11**).

3.4 Measurement of a Flavoprotein Oxidative Half Reaction

1. Take a tonometer bearing a side arm (Fig. 1), fill its main chamber with 4–5 ml of the flavoenzyme at a known concentration (a solution absorbing 0.2–0.4 at the flavin Band-I is recommended). Then, fill carefully its side arm with a small volume of a concentrated solution of reductant enough to fully reduce the flavoenzyme (~1.3-fold the concentration of protein).
2. Follow Subheading 3.1 to make the system anaerobic. When you consider your tonometer is anaerobic, turn it to pour the side-chain arm content on to the flavoprotein solution (*see* **Note 12**). You will detect that your flavoenzyme is reduced because it loses its yellowish color.
3. Follow Subheading 3.1 to prepare a tonometer containing around 4–5 ml of an anaerobic solution of the oxidant substrate at a known given concentration. Start with at least a molar 1:1 ratio regarding the flavoenzyme.

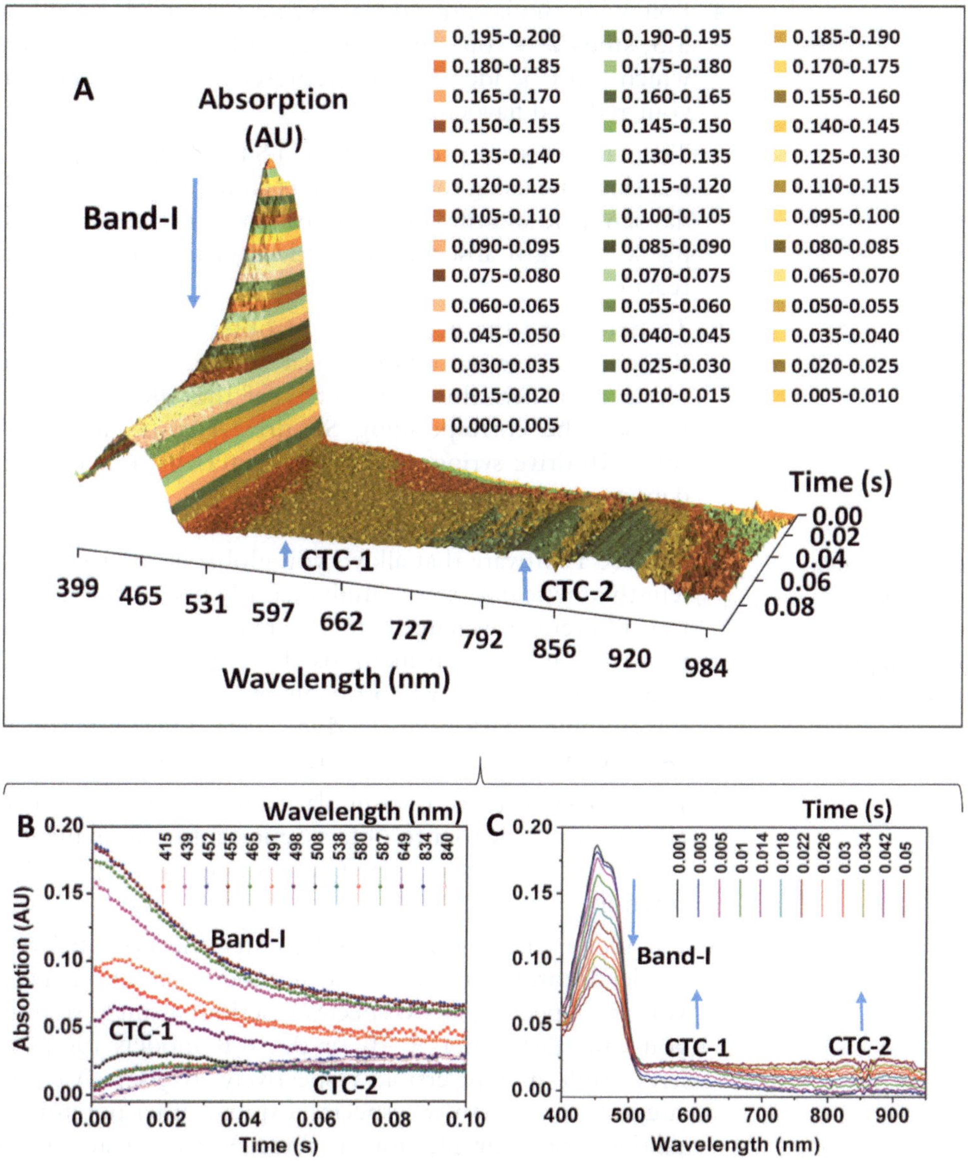

Fig. 4 Spectral evolution of the reduction of a bacterial ferredoxin-NADP$^+$ reductase (FNR) by a 1.5-fold excess of NADPH after their mixing in a SX-17MV stopped-flow equipment (*Applied Photophysics Ltd., Cambridge, UK*) coupled to a PDA detector with the X-SCAN software. (**a**) 3D plot showing the evolution of the protein spectral characteristics as a function of time and wavelength upon the processes of interaction with the NADPH and hydride transfer from the coenzyme to the FAD cofactor. (**b**) 2D plot showing the evolution of selected wavelengths along the reaction. (**c**) 2D plot showing the spectra of the reaction mixture at selected times. Data collected in 25 mM Tris/HCl pH 7.4 at 6 °C. The initial absorption increase in the 580 region envisages formation of an intermediate species that in this particular case correlates with the formation of a FNR_{ox}:NADPH charge transfer complex (CTC-1). Subsequently it is observed the absorption decrease in the Band-I (452 nm) characteristic of oxidized FAD, concomitantly with the appearance of a new broad band in the 800–900 nm region characteristic of a FNR_{rd}:NADP$^+$ charge transfer complex (CTC-2). Experimental Data taken from [35]

4. Following similar procedures to those reported in Subheading 3.3, **steps 2–4** you can obtain a set of spectra along the time, which must be identical and represent the initial spectrum of your reduced flavoenzyme.
5. Then, follow same procedures as in Subheading 3.3, **steps 5–8**, but this time using the tonometers containing one the reduced flavoenzyme and the other the oxidant agent, to acquire multiple wavelength absorption kinetic data for the oxidative half reaction.
6. To determine the dependence of observed rate constants on the oxidant concentration repeat the same procedure using reduced flavoenzyme tonometers of the same concentration to load the corresponding SF drive syringe, and load the other SF drive syringe using tonometers with increasing oxidant agent concentrations.

3.5 Determination of Apparent Kinetic Parameters from Multiwavelength Kinetic Data Sets

1. Choose a software that allows you global fitting to multiwavelength kinetic data sets (simultaneously fitting kinetic traces at the different wavelengths to the reaction model), as well as numerical integration methods. In general, the trademark of your own SF system will probably provide its own software, as for example Pro-Kineticist (*App. Photo. Ltd.*), but you might also use other software as MatLab (*Mathworks*).
2. It is advisable to run a singular-value decomposition (SVD) analysis of the dataset to obtain a model-free assessment of the number of spectrally distinct reaction species expected in the reaction.
3. For global analysis of data sets, enter the potential reaction model according the expected number of reaction species. If you are expecting a two species reaction, a single step model, you can start using A→B or A ⇆ B models for processes irreversible and reversible respectively. If your SVD analysis suggests more than two species, explore other potential possibilities: for example for a three species reaction model A ⇆ B ⇆ C, A ⇆ B→C, A→B ⇆ C or A→B→C (*see* **Note 13**) (Fig. 5).
4. Fitting routines will provide you concentration profiles, calculated spectra of reacting species (*see* **Note 14**) as well as the k_{obs} for the transformation between species, namely, $k_{A\to B}$, $k_{B\to C}$, $k_{B\to A}$, $k_{C\to B}$, . . ., according with the reaction model selected for multiwavelength data fitting and under the assayed conditions (Fig. 5) (*see* **Notes 15** and **16**).
5. Assess the quality of fitted data by the lack of systematic deviations from residual plots at different wavelengths and times (Fig. 5d), inspection of calculated spectra (Fig. 5a) and

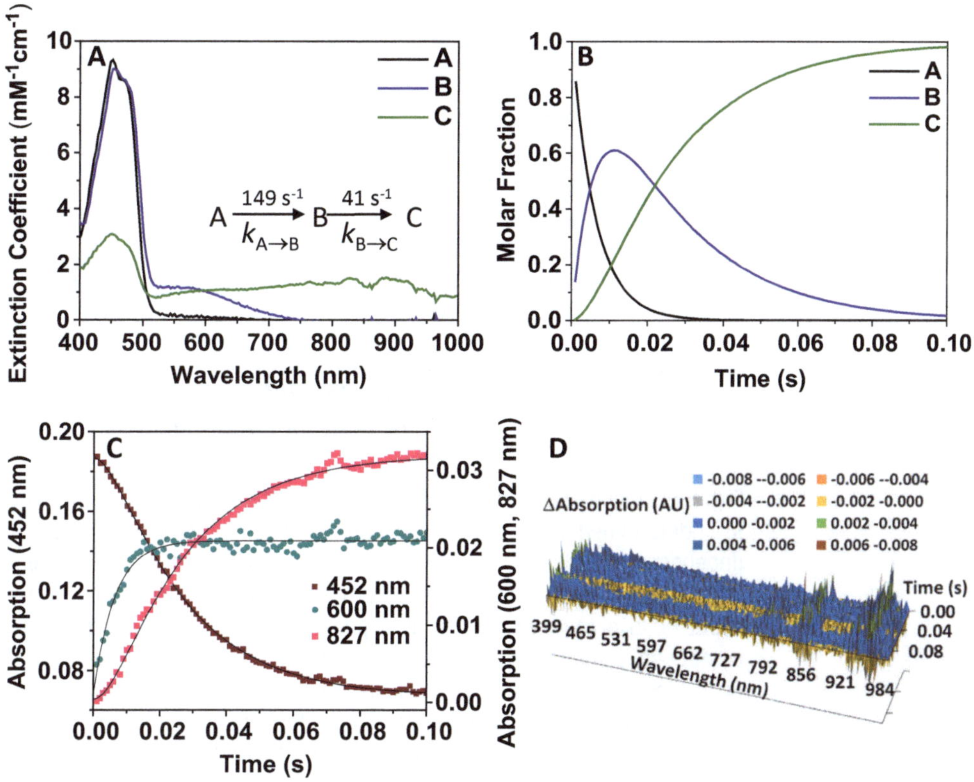

Fig. 5 Global-fit analysis of the data presented in Fig. 4a. (**a**) Absorbance spectra for the three pre–steady-state kinetically distinguishable spectroscopic species obtained by the global-fitting analysis of the reaction to a "A→B→C" kinetic model. (**b**) Corresponding evolution of A, B and C species along the time. (**c**) Experimental absorption kinetic transients obtained at the FAD Band-I (452 nm), the FNR_{ox}:NADPH CTC-1 band (600 nm) and the FNR_{rd}:$NADP^{+}$ CTC-2 band (827 nm). Symbols correspond to the experimental kinetic data, while lines show the corresponding global-fits to the "A→B→C" kinetic model. (**d**) Fitting residual plots as a function of both time and wavelength

consistence among the number of SVD species with the final fit model.

6. Time-dependent spectral simulations can be also performed to validate the results using the same software and the proposed reaction mechanism together with the experimentally determined interaction and limiting kinetics constants obtained after following procedures reported in Subheading 3.6.

3.6 Using Observed Rate Constants to Determine Binding Constants and Limiting Reaction Rates

1. Follow procedures in Subheading 3.3 or Subheading 3.4 to collect data at different concentrations of reductive or oxidative substrate agents, respectively.
2. For each reagent concentration determine k_{obs} values for the different reaction steps as explained in Subheading 3.5.

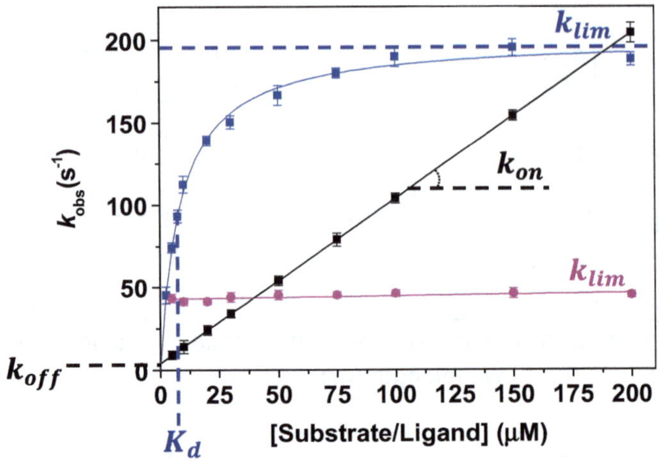

Fig. 6 Evolution of observed rate constants on the substrate concentrations. The k_{obs} determined by stopped-flow can be linearly dependent, hyperbolically dependent or independent on the substrate concentration. Lineal dependent data are indicative of the observed process either corresponding to a binding event or to a reactive process that does not require previous formation of a competent complex of the reactants. In the first case, fitting of these data to Eq. 1 would provide the ligand association rate, k_{on}, as the slope of the line and the corresponding dissociation rate, k_{off}, as the independent term. In the second case, the slope of the line will represent the second-order rate constant of the reactive process, k_2. Data showing a saturation profile are suggestive of ligand association followed by a process consequence of such interaction (as for example electron transfer process or reorganization process). Nonlinear data fitting to Eqs. 2–5 will allow to determine the dissociation constant of the association process, K_d, the rate constant for the subsequent reaction limiting rate constant, k_{lim}, and, depending on the best fitting equations, even reverse rate constants or inhibition constants. Finally, ligand concentration independent k_{obs} values directly relate the determined parameter with the limiting rate constant of the measured process

3. For each set of experiments, plot k_{obs} of each a particular reaction step as a function of the reactant concentration (Fig. 6).
4. When a linear profile of the k_{obs} on the substrate (S) concentration is observed (Fig. 6 black line) and you consider this reaction step corresponds to the binding of the reagent to the flavoprotein (FLV), as for example

$$\mathrm{FLV} + \mathrm{S} \underset{k_{off}}{\overset{k_{on}}{\rightleftharpoons}} \mathrm{FLV:S}$$

you can use the Eq. 1

$$k_{obs} = k_{on}[\mathrm{S}] + k_{off} \quad (1)$$

to fit your data, where k_{on} and k_{off} would account for the limiting rate constants for complex formation and for complex dissociation, and the equilibrium dissociation constant can be determined as $K_d = \frac{k_{off}}{k_{on}}$. If on the contrary you consider that this process rather relates to a reactive event, you will determine its second-order rate constant (k_2) from the slope of the line.

5. When a k_{obs} saturation profile on the substrate concentration is observed (Fig. 6 blue line), you can consider that the observed process includes a binding/reorganization step followed by a subsequent process, as for example for a half reductive flavoprotein redox reaction.

$$FLV_{ox} + S_{rd} \underset{k_{off}}{\overset{k_{on}}{\rightleftharpoons}} FLV_{ox} : S_{rd} \overset{k_{lim}}{\rightarrow} FLV_{rd} : S_{ox}$$

$$FLV_{ox} + S_{rd} \underset{k_{off}}{\overset{k_{on}}{\rightleftharpoons}} FLV_{ox} : S_{rd} \underset{k_{off-reorg}}{\overset{k_{on-reorg}}{\rightleftharpoons}} [FLV_{ox} : S_{rd}]^* \overset{k_{lim}}{\rightarrow} FLV_{rd} : S_{ox}$$

Them you can fit your data to Eq. 2

$$k_{obs} = \frac{[S_{rd}]k_{lim}}{[S_{rd}] + K_d^{S_{rd}}} \tag{2}$$

where k_{lim} is the limiting rate constant for the electron transfer process (reduction rate constant (k_{red}) or reoxidation rate constant (k_{ox}) for the reductive half and oxidative half reactions, respectively), while K_d might account for either the dissociation constant or the equilibrium reorganization constant to achieve the competent electron transfer complex.

6. Sometimes, when a k_{obs} saturation profile on the coenzyme concentration is observed and you would expect the electron transfer process to be reversible, you must consider that the k_{obs} might include the rates of the forward (k_{red}) and reverse (k_{ox}) reactive processes at equilibrium [13], as for example.

$$FLV_{ox} + S_{rd} \underset{k_{off}}{\overset{k_{on}}{\rightleftharpoons}} FLV_{ox} : S_{rd} \underset{k_{off-reorg}}{\overset{k_{on-reorg}}{\rightleftharpoons}} FLV_{rd} : S_{ox} \underset{k_{lim-1}}{\overset{k_{lim}}{\rightleftharpoons}} FLV_{rd} + S_{ox}$$

In the presence of excess of S_{rd}, the dependence of k_{red} and k_{ox} on S concentration is given by standard functions for substrate saturation and competitive inhibition (inhibition constant of the product, K_i), and Eq. 3 will apply

$$k_{obs} = \frac{[S_{rd}]k_{red}}{[S_{rd}] + K_d^{S_{rd}}} + \frac{[S_{ox}]k_{ox}}{[S_{ox}] + K_d^{S_{ox}}.(1 + [S_{rd}]/K_i^{S_{ox}})} \tag{3}$$

where the concentration of S_{ox} formed at equilibrium can also be estimated from the difference in the midpoint reduction potentials for S_{ox}/S_{rd} and $FLV_{ox/hq}$ redox couples (ΔE_m) and the total concentration of flavoprotein (FLV_T) by using Eq. 4.

$$[S_{ox}] = [S_{rd}]\frac{\left(1 + 4FLV_T\tilde{n}10^{\Delta E_m/29.5}/[S_{rd}]\right)^{1/2} - 1}{2\tilde{n}10^{\Delta E_m/29.5}} \quad (4)$$

Nonetheless, in this particular case you can also use Eq. 5 as a simplified option

$$k_{obs} = \frac{[S_{rd}]k_{red}}{[S_{rd}] + K_d^{S_{rd}}} + k_{rev} \quad (5)$$

that includes in a single reverse rate constant, k_{rev}, all the effects of the second term in Eq. 3.

7. When k_{obs} is independent on the substrate concentration (Fig. 6 pink line), k_{obs} values directly relate to the limiting rate constant of the process.

4 Notes

1. Anaerobic conditions might be also achieved by using a glovebox to remove oxygen. In this case, buffer, samples and equipment must be all inside the glovebox in an oxygen free atmosphere before the experiment starts. As a third option anaerobic conditions can also be obtained by using anaerobic buffers containing glucose (50 mM) and glucose oxidase (0.5–1 U/ml). This later possibility can be also used in combination with the others.
2. It is advisable to connect a flowmeter at the exit of the regulator valve to better control the anaerobic gas flow during system operation.
3. SF syringes might be of different volume if mixing of reactants in proportions different from 1:1 is required.
4. To activate the BASF catalyst purge it by flowing Ar with a 100 cc/min flow rate for at least 15 min. Then heat it up at 180–200 °C and continue flowing Ar for 3 h. Switch the purging gas from Ar to a mixture of 5% H_2 + 95% Ar, adjust the flow rate to 200 cc/min and purge the catalyst for 4 h maintaining the temperature. Finally, switch the purging gas back to Ar and turn off the heating tape. Let the catalyst cool down under Ar to either the working temperature (if you are using it immediately) or to room temperature.
5. For samples unstable at room temperature keep the tonometer on ice or protect it with freezer ice blocks.

6. Short times, 50 ms to 2 s, can be used here because the aim of this step is just to remove oxygen from the pipeline. No reaction is here followed and acquisition time is not relevant.
7. Considering the drive syringes having a volume around 2 ml and the system requiring a minimum of 180 μl by shoot to fill the pipeline (this will depend on the equipment), each refill of the syringes would allow at least 20 shoots.
8. It is recommended to discard the results from the three first shoots to ensure replacement of the pipelines with new solutions, as well as bubbles formed in the change of solutions.
9. To ensure adequate data fitting, enough time-dependent data must be collected representing a particular feature within the first half of the measuring time. If the measured process contains events occurring in different timescales, data should be collected along different temporal times to accurately fit each event in its respective timescale.
10. When doing sequential reloads of the drive syringes do the experiments in an increasing reactant concentration order, and ensure that the system maintains anaerobic conditions by avoiding introduction of air when changing the tonometers or loading the SF drive syringes.
11. A set of SF experiments can in addition be used to show the effect of other parameters such as temperature, pH and isotopic effects on the kinetics of the reaction.
12. Alternatively, you can also produce the hydroquinone form of your flavoenzyme by photoreduction within the tonometer in the presence of 3 mM EDTA and 3 μM 5-deazariboflavin [19].
13. Note that in general, A, B, and C are spectral species, reflecting a distribution of enzyme intermediates (reactants, CTCs, products, Michaelis Complexes, ...) at a certain point along the reaction time course and do not necessarily represent a single distinct reaction intermediate. Moreover, in general none of them represents individual species, and therefore their spectra cannot be included as fixed values in the global fitting.
14. Various concentration dependent multiwavelength data sets can also be used collectively for estimation of absorption spectra of intermediate species. In this case you must use a software that allows a simultaneous analysis of a series of kinetic traces according to a proposed reaction scheme, such as Pro-Kineticist II (*Applied Photophysics Ltd.*). Load your data, the proposed reaction scheme for the overall process under study and the spectra of the known oxidized and reduced reactants pure species. Use the fitting routine and the software will provide the best-fit parameter set that accommodates the

behavior of all kinetics across the entire concentration range, as well as the spectra of all the intermediate species involved.

15. The kinetic parameters obtained in this way satisfy all the wavelength data at each time point and therefore provide a robust fit that allows for the study of complex reactions involving spectroscopic intermediate species.
16. Errors in the determination of intermediate species spectra and extinction coefficients highly depend on the maximal amount stabilized and the lifetime of these species for each one of the analyzed processes.

References

1. Eccleston JF, Hutchinson JP, White HD (2001) Stopped-flow techniques. In: Harding SE, Chowdhry BZ (eds) Protein-ligand interactions. Structure and spectroscopy: a practical approach. Oxford University Press, Oxford, pp 201–237
2. DeSa RJ, Gibson QH (1969) A practical automatic data acquisition system for stopped-flow spectrophotometry. Comput Biomed Res 2:494–505
3. Gibson QH, Milnes L (1964) Apparatus for rapid and sensitive spectrophotometry. Biochem J 91:161–171
4. Massey V (2000) The chemical and biological versatility of riboflavin. Biochem Soc Trans 28:283–296
5. Massey V (1995) Introduction: flavoprotein structure and mechanism. FASEB J 9:473–475
6. Ghisla S, Massey V (1989) Mechanisms of flavoprotein-catalyzed reactions. Eur J Biochem 181:1–17
7. Müller F (1991) Free flavins: synthesis, chemical and physical properties. In: Müller F (ed) Chemistry and biochemistry of flavoenzymes. CRC Press, Boca Raton, FL, pp 1–71
8. Walsh CT, Wencewicz TA (2013) Flavoenzymes: versatile catalysts in biosynthetic pathways. Nat Prod Rep 30:175–200
9. Fraaije MW, Mattevi A (2000) Flavoenzymes: diverse catalysts with recurrent features. Trends Biochem Sci 25:126–132
10. Macheroux P (1999) UV-visible spectroscopy as a tool to study flavoproteins. Methods Mol Biol 131:1–7
11. Massey V, Palmer G (1962) Charge transfer complexes of lipoyl dehydrogenase and free flavins. J Biol Chem 237:2347–2358
12. Tejero J et al (2007) Catalytic mechanism of hydride transfer between NADP+/H and ferredoxin-NADP+ reductase from Anabaena PCC 7119. Arch Biochem Biophys 459:79–90
13. Daff S (2004) An appraisal of multiple NADPH binding-site models proposed for cytochrome P450 reductase, NO synthase, and related diflavin reductase systems. Biochemistry 43:3929–3932
14. Chapman SK, Reid GA (1999) Flavoprotein protocols. Humana Press, Totowa, NJ
15. Pimviriyakul P, Surawatanawong P, Chaiyen P (2018) Oxidative dehalogenation and denitration by a flavin-dependent monooxygenase is controlled by substrate deprotonation. Chem Sci 9:7468–7482
16. Chaiyen P et al (2004) Use of 8-substituted-FAD analogues to investigate the hydroxylation mechanism of the flavoprotein 2-methyl-3-hydroxypyridine-5-carboxylic acid oxygenase. Biochemistry 43:3933–3943
17. Batie CJ, Kamin H (1986) Association of ferredoxin-NADP+ reductase with NADP (H) specificity and oxidation-reduction properties. J Biol Chem 261:11214–11223
18. Pollegioni L, Fukui K, Massey V (1994) Studies on the kinetic mechanism of pig kidney D-amino acid oxidase by site-directed mutagenesis of tyrosine 224 and tyrosine 228. J Biol Chem 269:31666–31673
19. Medina M, Martínez-Júlvez M, Hurley JK, Tollin G, Gómez-Moreno C (1998) Involvement of glutamic acid 301 in the catalytic mechanism of ferredoxin-NADP+ reductase from Anabaena PCC 7119. Biochemistry 37:2715–2728
20. Batie CJ, Kamin H (1984) Electron transfer by ferredoxin:NADP+ reductase. Rapid-reaction evidence for participation of a ternary complex. J Biol Chem 259:11976–11985
21. Gorelick RJ, Schopfer LM, Ballou DP, Massey V, Thorpe C (1985) Interflavin oxidation-reduction reactions between pig

kidney general acyl-CoA dehydrogenase and electron-transferring flavoprotein. Biochemistry 24:6830–6839

22. Reis RA, Ferreira P, Medina M, Nonato MC (2016) The mechanistic study of *Leishmania major* dihydro-orotate dehydrogenase based on steady- and pre-steady-state kinetic analysis. Biochem J 473:651–660
23. Ferreira P et al (2015) Aromatic stacking interactions govern catalysis in aryl-alcohol oxidase. FEBS J 282:3091–3106
24. Seo D, Soeta T, Sakurai H, Sétif P, Sakurai T (2016) Pre-steady-state kinetic studies of redox reactions catalysed by Bacillus subtilis ferredoxin-NADP(+) oxidoreductase with NADP(+)/NADPH and ferredoxin. Biochim Biophys Acta 1857:678–687
25. Hedison TM, Hay S, Scrutton NS (2015) Real-time analysis of conformational control in electron transfer reactions of human cytochrome P450 reductase with cytochrome c. FEBS J 282:4357–4375
26. Meints CE, Gustafsson FS, Scrutton NS, Wolthers KR (2011) Tryptophan 697 modulates hydride and interflavin electron transfer in human methionine synthase reductase. Biochemistry 50:11131–11142
27. Marshall KR et al (2005) The human apoptosis-inducing protein AMID is an oxidoreductase with a modified flavin cofactor and DNA binding activity. J Biol Chem 280:30735–30740
28. Belcher J et al (2014) Structure and biochemical properties of the alkene producing cytochrome P450 OleTJE (CYP152L1) from the Jeotgalicoccus sp. 8456 bacterium. J Biol Chem 289:6535–6550
29. Dai Y, Haque MM, Stuehr DJ (2017) Restricting the conformational freedom of the neuronal nitric-oxide synthase flavoprotein domain reveals impact on electron transfer and catalysis. J Biol Chem 292:6753–6764
30. Chenprakhon P, Trisrivirat D, Thotsaporn K, Sucharitakul J, Chaiyen P (2014) Control of C4a-hydroperoxyflavin protonation in the oxygenase component of p-hydroxyphenylacetate-3-hydroxylase. Biochemistry 53:4084–4086
31. Pennati A et al (2006) Role of the His57-Glu214 ionic couple located in the active site of Mycobacterium tuberculosis FprA. Biochemistry 45:8712–8720
32. Argyrou A, Sun G, Palfey BA, Blanchard JS (2003) Catalysis of diaphorase reactions by Mycobacterium tuberculosis lipoamide dehydrogenase occurs at the EH4 level. Biochemistry 42:2218–2228
33. Fitzpatrick TB et al (2001) Chorismate synthase from the hyperthermophile Thermotoga maritima combines thermostability and increased rigidity with catalytic and spectral properties similar to mesophilic counterparts. J Biol Chem 276:18052–18059
34. Fraaije MW, van den Heuvel RH, van Berkel WJ, Mattevi A (1999) Covalent flavinylation is essential for efficient redox catalysis in vanillyl-alcohol oxidase. J Biol Chem 274:35514–35520
35. Pérez-Amigot D et al (2019) Towards the competent conformation for catalysis in the ferredoxin-NADP+ reductase from the Brucella ovis pathogen. BBA-Bioenergetics 1860:148058

Chapter 10

Atomic Force Microscopy: Single-Molecule Imaging and Force Spectroscopy in the Study of Flavoproteins Ligand Binding and Reaction Mechanisms

Anabel Lostao and Milagros Medina

Abstract

Atomic force microscopy (AFM) is one of the most versatile tools currently used in nanoscience. AFM allows for performing nondestructive imaging of almost any sample in either air or liquid, regardless whether the specimen is insulating, conductive, transparent, or opaque. It also allows for measuring interaction forces between a sharp probe and a sample surface, therefore allowing to probe nanomechanical properties of the specimen by either applying a controlled force or pulling the sample. It can provide topography, mechanical, magnetic, and conductive maps for very different type of samples. Transferred to the field of biology, today, AFM is the only microscopy technique able to produce images from biomolecules to bacteria and cells with nanometric resolution in aqueous media. Here, we will focus on the biological applications of AFM to flavoproteins. Despite references in the literature are scarce in this particular field, here it is described how imaging with AFM can contribute to describe catalysis mechanisms of some flavoenzymes, how oxidation states or binding of relevant ligands influence the association state of molecules, the dynamics of functional quaternary assemblies, and even visualize structural differences of individual protein molecules. Furthermore, we will show how force spectroscopy can be used to obtain the kinetic parameters, the dissociation landscape and the mechanical forces that maintain flavoprotein complexes, including the possibility to specifically detect particular flavoproteins on a sample.

Key words Atomic force microscopy, Forces, Quaternary assemblies, Interactions, Dissociation landscape, Single molecule, Ligand binding

1 Introduction

The development of techniques able to manipulate and measure the signals of individual molecules has revolutionized the study of biological processes at the molecular level [1]. These "single-molecule techniques" offer the possibility to directly sample distributions of molecular properties, allowing for identification of subpopulations. This capability is not viable with bulk biochemical or biophysical assays where provided values are the average of the signals, canceling their variability, assuming that all the molecules

Maria Barile (ed.), *Flavins and Flavoproteins: Methods and Protocols*, Methods in Molecular Biology, vol. 2280,
https://doi.org/10.1007/978-1-0716-1286-6_10,

behave the same. Among the current single molecule methods stand out those designed to measure forces between molecules or inside an individual molecule [2]. The most broadly used technique in this area is atomic force microscopy (AFM) [3] in the force spectroscopy mode [4], but other techniques are also available, as the biomembrane force probe [5], the laminar flow chambers [6], and optical [7] and magnetic tweezers [8]. All of them can manipulate, follow motions at the nm level and apply forces in a large range depending on the approach used. The magnetic tweezers may even introduce and measure rotations as a function of the force, which allows to study rotary motors or specific enzymes [9]. This area of research tackles the complex relation between force, lifetime, and chemistry in single molecular bonds, known as dynamic force spectroscopy (DFS), used to analyze the dynamics of the process observed, specifically, by measuring force as a function of the loading rate or velocity at which the force is applied. Processes usually studied are dissociation of ligand–receptor complexes under force [4], molecular stretching of a protein to analyze unfolding [10], stretching of nucleic acids [11], and even unraveling of enzyme catalysis mechanisms [9]. However, while the main advantage of the other cited methods, specially both type of tweezers, is that they can measure tremendously small forces, reaching the hundredth of the pN, the AFM allows, in addition, to measure forces in the range of $10–10^4$ pN, visualizing samples with subnanometer resolution.

AFM belongs to the scanning probe microscopy (SPM) family where a very sharp tip is set in (weak) interaction with a sample and scanned on it while interaction is measured and controlled. The tip tracks surface morphology while its XYZ position is registered by the electronics to compose a 3D map of the sample surface. Since the interaction can be controlled and limited to very low values, this kind of imaging is usually nondestructive. Depending on the type of interaction measured, SPMs take different names, such as scanning tunneling (STM; current between tip and samples), atomic force (AFM; force between last part of the tip and sample), magnetic probe (MFM; magnetic force between tip and samples), and scanning near field optical (SNOM; optical coupling). Today, AFM [3] has become a powerful and versatile technique being unique at imaging specimen of almost any type, not only in air or vacuum, but also in aqueous media at subnanometer resolution. It can image from biomolecules to cells and bacteria without the use of contrast enhancement, but requiring immobilization onto an ultraflat surface to prevent from drag phenomena. AFM can provide maps on the topology, adhesion, elasticity, dynamics, and other properties of biological samples in physiologically relevant buffers [12]. Figure 1a shows the main components of an AFM: a very sharp tip at the end of a flexible cantilever, which behaves according to Hooke's law (Fig. 1b), a sensitive deflection optical sensing

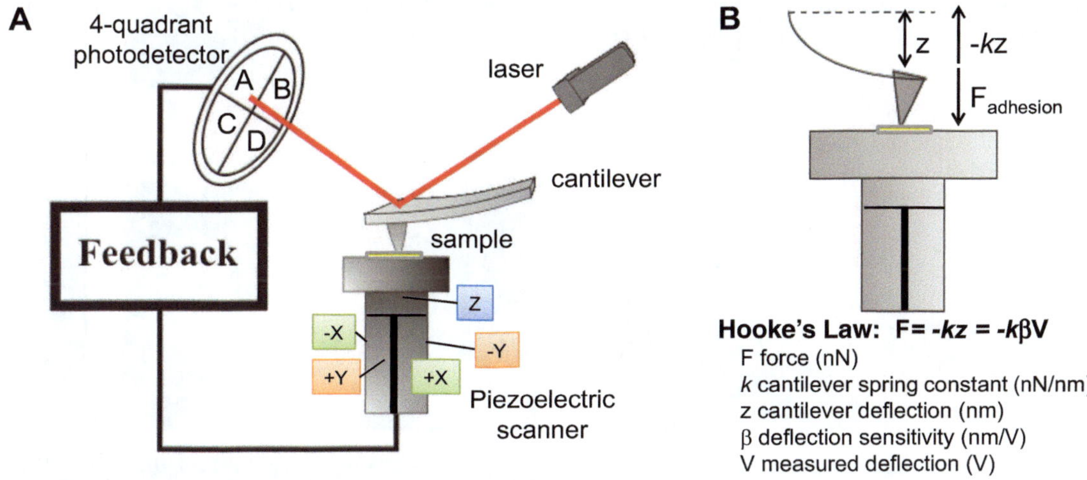

Fig. 1 The AFM equipment. (**a**) Schematic diagram of an AFM, of the scanned-sample type, where the sample is located on the piezoelectric scanner, and based on an optical beam deflection system. In the case of scanned probe type, it is the tip that is scanned instead of the sample. The scanner is the nanopositioning element allowing movement with subnanometer precision: it works by applying opposite voltages to $\pm X$ and $\pm Y$ sectors to move the sample in X and Y directions, respectively; a Z sector moves the sample in the vertical direction. A laser beam strikes on the reverse face of the lever and deflects toward a segmented photodiode. The resulting electrical signal that is proportional to the cantilever deflection is fed into a signal processor and compared with the desired deflection (set point or feedback parameter). When a correction is needed the voltage to the z-piezo is changed to correct the height of the cantilever with respect to the sample. (**b**) The cantilever acts as a small spring and the deflection can be converted into a force F acting on the cantilever using the Hooke's law. As signals are measured in volts, estimation of cantilever deflection sensitivity is required (*see* Subheading 3.2, **step 8**)

system capable of controlling with high accuracy the tip–sample relative position, and a 3D positioning system, called piezoelectric scanner, to adjust tip–sample relative position according to a set-point parameter through a feedback electronics.

Regarding flavoproteins, studies using single molecule techniques are still scarce [13], but recent reports show the enormous potential of AFM in the study of their mechanisms of action. Two different approaches can be found. The first based on the visualization of the influence of the formation of flavoprotein–ligand complexes at relevant buffered conditions in the conformation, association degree and/or topology flavoprotein conformational. This task is favored by using ultrasharp probes that allow to distinguish even among different domains, something that allows us to unequivocally identify their quaternary organization as well as to verify that the assemblies preserve their catalytic activity when adsorbed on mica and when desorbed after imaging [14]. That is the case of the organized oligomeric structures found in the catalytic pathway of the bifunctional FAD synthase (FADS) from *C. ammoniagenes*, which envisage (1) how reactants and products

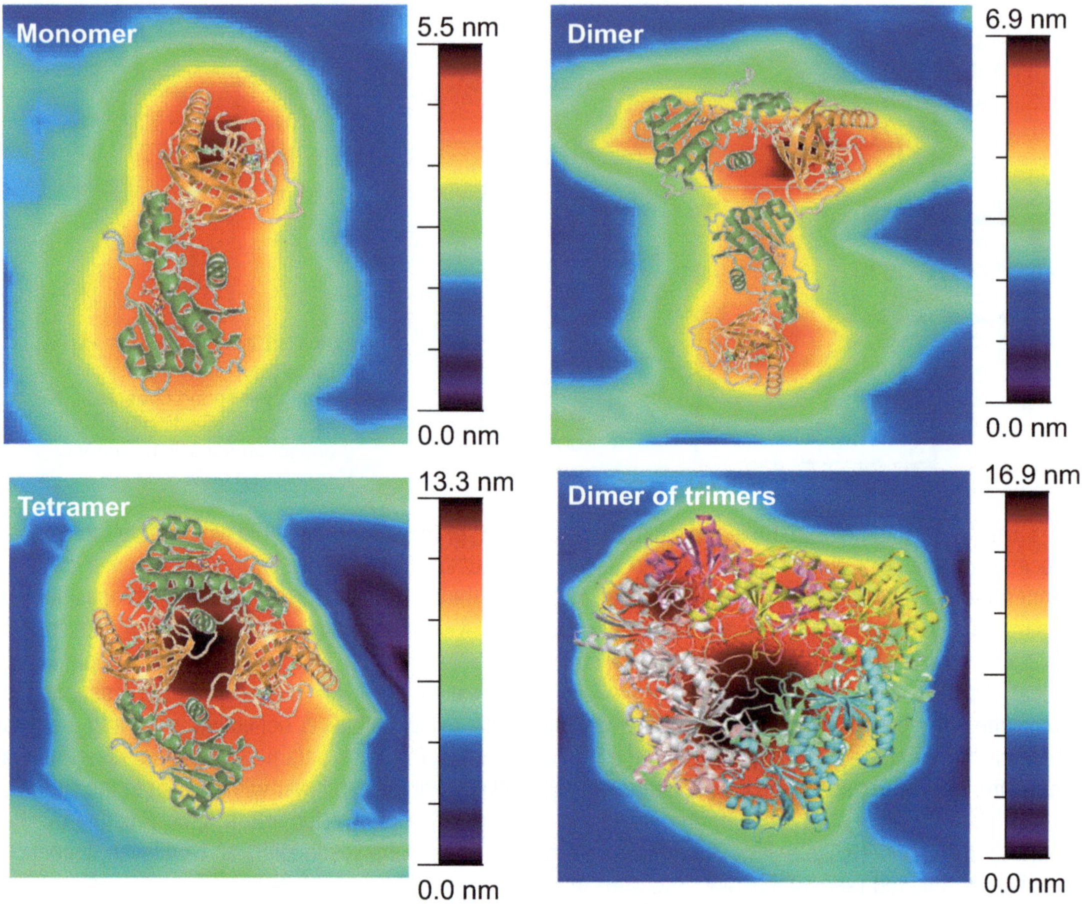

Fig. 2 AFM topography images of FAD synthase from *Corynebacterium ammoniagenes* individual associates upon incubation with ligands. The oligomerization pattern observed at the images can be estimated and translated into a detailed table to unravel the stoichiometry and other mechanistic aspects of this bifunctional enzyme involved in the production of FMN and FAD in prokaryotes. The structure of the protein is superposed on the images to clarify the identification of the individual protomers. From left to right and up to down: Monomer obtained when incubating with FMN, it is possible to distinguish among the RFK (brown) and the FMNAT (green) domains. Dimers obtained in different conditions showed a head–tail interaction between the N-terminal and C-terminal modules of two protomers, making a ~90° angle. Tetramers observed upon incubation with RF. Dimer of trimers imaged upon simultaneous incubation with the products of the RFK reaction, ADP:Mg^{2+} and FMN. All measurements in 20 mM PIPES, 2 mM DTT, pH 6.0. These measurements and results were described previously [14]. The scale of the images in the XY plane has been adapted to maximize observation of single oligomers

can transfer among domains within a dimer of trimers assembly (Fig. 2) [14, 15], as well as (2) how mutants affecting the RFK activity promote other quaternary organizations [16]. In addition, AFM has allowed to describe these organizations as species-specific, showing that the FADS from *Streptococcus pneumoniae* forms much less and different quaternary assemblies [17]. Other interesting case is the analysis of the human apoptosis-inducing factor (hAIF) where AFM imaging demonstrated clearly the displacement of the monomer–dimer equilibrium toward the dimer after its flavin ring reduction by NADH while a variant affected in dimerization did not; these and other results suggested that both the mitochondrial and apoptotic functions of hAIF are interconnected and coenzyme controlled [18]. Furthermore, AFM imaging of hAIF variants at the flavoenzyme active site at positions contributing to either binding of the FAD cofactor and/or the NADH coenzyme further contributed to understand the redox dependence of the monomer–dimer conformational transition of hAIF [19]. The second approach to the study of flavoenzymes is based on the use of DFS described above. This method requires covalent immobilization of the flavoenzyme on a nanoflat substrate, and of the ligand at the AFM tip, preferably through a flexible linker that constitutes a fingerprint of the specific rupture force peaks attributed to the complex bonds. In this case the exhaustive study of the ferredoxin-NADP$^+$ reductase (FNR):ferredoxin (Fd)/flavodoxin (Fld) complexes from *Anabaena* constitutes a good example [20, 21]. The interaction of the reductase with the natural electron donor, Fd, was much stronger and its lifetime longer and more specific than that with the substitute under iron-deficient conditions, Fld. A new method for the covalent attachment of FNR optimizing the recognition ability was developed [20]. The differences in the recognition efficiency of this orienting method regarding a random attachment procedure, together with the nanomechanical results, showed two binding models for these systems [22]. The higher bond probability and two possible dissociation pathways in Fld binding to FNR were attributed to the nature of this complex, closer to a dynamic ensemble model. This was in contrast with the one step dissociation kinetics observed for the FNR–Fd complex, in agreement with the specific interactions described for this complex. The force data of this analysis was also used to develop a new theoretical framework that proposes a new free energy profile for the dissociation process characterized by two key magnitudes, a free energy barrier $\Delta G^{\dagger}$ and the dissociation free energy ΔG^0. $\Delta G^{\dagger}$ can be estimated from the dependence of the rupture force with the loading rate, related by the dynamic force spectroscopy theory [23]; and ΔG^0 can be calculated from measured irreversible work between two equilibrium states as the integral of the rupture force peaks, which converges with the calorimetry-determined values for ΔG^0 at low loading rates [24], bridging the gap between bulk and single molecule techniques [25, 26]. Finally, comment that the use of ligand-functionalized tips can be used for the detection or localization of a particular flavoprotein on a substrate or directly on a membrane through adhesion force maps [20, 27].

2 Materials

Prepare all solutions using ultrapure milli-Q water.

2.1 AFM Setup

1. Cervantes FullMode SPM (Nanotec Electronica, S.L) or other commercial AFM system.
2. AFM liquid cell for cantilever holder in liquid operation.
3. AFM cantilevers for imaging in fluid. V-shaped silicon nitride cantilevers with integrated pyramidal 2-nm final radius tips with spring constants from 0.01 to 0.06 N/m and nominal resonance frequencies of 7–18 kHz (Bruker Probes, SNL and MSNL Lever Probes). These probes are optimum for the Jumping Mode (Nanotec Electronica, S.L), the amplitude-modulation modes and the other intermittent force modes operated in fluid, which are described in this chapter.
4. AFM functionalized cantilevers for force spectroscopy measurements. Maleimide-terminated flexible polyethylene glycol (PEG) linker silicon nitride AFM cantilevers (CT.PEG.MAL, MW 3400; Novascan Technologies Inc.) (Fig. 3c). Two different types of probes are recommended, one with triangular shape and spring constant of 0.06 N/m and other with rectangular shape and spring constant of 0.02 N/m.
5. WSxM freeware for data analysis (Nanotec Electrónica, S.L., http://www.wsxm.es/download.html), or other suitable software.
6. 20% isopropanol and Milli-Q quality water.

2.2 Preparation of Flavoprotein Samples for Imaging Analysis

1. V-5 quality muscovite mica (Electron Microscopy Sciences).
2. Invisible adhesive tape (19 × 33 mm).
3. Thin hooked tip tweezers (72750-D, Dumoxel Alloy, Electron Microscopy Sciences).
4. Blunt-tip scissors (histological quality).
5. 0.5 μM solutions of purified flavoprotein (*see* **Note 1**).
6. 20 mM PIPES, pH 6.0 (buffer and pH selection depends on enzyme isoelectric point and stability; *see* **Note 1**).
7. 50–250 μM solutions of flavoprotein ligands; (1) small molecules acting as substrates, products, or modulators, such as ATP, ADP and PPi, RF, FMN and FAD, NADH, NADPH, NAD^+, and $NADP^+$, or (2) protein partners (*see* **Note 2**).
8. Freshly prepared 2 mM dithiothreitol (DTT) or solutions including cations at different concentrations, if required to avoid protein oxidation or to produce ligand binding (*see* **Note 2**).

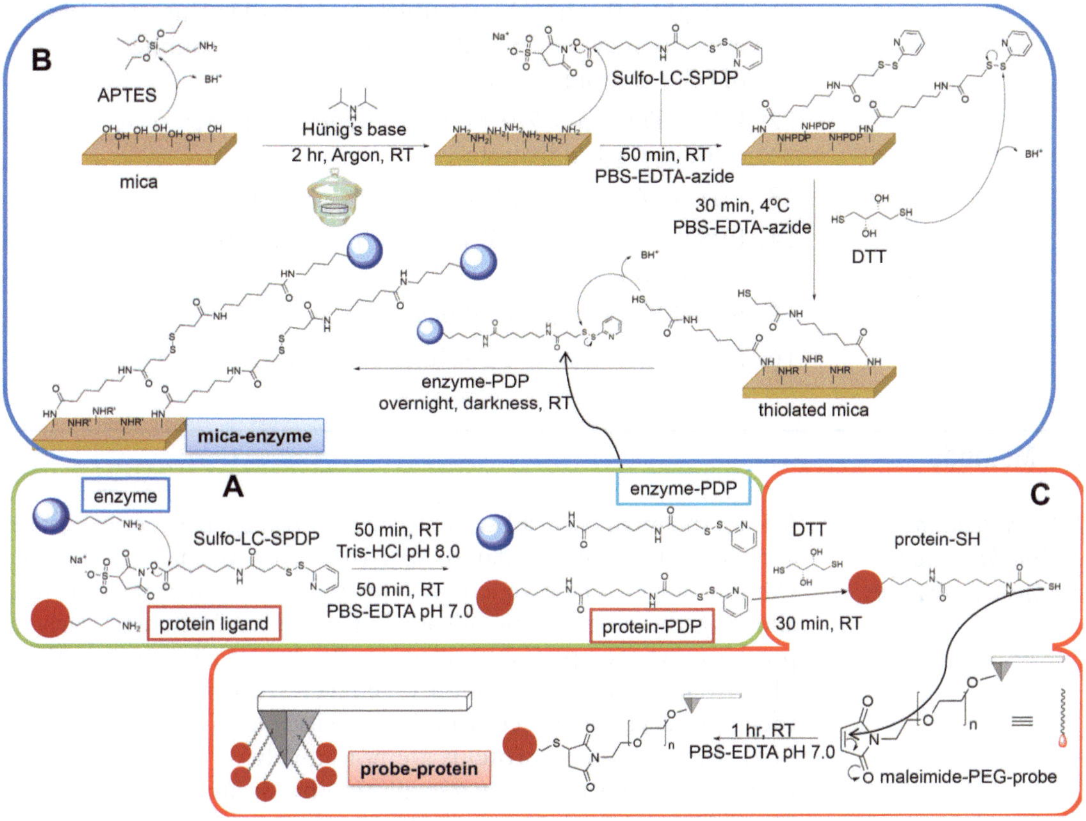

Fig. 3 Schemes for sample functionalization. Labeling of the enzyme and the protein ligand with a bifunctional chemical cross-linker (**a**). Main functionalization steps to prepare enzymatic samples covalently attached to mica sheets (**b**), and AFM probe tips functionalized with proteins through PEG linkers (**c**). These immobilization methods are required as previous steps for FS. *See* [20] for more details

2.3 Preparation of Flavoprotein Samples for Force Spectroscopy (FS) Analysis

1. **Items 1–4** as in Subheading 2.2.
2. 0.5 μM solution of purified flavoenzyme and 0.5 mM solution of protein ligand (*see* **Note 1**).
3. Solutions of 20 mM sulfosuccinimidyl 6-(3′-[2-pyridyldithio] propionamido) hexanoate (Sulfo-LC-SPDP; Pierce) in either 50 mM Tris-HCl, pH 8.0 or milli-Q water (Fig. 3).
4. PD MiniTrap Sephadex G-25 desalting columns for size exclusion separation (GE Healthcare).
5. 50 mM Tris-HCl, pH 8.0, or any other buffer where the studied flavoprotein is stable and functional.
6. UV-Vis spectrophotometer.
7. Petri dishes, watch glasses, vacuum grease.
8. Desiccator with an input connectable to Argon.
9. Aminopropyl triethoxysilane (APTES) (Fig. 3b).
10. N,N-diisopropylethylamine (Hünig's base) (Fig. 3b).

11. PBS pH 7.0 (phosphate-buffered saline; pH values in Subheading 3.6 may be in 7.0–7.5 range, but in Subheading 3.7, pH 7.0 is more strictly required).
12. PBS–EDTA pH 7.0 (1 mM ethylenediaminetetraacetic acid, disodium salt dihydrate).
13. PBS–EDTA–azide pH 7.0 (0.02% sodium azide).
14. PBST–SDS pH 7.0 (PBS–0.2% Tween 20–0.1% sodium dodecyl sulfate).
15. 6-well ELISA Sterile *Nunclon Surface* plates.
16. 50 and 150 mM DTT solutions in PBS–EDTA–azide.

3 Methods

3.1 Preparation of Protein Samples for Imaging Analysis

Mica is the most used substrate to image proteins because it exposes a clean and atomically flat surface that after mechanical cleavage exhibits a net negative charge, attributed to hydroxyl groups, allowing the electrostatic adsorption of molecules, particularly of proteins whose pI is similar or higher than the working pH. Evaluation of adsorption of enzymes on mica showed that they preserve their enzymatic activity, even after desorbed [14]. In cases where proteins are not adsorbed to the mica at the working pH, it is necessary to explore other strategies (*see* **Note 1**).

1. Prepare 150 μl of each type of flavoprotein solution at 0.5 μM in 20 mM PIPES, pH 6.0 (*see* **Note 1**). Check that this flavoprotein concentration allows producing images that individually resolve molecular species as shown in Fig. 2 (otherwise, *see* **Note 1**).
2. If you wish to evaluate the effect of ligand binding, as well as of subsequent catalytic process in the case of enzymes, changes in flavoprotein conformation or association patterns, prepare sets of protein samples incubated with different combination of ligands and/or at different ratios (including substrates, substrates analogs, and/or products). To favor formation of complexes mixtures can be preincubated at 4 °C for 10 min under mild stirring. To prevent from formation of intramolecular and intermolecular ***disulfide bonds*** add to your sample 2 mM of freshly prepared DTT. When studying binding of nucleotide ligands dependent of Mg^{2+} (such as ATP) supplement your sample with $MgCl_2$ (Fig. 2) (*see* **Note 2**).
3. To evaluate the incubation time influence on the proportion of observed features, incubate samples at different times, from a few seconds to overnight, before imaging. Try relevant flavoprotein ligands such as substrates, flavins, or coenzymes (Fig. 2) (*see* **Note 2**). Samples containing free flavins are protected from light by using opaque eppendorf tubes to minimize side photochemical reactions.

4. Cut a mica sheet into pieces of around 1 cm^2 using blunt-tip scissors, exfoliate twice a piece using invisible adhesive tape, add 150 μl of the flavoprotein sample, free (control) or incubated with ligands, and incubate for 10 min at room temperature. Wash the mica sheet three times with 20 mM PIPES, pH 6.0 to remove weakly joined molecules. Cover the sample with the same buffer.

3.2 AFM Imaging in Fluid

Standard AFMs use an optical detection system based on a laser beam deflection method to monitor the deflection of the cantilever during scanning (*see* Fig. 1a).

1. Connect to the system the suitable piezoelectric scanner with the best resolution to visualize biomolecules. It is usually the smallest one, with a range of about 10 μm in the XY plane, and also compatible with the measurement in liquid.
2. Turn on the AFM electronics module for at least 15 min to warm it up, and run the acquisition software. Select the SPM mode (in general, these systems offer the possibility of working with other SPM, e.g., STM, MFM, and KPM), AFM, and then the correct scanner (these systems offer the possibility of working with short, around 10 μm^2, and long, around 100 μm^2, scanners, apart from other special scanners for spectroscopy, temperature control, etc.) to introduce the corresponding calibration parameters.
3. Wash three times the liquid cell and the cantilever holder with ultrapure water, then rinse them with 20% isopropanol and finally dry them with argon or nitrogen.
4. Using clean tweezers insert the selected cantilever chip onto the cantilever probe holder and load it into the AFM head.
5. Align the laser beam to focus it on the cantilever end, particularly on the back of the tip, assisted by the optical microscope mounted on the system.
6. Adjust the photodiode position maximizing the total intensity of light.
7. Place the wet sample mica piece onto the AFM scanner plate using the tweezers—most systems move the sample relative to the probe (*see* Fig. 1a), but certain models move the probe attached to the scanner-. Cover the sample with 20 mM PIPES, pH 6.0, and place a 50 μL drop of the same buffer on the cantilever chip. Take care to prevent sample from getting dry, consider you might require adding buffer to the sample after some time. Return the AFM head to the acquisition position. Adjust the height of the stage so that the sample comes into contact with the buffer suspended on the cantilever (be careful not to get too close, crashing will break the cantilever tip). The presence of liquid produces displacement of the laser position

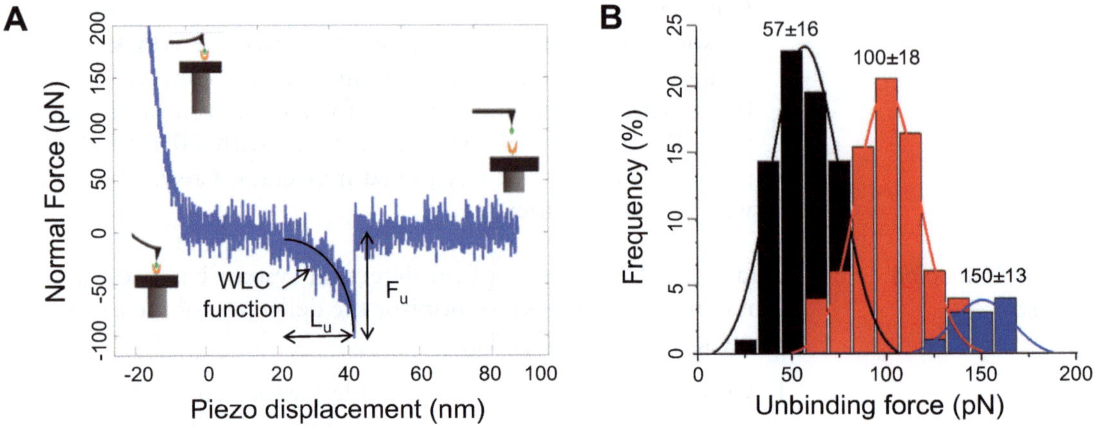

Fig. 4 Force determinations (**a**) representative experimental retraction force curve showing a specific unbinding event corresponding to a single FNR–Fd complex [21]. The Fz scan does not show nonspecific adhesion peaks that usually follow the slope of the retraction. The rupture event of force F_u occurs at the unbinding length L_u, or tip–sample separation that is close to the length of the stretched linker, around 20 nm in these measurements, given by the piezo displacement encompassing the non-linear portion of the retraction curve before the rupture. The superposed black line represents the corresponding stretch of PEG according to the WLC function. The shape of the force peak and the distance at which it occurs add certainty to measured rupture forces coming from recognition events under study and not from artifacts or nonspecific tip–sample adhesions. (**b**) Force histogram distribution for the FNR–Fd complex operating at R 10 nN. s^{-1}. It is possible to assign the grouped smaller values of the first fitting to the rupture of single complexes, the result is known as "the most probable rupture force," 57 pN, and subsequently attribute greater values to the simultaneous rupture of two, 100 pN, or three complexes, 150 pN [21]

on the cantilever in the longitudinal direction, realign it. Readjust also the photodiode detector position using the AFM software; maximize the total signal (A + B + C + D of the quadrant photodetector), and reduce the difference signal (A + B − C − D)/(A + B + C + D) so that it is zero (*see* Fig. 1a). Monitoring the laser light on a small piece of paper held between the photodetector and the optical lever path can help to ensure that the light beam concentrates into a small point of the lever and not outside it.

8. Record a force-distance curve (Fz) (Fig. 4a) and, aided by the cursors, select a straight section and estimate the slope in the retraction. Data are stored by the system during the measurement. These data provide the sensitivity to deflection (in nm. V^{-1}), that is, the relation between the distance the cantilever moves (in nm) and the output it generates at the detection system (in V) (*see* Fig. 1b).
9. Then calibrate the cantilever real spring constant. If your AFM control software includes a calibration method, follow the manufacturer's instructions. Otherwise, you can use the commonest thermal method (*see* **Note 3**).

10. Select the Jumping Mode operation (JM) or other mode compatible with soft samples in fluid (*see* **Note 4**). JM is a suitable nonintrusive scanning method for soft samples and especially for biomolecules [28]. JM is an intermittent contact mode and works by performing a force curve at each point of the sample surface with a feedback time in between, the feedback signal being the applied normal force. At maximum tip–sample separation, the tip is moved laterally to the next point, avoiding lateral forces. Since the tip is moving in and out of contact, an algorithm is used to refresh the zero force level, ensuring a constant set point. This feature is appropriate for a careful control of the forces applied to biomolecules. For each pixel of the image maximum tip–sample adhesion and height are registered and stored, allowing for quick production of simultaneous topography and adhesion images. To obtain quality images that allows visualizing biomolecules in liquid apply: (1) normal forces of 150 pN or less, corresponding with low 0.01–0.10 voltages for the set point (begin with higher voltages and then try to lower them as the measurement stabilizes), scan rates of around 1–2 lines per second, high proportional/integral (P/I) gains (e.g., 400/200), begin using a jump off—maximum distance between the tip and the point on every jump—of around 250 nm and try to lower progressively till 150, jump sample around 50—number of steps taken for every jump, and control cycles around 6—time the tip spends with the feedback closed after every jump; (2) it is advisable to turn the oscilloscope on—or place an external one if the microscope does not have one—to visualize Fzs that present hysteresis attributed to the dragging force due to liquid viscosity. Softer cantilevers, as the required here, produce bigger hysteresis; (3) operation in repulsive regime generates a better measurement, do an Fz and select a value inside the repulsive area to select a set point (*see* **Note 5**). Press "approach" trying to enter JM imaging range. If reached optimally, hysteresis should appear. Choose an area (e.g., 2 μm) and begin to scan. Depending on observed image quality, parameters might need to be changed until features attributed to the studied flavoprotein are clearly resolved.

11. Acquire images of different sizes and from different sample areas to get representative features. If false approaches occur or the piezo remains stretched despite continuing to push "approach" several times, change the set-point value (*see* **Note 6**). Once measurement is stable to achieve the best resolution images, that is, the measurement remains in image range and no artifacts appear, try to reduce the scan size to 400 × 400 nm, increment to 512 × 512 pixels the resolution.

3.3 Analysis of AFM Images to Characterize Protein Species Upon Ligand Binding

1. Images have to be analyzed in detail to characterize each feature or individual "particle" found, and to assign it to a single protein molecule or to an oligomer, distinguishing also among different conformers (*see* Fig.2). AFM images can be analyzed with the free WSxM software for SPM image processing [29], or any other suitable software (*see* **Note** 7).
2. If ligand binding produces characteristic features, make use of available structural information, as PDB 3D structure, to identify dimensions and shapes (*see* Fig. 2).
3. Percentages for different quaternary assemblies can be estimated by the identification of each isolated species using the zoom function (*see* Fig. 2). This allows observing each potential oligomer in detail, which complemented with their shape and height profiles can make possible to differentiate each monomer within the oligomer and, as consequence, to unequivocally distinguish the type of association. At least ten images of ten different areas of a 500 × 500 nm sample surface need to be analyzed, and a minimum of 250 features examined for each sample. Quantify only clearly identified features, discarding unclear elements. Determine percentages of each species as the existing number of this species with respect to the total number of species in the sample. Calculate error from the dispersion of results in the analysis of different AFM images corresponding to different areas of the sample.
4. The interaction of ligand to flavoproteins can produce structural conformational changes or formation of quaternary assemblies that can be observable by AFM, but in occasions the go beyond the discrimination capability of the technique (*see* **Note 8**).
5. In addition to the zooms of 2D and 3D topography maps, data can be shown in the form of a table. This allows to show the estimated percentage of each conformational form either relative to the total number of species found or to the total of molecules [14].

3.4 Labeling of Enzymes for Force Spectroscopy

Enzymes are labeled using the water-soluble heterobifunctional cross-linker sulfo-LC-SPDP. This cross-linker contains an amine-reactive end, an N-hydroxysuccinimide (NHS) ester group able to bind to the primary amines (-NH_2) at lateral chains of surface lysines. The other terminal group of the cross-linker, 2-pyridyldithiol (PDP), contains a sulfhydryl-reactive portion that optimally reacts with protein sulfhydryl groups resulting in displacement of the pyridine-2-thione molecule and the formation of a disulfide bond (*see* Fig. 3a).

1. To label the enzyme prepare a fresh 20 mM SPDP solution in Tris-HCl and add 15 μl of this solution by milligram of protein in the same buffer. Incubate during 50 min at room

temperature an under mild stirring. Tagged species of free enzyme carrying a C9-long arm ending in a 2-pyridyldithiol group (enzyme-PDP) will be produced.

2. Use a PD MiniTrap desalting column to isolate enzyme-PDP molecules from unreacted SPDP and other molecules, as well as to concentrate the labeled complex. Centrifugation is performed at 1000 × *g* during 2 min using the same buffer.
3. Use UV-Vis spectrophotometry and the Lambert–Beer law, taking the absorption at the Flavin band-I and the enzyme molar absorption coefficient at this wavelength, to quantify the enzyme-PDP concentration.
4. Store the sample at −20 °C until used.

3.5 Covalent Immobilization of Enzymes on Mica for Force Spectroscopy

Freshly cleaved square muscovite mica pieces of approximately 1 cm wide are used as the functionalization substrate for enzyme immobilization. Cleaved mica pieces are fixed on 6-well ELISA *Nunclon Surface* plates in order to proceed with the functionalization process following several steps. APTES is used in vapor phase to achieve a monolayer of amine groups on mica, contrary to what happens when is directly incubated in liquid phase. Contrary to the separated molecules suitable for topography analysis, dynamic force spectroscopy (DFS) measurements require saturated enzymatic layers to favor complex formation. The main steps can be followed in the schemes of Fig. 3b.

1. Place cleaved mica pieces separately on a clean container, for example in an open petri dish, inside a desiccator, and expose them to APTES and Hünig's Base vapors in a 3:1 volume ratio, both placed on different watch glasses, under argon atmosphere for 2 h. APTES will become vaporized and covalently bound to the hydroxyl groups on the mica surface. Store aminated mica pieces in the desiccator till required.
2. Dilute freshly prepared 20 mM SPDP in milli-Q water by 20 times in PBS–EDTA–azide. Add EDTA as chelator of heavy metal cations and sodium azide to prevent microorganism contamination. Attach each mica piece at the bottom of each plate well aided by a small portion of vacuum grease stuck with a pipette tip. Add 200 μl of the solution to each mica-NH_2 piece and incubated for 50 min at room temperature without stirring. After incubation, wash mica pieces three times under mild stirring (5 min each) using the same incubation buffer. Following this procedure, the NHS group of the SPDP will bind to the amines on the mica surface, leaving a reactive PDP group at the other end of the cross-linker.
3. Reduce the exposed PDP groups on the mica pieces by covering then with 3 ml of 150 mM DTT in PBS–EDTA–azide at 4 °C during 30 min under stirring. After incubation, wash mica pieces three times under mild stirring (5 min each) with PBS–EDTA.

4. Add 100 μg of enzyme-PDP in 3 ml of the working buffer to the wells where each thiolated mica piece is attached. Incubate overnight at room temperature, under mild stirring and in the absence of light to favor the disulfide bridge formation. You might require to assay different quantities of enzyme-PDP to achieve a monolayer (*see* **Note 9**).
5. Wash mica pieces three times for 15 min each with PBST–SDS under stirring to remove aggregated proteins. Kept in PBS at 4 °C and in the absence of light until used.
6. Determine if labeling and functionalization might have affected the enzyme performance by checking its functionality (e.g., using a kinetic assay) (*see* **Note 10**).

3.6 Functionalization of the AFM Cantilever Tip with a Protein

To perform force measurements, prefunctionalized commercial probes for ease ligand attachment can be used, but it is also possible to a la carte functionalize bare tips using linkers (*see* **Note 11**). These AFM probes are coated with a 30 nm-wide gold layer and their tips are modified with a maleimide terminal PEG moiety that is 20 nm long when fully stretched. This spacer constitutes a fingerprint that allows discerning peaks in the Fz curves due to specific interactions attributed to the rupture of enzyme–ligand complexes from those due to nonspecific cantilever–sample interactions (*see* Fig. 4a). The steps can be followed in Fig. 3c.

1. Follow **steps 1–3** of Subheading 3.4 to produce the protein-PDP sample. Use PBS–EDTA, pH 7.0 from **step 2**, to ensure the pH of the maleimide–sulfhydryl reaction in the 6.5–7.5 range. Under more alkaline conditions, the reaction favors primary amines and increases the rate of hydrolysis of the maleimide groups. Ligands different from proteins can also be used (*see* **Note 11**).
2. Treat the protein-PDP with 50 mM DTT for 30 min at room temperature to expose sulfhydryl groups on the protein surface.
3. Cover each maleimide-PEG-probe with 40 μM thiolated protein in PBS–EDTA pH 7.0 for 1 h to attach the ligand to the tip, and wash three times to remove the excess of reactants.

3.7 Dynamic Force Spectroscopy (DFS)

In the FS mode of AFM, cantilever deflection is recorded as a function of the vertical displacement of the piezoelectric scanner (*see* Fig. 1b). If a receptor is immobilized on a sample and a ligand is attached to the AFM tip, the force attributed to the interaction of single complexes can be measured. Figure 4a shows an Fz curve corresponding with a rupture event of an FNR–Fd complex [21]. The sample is pushed toward the tip and retracted. If during the approach an interaction between the two molecular partners occurs, a dissociation process will be induced when retracting the tip from the sample and we will be able to measure the corresponding rupture force. The cantilever deflection signal can be converted into a force introducing the spring constant of the system according to the Hooke's law (*see* Fig. 1b). Formation of biomolecular complexes is a stochastic process, so hundreds or thousands of approaches need to be done to collect Fz curves containing clear specific adhesion peaks (*see* Fig. 4a). Control experiments are needed to verify the specificity of the measurements. The easiest is to block the enzyme sites on the mica surface with an excess of ligand—or alternatively blocking the tip with enzyme. An important decrease in the number of rupture events should be observed, while for those remaining rupture force values must not change. To perform the FS, the first steps are similar to those for imaging.

1. Follow **steps 1–9** of Subheading 3.2 to prepare the AFM system.
2. Image different areas of the sample to evaluate its quality. Ideally, the sample should be homogeneous in covering and height.
3. Use the system in FS mode to obtain several hundreds of Fz cycles for ligand-cantilever/enzyme-mica approaches at different loading rate (R) values. Keep the loaded force between the tip and the sample constant in 1.25 nN—this value is indicative. Select a distance from which approach and withdraw traces in an Fz curve coincide in those parts where peak forces do not appear; this distance should not be very long because it will increase the recording time, in general 150–250 nm are enough. Obtain Fz curves by applying a voltage to the z-piezo at a tip-retraction velocity from 50 to 4000 nm/s. These data translate into R in the 3–80 nN/s range. The curves are collected as voltage versus distance scans (*see* Fig. 1b). Translate voltage values into force data by using the slope of the backward curve and the calibrated value of the spring constant of the functionalized used cantilever (*see* Subheading 3.2, **steps 8** and **9**). Perform measurements in PBS–EDTA pH 7.0 at room temperature. A minimum of 6 R values should be assayed to achieve good data quality settings.
4. For negative control experiments, block the available enzyme sites for interacting with the ligand by incubating the mica sample with excess of a 0.5 mM free ligand solution for 15 min. At least two different R of the total selected in **step 3** should be assayed.

3.8 Mechanostability Analysis

DFS is used to analyze the dynamics of the dissociation process, specifically, by measuring the rupture force as a function of the loading rate or velocity at which force is applied to the molecule.

1. Analyze individually each collected Fz curve with WSxM software (*see* **Note 7**). There have been some attempts to develop algorithms or software to analyze curves collectively, but to this day they are not reliable. Force histograms will be created by using only peak force data from Fz curves that meet the specificity requirements; those Fz that show a force peak produced at a distance coinciding with the length of the PEG spacer stretched (or multiples of this) that binds the ligand molecule to the tip, and present a parabolic shape typical of PEG stretch according to the worm-like chain model [30] (*see* Fig. 4a). In particular, those peaks that follow the retraction slope in the withdrawing Fz trace are attributed to tip–sample nonspecific forces. It is also necessary to discard all those signals that do not reflect specific interactions, as well as to consider those ambiguous as "false events." To facilitate the analysis, first correct the baseline of the Fz, and second align the *y*-axis to zero with the specific functions provided by your analysis program to get directly the peak data values. Change the strategy if nearly all curves lack of rupture events (*see* **Note 12**). Combine FS with scanning to further check the quality of tip and sample. In the Cervantes system, it can be operated using JM, in other microscopes force volume or quantitative nanomechanical methods can be used (*see* **Note 13**).
2. Construct force histograms for each R from specific data peak rupture forces. Start by grouping force values in small intervals. Typically, two or three different asymmetrical force groups will fit to Gaussian functions by using the least-squares method (*see* Fig. 4b). You can relate rupture events taking place at (1) the lowest force values to interactions that account for the participation of a single couple of proteins, (2) intermediate force values with the ones where two protein couples bind at the same time (the force required to break the interaction is approximately twice that required for a single event), and (3) higher force values in which three proteins interact simultaneously with their couples (the force required is around three times that for the single event). The value attributed to a single complex is known as "the most probable rupture force" value, as the value obtained from the Gaussian fitting of dark bars in Fig. 4b. Analyze similarly histograms from blocking control experiments.
3. Beyond the possibility to obtain the forces that maintain the two molecules bound (10 $nN.s^{-1}$ is in general the reference R at which the mechanostability data from different complexes are compared), you can also use the representation of the more probable unbinding forces versus R in the logarithm scale to obtain interesting kinetic data. Use the Evans–Ritchie expression [31]:

$$F^* = \left(\frac{k_B \cdot T}{x_\beta}\right) + \ln\left(\frac{R \cdot x_\beta}{k_{off} \cdot k_B \cdot T}\right)$$

and evaluate if one or several linear fitting regimes appear. The slope of each linear fitting is equal to $k_B T/x_b$, in which k_B is the Boltzmann constant, T is the temperature, and x_b (in Å) is the distance from the energy minimum to the transition state. Calculate the kinetic dissociation constant at zero force, k_{off}, from the extrapolation of the fitting to force zero. One or two linear regimes can be traced back to the presence of one and two intermediate states in the dissociation process, according to one or two energy barriers.

4 Notes

1. Proteins should be firmly attached to the mica surface to prevent them from being dragged while scanning. If proteins bind to the mica at pH 7.0, use 20 mM PBS; higher buffer concentrations increase the number of ions in the medium and can disturb probe–sample interactions. If proteins present a net negative charge and do not bind to the mica sheets, it is necessary to low the pH below the pI of the protein. In case this affects flavoprotein stability and/or functionality, check if addition of divalent cations such as Mg^{2+}, Ca^{2+}, Mn^{2+}, and Ni^{2+}, which electrostatically bridge the negatively charged surface of the mica with the negatively charged protein, improves the performance; for example, 5 μM $MnCl_2$ helped to immobilize an enzyme whose pI was below 6 [32]. If the presence of cations is not convenient for the measurement, try to use Mg^{2+}. It has been reported to immobilize negative DNA on the mica; good results have been reported by pretreatment of each mica piece for 5 min with 50 μl of a 200 mM $MgCl_2$ solution, followed by a rinse with the working buffer [33]. The concentration of the sample must be such that molecular species are individually visualized in the images, conveniently separated to be individually analyzed and in sufficient quantity to make a statistical estimation. If protein is observed in excess, reduce the incubation concentration. If little protein is observed, increase its concentration. Nonetheless, if you barely see improvement at higher concentrations, change pH, add divalent cations or make a pretreatment.
2. To analyze the influence of any ligand agent or medium condition on an enzymatic reaction, incubate both separately and altogether the different components of the reaction to analyze the whole "puzzle" of the catalytic process. Proportions reported as optimal in solution enzymatic assays must be used as starting point. Typically, it is of interest to analyze images

under different conditions such as presence of compounds that influence the flavoprotein redox state, or concentrations of ligands, metals, or other agents in the reaction media. If reaction time or pH is going to be analyzed, repeat the same protocols to prepare the reaction mixtures by modifying these parameters.

3. The "thermal method" is based on the cantilever's thermal distribution spectrum (square of the fluctuations in amplitude as a function of frequency). According to the equipartition theorem, the mean square amplitude of the cantilever's thermal fluctuation in the vertical direction (z^2) can be expressed as $(z^2) = (k_B\ T)/k_N$, where k_B is Boltzmann's constant, T is the temperature of the cantilever, and k_N is the cantilever spring constant [34].
4. JM is the most suitable operative mode in liquid when using a Cervantes system, but today there are many available standard or special developments for different models and companies. Modern AFMs include easy-to-use routines for amplitude-modulation modes, using the amplitude of the cantilever oscillation as the control parameter in a similar way that force is used in JM. They do not need parameter selection for tip–sample approach, but the user has to tune the cantilever and choose an amplitude slightly lower that the selected resonant peak for presenting a higher symmetry of those found. Sometimes this is not easy to achieve in fluid and several peaks have to be explored. This is due to fluid dampening the resonance capacity of the lever in liquid, causing the quality factor of the main resonance peak to decrease drastically. In addition, in most systems, the small piezo that generates the resonant vibration is located above the lever, making the entire holder to vibrate, which in turn decreases resolution in the measurement. These modes are more suitable for working in air, but they allow working in liquid. In addition, many companies have developed their own intermittent force modes for liquids, as is the case of the Nanotec JM itself or the Peak Force Tapping by Bruker, able to control the applied force.
5. When JM is operated in repulsive electrical double layer regime (REDL), operation can be very stable. The tip is very close to the sample but in noncontact, being therefore nonintrusive and allowing to resolve with high quality the features. It is necessary to choose a reference set-point value at a lower force before the jump to contact position in an Fz within the electrostatic repulsion region, in such a way that no contact between the tip and the sample is ensured [35].
6. It might happen that at the beginning or during a measurement the piezo scanner fully extends and does not reach the imaging range or once reached gets out of it. If this happens,

perform a coarse approach using the "motor steps" until the vertical scanner position is within its elongation range. Change the set point to extend the piezo until the tip gets in imaging range or repeat the coarse approach procedure. One way to check if the AFM is in imaging range is by performing an Fz curve. If the tip is in imaging range, the force should increase with tip–sample distance and beyond this point, the force remains constant.

7. Alternative Software for scanning probe microscopy data visualization and analysis can be used, as freeware Gwyddion (http://gwyddion.net/) or other payment software as SPIP (Image Metrology, https://www.imagemet.com/products/spip/) and NanoScope Analysis (Bruker).
8. Ligand binding might not influence in general the quaternary organization of flavoproteins, but it can induce conformational changes. Usually these changes are very small to be observed by AFM, but if they are large enough different conformational species might be visualized as consequence of ligand binding and/or subsequent catalysis. In these cases AFM can allow to clearly visualize the different compaction degree and aperture of different protein domains [36].
9. To determine the proper amount of protein needed for obtaining a homogeneous layer a study can be done. In the attempt to saturate the mica by directly adding a high excess of protein, a high degree of aggregation on the covalent layer can be produced, being difficult to remove it by using detergent (*see* Subheading 3.5, **step 5**). This will produce problems during FS. Therefore, it is of interest to relate the enzyme-PDP incubation concentration with both the degree of coverage that may be analyzed through AFM topography imaging and the enzymatic activity on surface [20].
10. Functionality of the tagged enzymes, both immobilized and in solution, can be verified as control by using specific enzymatic tests [20] (*see* Subheading 3.4).
11. Bare probes can be functionalized to achieve maleimide-PEG-probes or PEG-probes ended in other reactive groups. PEG lengths from 4 to 12 repetitive units can be acquired (Polypure AS). Different chemistries have been probed [37], including the use of modified antibodies that introduce an additional length [38]. If the ligand is not a protein other protocols should be designed specifically including a spacer that introduces a fingerprint to recognize specific events [39].
12. Nominally there should be an event every ten pulls, thus, most recording events being almost empty suggests too few molecules on the surface. In this case, start again the experiment by increasing the concentration of the protein sample by tenfold.

Tip functionalization using a high protein excess guarantees the probability of encounter. If increasing flavoenzyme concentration in the incubation step does not work, a procedure able to orientate the binding site to the ligand at the tip can be used. Try to label the flavoenzyme–ligand complex instead of the free protein. This will avoid labeling at the interaction surface and will prevent the subsequent immobilization of the enzyme through the interactions surface area, as a consequence your sample will be oriented. This methodology has been shown to efficiently increase the percentage of successful complex formation events compared to the total number of approaches [21]. Other strategies improving the binding sites relative orientation can be designed.

13. An alternative interesting control to check the quality of functionalized tip and sample is to combine FS with scanning. In the Cervantes system, this can be operated by using JM, in other microscopes other methods, as force-volume or TREC, can be used. Force-volume modes collect the complete Fz curves so they store a huge amount of data and become very slow. A second strategy uses a dynamic mode, where the tip is oscillating close to its resonant frequency and allows the simultaneous production of topography and recognition images (TREC). Here receptor sites in the recognition image can be correlated with definite features of the topography image thanks to the amplitude reduction produced by tip–sample specific interactions [40]. Herein we propose JM operated in repulsion REDL regime (*see* **Note 5**). By operating in this regime lateral unspecific forces are avoided because there is no tip–sample mechanical contact and the adhesion map exclusively shows molecular recognition events. Additionally, the adhesion maps are not only qualitative but also quantitative maps [39]. This capability can be applied to design specific detection methods based on adhesion maps.

Acknowledgments

We thank MINECO (BIO2016-75183-P) and Gobierno de Aragón (E35_17R and LMP58_18) with FEDER (2014-2020) funds for "Building Europe from Aragón" for financial support. A.L. acknowledges support from ARAID. The authors also thank Prof. Carlos Gómez-Moreno for introducing the authors in the flavoproteins field and I. Echániz for technical support.

References

1. Miller H, Zhou Z, Shepherd J, Wollman AJM, Leake MC (2017) Single-molecule techniques in biophysics: a review of the progress in methods and applications. Rep Prog Phys 81:24601
2. Neuman KC, Nagy A (2008) Single-molecule force spectroscopy: optical tweezers, magnetic tweezers and atomic force microscopy. Nat Methods 5:491–505
3. Binnig G, Quate CF, Gerber CH (1986) Atomic force microscope. Phys Rev Lett 56:930
4. Florin EL, Moy VT, Gaub HE (1994) Adhesion forces between individual ligand-receptor pairs. Science 264:415–417
5. Merkel R, Nassoy P, Leung L, Ritchie K, Evans E (1999) Energy landscapes of receptor-ligand bonds explored with dynamic force spectroscopy. Nature 397:50–53
6. Pierres A, Touchard D, Benoliel AM, Bongrand P (2002) Dissecting streptavidin-biotin interaction with a laminar flow chamber. Biophys J 82:3214–3223
7. Capitanio M, Pavone FS (2013) Interrogating biology with force: single molecule high-resolution measurements with optical tweezers. Biophys J 105:1293–1303
8. Sarkar R, Rybenkov VV (2016) A guide to magnetic tweezers and their applications. Front Phys 4(48)
9. Alegre-Cebollada J, Perez-Jimenez R, Kosuri P, Fernandez JM (2010) Single-molecule force spectroscopy approach to enzyme catalysis. J Biol Chem 285:18961–18966
10. Puchner EM, Gaub HE (2009) Force and function: probing proteins with AFM-based force spectroscopy. Curr Opin Struct Biol 19:605–614
11. Williams MC, Rouzina I (2002) Force spectroscopy of single DNA and RNA molecules. Curr Opin Struct Biol 12:330–336
12. Muller DJ, Dufrene YF (2008) Atomic force microscopy as a multifunctional molecular toolbox in nanobiotechnology. Nat Nanotech 3:261–269
13. Gomez-Moreno C, Lostao A (2013) Single molecule methods to study flavoproteins. In: Hille R, Miller S, Palfey B (eds) Complex flavoproteins, dehydrogenases and physical methods, vol 2. De Gruyter, Boston, pp 277–298
14. Marcuello C, Arilla-Luna S, Medina M, Lostao A (2013) Detection of a quaternary organization into dimer of trimers of *Corynebacterium ammoniagenes* FAD synthetase at the single-molecule level and at the in cell level. Biochim Biophys Acta 1834:665–676
15. Marcuello C, Arilla-Luna S, Medina M, Lostao A (2012) Atomic force microscopy reveals a dimer of trimers organization in *Corynebacterium ammoniagenes* FAD synthetase. FEBS J 279:513–514
16. Ana S, Sebastian M, Arilla-Luna S, Baquedano PMC, Lostao A, Herguedas B, Velázquez-Campoy A, Martínez-Júlvez M, Medina M (2015) Quaternary organization in a bifunctional prokaryotic FAD synthetase: involvement of an arginine at its adenylyltransferase module on the riboflavin kinase activity. Biochim Biophys Acta 1854:897–906
17. Sebastián M, Lira-Navarrete E, Serrano A, Marcuello C, Velázquez-Campoy A, Lostao A, Hurtado-Guerrero R, Medina M, Martínez-Júlvez M (2017) The FAD synthetase from the human pathogen Streptococcus pneumoniae: a bifunctional enzyme exhibiting activity-dependent redox requirements. Sci Rep 7:7609
18. Ferreira P, Villanueva R, Martinez-Julvez M, Herguedas B, Marcuello C, Fernandez-Silva P, Cabon L, Hermoso JA, Lostao A, Susin SA, Medina M (2014) Structural insights into the coenzyme mediated monomer-dimer transition of the pro-apoptotic apoptosis inducing factor. Biochemistry 53:4204–4215
19. Villanueva R, Ferreira P, Marcuello C, Usón A, Miramar MD, Peleato ML, Lostao A, Susin SA, Medina M (2015) Key residues regulating the reductase activity of the human mitochondrial apoptosis inducing factor. Biochemistry 54:5175–5184
20. Marcuello C, de Miguel R, Gómez-Moreno C, Martínez-Júlvez M, Lostao A (2012) An efficient method for enzyme immobilization evidenced by atomic force microscopy. Protein Eng Des Sel 25:715–723
21. Marcuello C, de Miguel R, Martínez-Júlvez M, Gómez-Moreno C, Lostao A (2015) Mechanostability of the single electron-transfer complexes of *Anabaena* ferredoxin–$NADP^+$ reductase. ChemPhysChem 16:3161–3169
22. Marcuello C, de Miguel R, Martínez-Júlvez M, Gómez-Moreno C, Lostao A (2015) Mechanostability of the single electron-transfer complexes of *Anabaena* ferredoxin–$NADP^+$ reductase. ChemPhysChem 16(issue 15):3120
23. Dudko OK, Hummer G, Szabo A (2006) Intrinsic rates and activation free energies from single-molecule pulling experiments. Phys Rev Lett 96:108101

24. Jarzynski C (1997) Nonequilibrium equality for free energy differences. Phys Rev Lett 78:2690
25. Tapia-Rojo R, Marcuello C, Gomez-Moreno-C, Lostao A, Gómez-Moreno C, Mazo J, Falo F (2015) Characterizing the mechanical dissociation of biological complexes: from forces to free energies. Eur Biophys J 44:130
26. Tapia-Rojo R, Marcuello C, Lostao A, Gómez-Moreno C, Mazo J, Falo F (2017) A physical picture for mechanical dissociation of biological complexes: from forces to free energies. Phys Chem Chem Phys 19:4567–4575
27. Marcuello C, de Miguel R, Gomez-Moreno C, Lostao A (2013) Discrimination of protein receptors through quantitative adhesion force maps. Eur Biophys J 42:201
28. de Pablo PJ, Colchero J, Gomez-Herrero J, Baro AM (1998) Jumping mode scanning force microscopy. Appl Phys Lett 73:3300–3302
29. Horcas I, Fernández R, Gómez-Rodríguez JM, Colchero J, Gómez-Herrero J, Baro AM (2007) WSxM: a software for scanning probe microscopy and a tool for nanotechnology. Rev Sci Instrum 78:013705
30. Bouchiat C, Wang MD, Allemand J, Strick T, Block SM, Croquette V (1999) Estimating the persistence length of a worm-like chain molecule from force-extension measurements. Biophys J 76:409–413
31. Evans E, Ritchie K (1997) Dynamic strength of molecular adhesion bonds. Biophys J 72:1541–1555
32. Valero-González J, Leonhard-Melief C, Lira-Navarrete E, Jiménez-Osés G, Hernández-Ruiz C, Pallarés MC, Yruela I, Vasudevan D, Lostao A, Corzana F, Takeuchi H, Haltiwanger RS, Hurtado-Guerrero R (2016) A proactive role of water molecules in acceptor recognition by protein O-fucosyltransferase 2. Nat Chem Biol 12:240–246
33. Pallarés MC, Marcuello C, Botello-Morte L, González A, Fillat MF, Lostao A (2014) Sequential binding of FurA from Anabaena sp. PCC 7120 to iron boxes: exploring regulation at the nanoscale. Biochim Biophys Acta 1844:623–631
34. Butt HJ, Jaschke M (1995) Calculation of thermal noise in atomic force microscopy. Nanotechnology 6:1–7
35. Sotres J, Lostao A, Gómez-Moreno C, Baró AM (2007) Jumping mode AFM imaging of biomolecules in the repulsive electrical double layer. Ultramicroscopy 107:1207–1212
36. Lira-Navarrete E, de Las Rivas M, Compañón I, Pallarés M, Kong Y, Iglesias-Fernández J, Bernardes P, Rovira B, Bruscolini C, Lostao C, Hurtado-Guerrero R (2015) Dynamic interplay between catalytic and lectin domains of GalNAc-transferases modulates protein O-glycosylation. Nat Commun 6:6937
37. Wildling L, Unterauer B, Zhu R, Rupprecht A, Haselgrübler T, Rankl C, Ebner A, Vater D, Pollheimer P, Pohl EE, Hinterdorfer P, Gruber HJ (2011) Linking of sensor molecules with amino groups to amino-functionalized AFM tips. Bioconjug Chem 22:1239–1248
38. Sotres J, Lostao A, Wildling L, Ebner A, Gómez-Moreno C, Gruber HJ, Hinterdorfer P, Baró AM (2008) Unbinding molecular recognition force maps of localized single receptor molecules by atomic force microscopy. ChemPhysChem 9:590–599
39. Akhremitchev BB (2012) Immobilization and interaction strategies in DFS of biomolecular partners. In: Bizarri AR, Cannistraro S (eds) Dynamic force spectroscopy and biomolecular recognition. CRC Press, Boca Ratón, pp 133–162
40. Stroh C, Wang H, Bash R, Ashcroft B, Nelson J, Gruber H, Lohr D, Lindsay SM, Hinterdorfer P (2004) Single-molecule recognition imaging microscopy. PNAS 101:12503–12507

Chapter 11

Ligand Binding in Allosteric Flavoproteins: Part 1. Quantitative Analysis of the Interaction with NAD^+ of the Apoptosis Inducing Factor (AIF) Harboring FAD in the Reduced State

Paolo Cocomazzi, Luca Sorrentino, Federica Cossu, and Alessandro Aliverti

Abstract

To perform their action, flavoproteins usually interact with a variety of low molecular weight partners, including electron transporters, yielding transient complexes whose tightness is often controlled by the redox state of the bound flavin cofactor. As a case study, here we describe the quantitative analysis of the redox-dependent interaction of the mammalian apoptosis inducing factor (AIF) with its NAD^+ ligand. In particular, we report a protocol for the spectrophotometric titration of AIF in its reduced state under anaerobic conditions with NAD^+, in order to determine the dissociation constant of the resulting complex.

Key words Protein–ligand interaction, Dissociation constant, Electron carrier, Charge-transfer complex, Anaerobiosis, Photoreduction, Spectrophotometric titration

1 Introduction

In flavoenzymes, allostery is often displayed as the modulation of the affinity of the ligand binding sites of the protein determined by the redox state of the bound flavin cofactor. The apoptosis-inducing factor (AIF), a mitochondrial intermembrane-space FAD-containing protein, is paradigmatic in such behavior [1]. AIF possesses a three-domain organization, comprising FAD-binding, NAD-binding, and C-terminal domains [2]. FAD reduction dramatically increases AIF affinity for NAD^+, whose binding yield a very tight $FADH^-$-NAD^+ charge-transfer (CT) complex, extremely resistant to oxidation by O_2. In addition to lock the protein in the reduced state, CT formation triggers a general rearrangement of the protein tertiary structure, which induce a change also in its quaternary level of organization [1, 2]. Indeed, FAD reduction is strictly coupled to the transition

Maria Barile (ed.), *Flavins and Flavoproteins: Methods and Protocols*, Methods in Molecular Biology, vol. 2280, https://doi.org/10.1007/978-1-0716-1286-6_11,

from the monomeric to the dimeric state of AIF [3]. The ability of AIF to adopt a CT/dimeric state is crucial for its functions and, in particular, for its role in assisting the biogenesis and maintenance of respiratory complexes. This is testified by the observation that pathogenic variants of human AIF, hampered in undergoing such transition, are the cause of mitochondriopathies with often severe neurological and muscular symptoms [4].

The ability to adopt the CT/dimeric state has been recently shown to be crucial for the role of AIF, in conjunction with its CHCHD4/MIA40 [5] partner, in assisting the import and folding of nuclear-encoded mitochondrial-targeted proteins, as well as their maturation by the formation of disulfide bonds [6].

Thus, the study of the interaction of AIF in different redox states with NAD^+/H is of paramount importance to investigate its action at the molecular level and to interpret the pathological effects of some of its allelic variants. Here, we describe a protocol for the characterization on a quantitative basis of the complex between the reduced form of AIF and NAD^+. Usually, flavin nucleotides and flavoproteins are reactive toward molecular oxygen, so that, to preserve their reduced state during in vitro experiments, anaerobic conditions are required. Thus, a specifically designed anaerobiosis apparatus and vacuum glassware items are required to apply this method. The experimental procedure here described to analyze AIF–NAD^+ interaction, with the appropriate modifications, could find applications in the study of other flavoproteins. In Chapter 12 we report a titration procedure to determine the dissociation constants of the complexes formed by AIF under different redox states with its protein partner CHCHD4/MIA40.

2 Materials

NAD^+ and NADH are purchased from Sigma-Aldrich. All other chemicals are of analytical grade. Water used for the preparation of all solutions is purified using Elix and Milli-Q systems connected in series to attain a final conductivity of about 18 MΩ × cm at 25 °C. All solutions are filtered through low-extractable 0.22 μm filters (*see* **Note 1**).

2.1 Instrumentation and Other Devices

1. Desalting cartridge: 10 ml PD10 cartridge (GE Healthcare).
2. Spectrophotometer: any UV-Vis spectrophotometer suitable to accommodate an anaerobiosis cuvette may be used. In our case, a diode-array 8453 spectrophotometer (Agilent) is used to record all spectra during the entire procedure.
3. Anaerobiosis apparatus: vacuum-nitrogen manifold system.
4. Anaerobiosis cuvette: in our case, a customized spectrophotometric anaerobic cuvette (Fig. 1a), purchased from a local

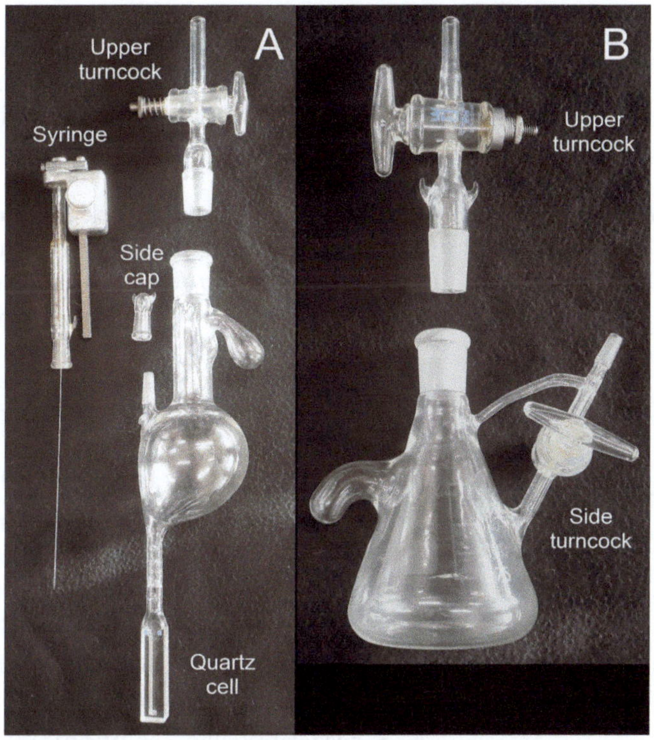

Fig. 1 Special glassware for spectrophotometric measurements and related liquid handling under anaerobiosis conditions. (**a**) Set for anaerobic titration, including anaerobiosis cuvette and gastight microsyringe. The syringe can be connected to the cuvette body in place of the side cap. (**b**) Vacuum flask. Solutions made anaerobic within the flask can be withdrawn trough the side opening using a gastight syringe

scientific glassblower (Arbore Cataldo, Milano, Italy), based on a quartz 1 ml optical cell, is used.

5. Vacuum flask: in our case, a 50 ml vacuum flask (Fig. 1b) was obtained from the above provider, specifically designed to be suitable to make the titrant solution anaerobic and to transfer it to a graduated microsyringe (*see* **Note 2**).
6. Gastight microsyringe: 100 μl gastight graduated microsyringe with a glass adaptor for the connection to the side arm of the anaerobiosis cuvette (Fig. 1a).
7. Vacuum grease: high-vacuum grease (Apiezon, type N) to lubricate glass-to-glass connections.
8. Infrared light filtration device: in our case, a thin-layer-chromatography glass container about 10 cm wide filled with water is used.
9. Light source: in our case, a conventional slide projector is used.

10. Data analysis and graphing software: any scientific data analysis software able to perform nonlinear regression using equations provided by the user can be used. In our case, GraFit 5.0 (Erithacus Software Ltd., Wilmington, Sussex, Great Britain) for Microsoft Windows is adopted.

2.2 Solutions

1. Target flavoprotein stock solution: in our case, recombinant mouse AIFΔ1-101 (hereafter indicated as AIF) is overproduced in *Escherichia coli* as a fusion with a C-terminal His-tag, using a plasmid based on the pKK223-3 vector. Expression and purification of AIF are reported elsewhere [3]. The protein is typically obtained at the final concentration of about 0.5 mM, determined spectrophotometrically on the basis of $\varepsilon_{452} = 12.84\ \text{mM}^{-1}\ \text{cm}^{-1}$ and stored at −20 °C in 50 mM Tris-HCl, pH 7.4 (at 25 °C), 10% glycerol (*see* **Note 1**).
2. Protein medium: the composition of the protein solution medium depends on the physicochemical feature of the target protein. For AIF, 2× protein medium is 100 mM Na-phosphate, pH 7.5 (*see* **Note 1**).
3. EDTA solution: 150 mM EDTA, pH 7.5. The pH of the solution should match that of the protein medium.
4. 5-deazariboflavin solution: 70 mM 5-deazariboflavin in H_2O. The chemical is not available from commercial suppliers: in our case, it was a gift of Dr. Sandro Ghisla. The compound can be quantified spectrophotometrically using $\varepsilon_{390} = 12.39\ \text{mM}^{-1}\ \text{cm}^{-1}$.
5. Titrant solution: the chemical nature and concertation of the titrant depends of the protein and the ligand under study. In the case of AIF, the solution is 400 μM NAD^+ in H_2O. The pH of the solution should be adjusted to match that of the protein medium (*see* **Note 3**).

3 Methods

3.1 Spectrophotometric Titration with NAD^+ of the Reduced Form of AIF Under Anaerobic Conditions: Theoretical Background

Reaction of AIF with NADH occurs through an equilibrium strongly shifted towards FAD reduction generating a particularly stable $FADH^-$-NAD^+ CT complex. Despite its virtually irreversible nature, the reaction is particularly slow, requiring up to one hour at low NADH concentration [2, 3]. We found that reaction of AIF bearing reduced FAD and NAD^+ generates the same end-product at a considerably higher rate [3], thus requiring much lower reaction time. Based on these observations, we devised a titration procedure to analyze the interaction of AIF with its dinucleotide ligand, suitable for the determination of the dissociation constant of the resulting CT complex. The protocol consists in the

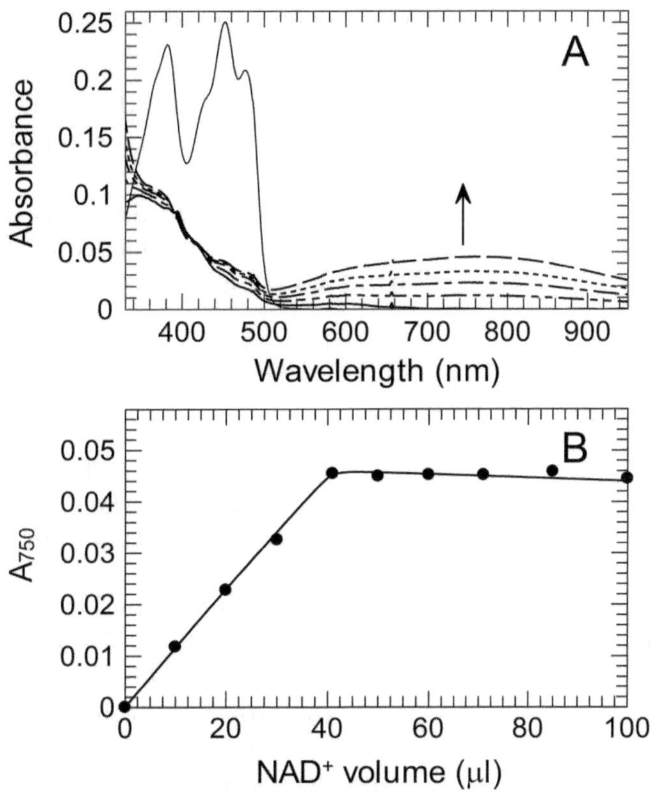

Fig. 2 Representative titration of reduced AIF with NAD^+, under anaerobiosis. (**a**) Selected absorption spectra, recorded as equilibrium conditions were established, before and after each stepwise titrant solution addition. The spectrum of the oxidized protein is shown for comparison (thin solid line). The arrow indicates the direction of the absorbance change upon increase of NAD^+ concentration. (**b**) Values of A_{750}, reflecting CT complex concentration, as a function of total titrant volume. Dots represent experimental data, while the solid line is the best-fitting curve corresponding to the interpolating equation

photochemical reduction of the AIF-bound FAD under anaerobic conditions, using the procedure devised by Massey and Hemmerich [7], followed by the stepwise addition of an NAD^+ solution up to fully conversion of the protein to the corresponding CT-transfer complex. The procedure is carried out in a sealed anaerobic cuvette and the titration is monitored spectrophotometrically (Fig. 2a), essentially as previously reported elsewhere [8].

In the case of AIF, protein–ligand complex formation is assumed to occur via the following reaction:

$$\mathrm{AIF(FADH^-)} + \mathrm{NAD^+} = \mathrm{AIF(FADH^- {-} NAD^+)_{CT}}, \qquad (1)$$

where $\mathrm{AIF(FADH^-)}$ indicates AIF harboring the reduced form of FAD, and $\mathrm{AIF(FADH^- \text{-} NAD^+)_{CT}}$ its CT complex with NAD^+, and whose equilibrium is described by the dissociation constant K_d, as follows:

$$K_d = \frac{[AIF(FADH^-)] \times [NAD^+]}{[AIF(FADH^- \text{-} NAD^+)_{CT}]} \tag{2}$$

The titration is predicted to proceed as a classical protein–ligand process, which under equilibrium follows the general equation of a 1:1 complex dissociation:

$$[PL] = \frac{[P]_T + [L]_T + K_d - \sqrt{([P]_T + [L]_T + K_d)^2 - 4[P]_T[L]_T}}{2}, \tag{3}$$

where [PL] here represents the concentration of the CT complex, $[P]_T$ that of total AIF(FADH$^-$)—both free and ligand bound—and [L] that of total NAD$^+$—both free and ligand bound.

3.2 Titration Procedure

1. Desalt an aliquot of purified target protein by gel filtration on the PD10 cartridge to change its solvent to the selected protein medium (*see* **Note 1**).
2. Mix the protein with the appropriate amounts of stock solutions of EDTA, 5-deazariboflavin, and protein medium within the main body of the anaerobic cuvette (Fig. 1a) in order to obtain a total volume of 1.2 ml of a solution containing ca. 20 μM AIF, 15 mM EDTA, 1.5 mM 5-deazariboflavin, and H_2O in 1× protein medium.
3. Cover the side opening of the anaerobiosis cuvette with the glass stopper and close its upper turncock.
4. Connect the cuvette to the vacuum-nitrogen manifold system through a rubber tubing, open the upper turncock, and make the content of the cuvette anaerobic operating several cycles of vacuum application and nitrogen flushing.
5. Close the upper turncock of the cuvette.
6. Place the NAD$^+$ solution (*see* **Notes 3** and **4**) into the vacuum flask (Fig. 1b), connect it to the vacuum-nitrogen manifold system and flush the content of the flask with nitrogen for ca. 30 min to make it anaerobic.
7. Repeatedly rinse a gastight 100 μl glass microsyringe with the anaerobic NAD$^+$ solution and then fill up the syringe with this solution, while keeping the solution under nitrogen flux.
8. Assess the actual NAD$^+$ concentration spectrophotometrically on another aliquot of the solution, on the basis of $\varepsilon_{260} = 17.6\ mM^{-1}\ cm^{-1}$.
9. Open the upper turncock of the anaerobiosis cuvette, flush it with nitrogen, and open the cuvette side arm.

10. Connect the microsyringe, filled with the titrant, to the side opening of the anaerobiosis cuvette through its terminal conic glass fitting, while keeping the cuvette content under constant nitrogen stream.
11. Close the upper turncock of the cuvette and disconnect it from the vacuum-nitrogen manifold system.
12. Place the cuvette–syringe assembly in the thermostatic cell holder of the spectrophotometer, and record the absorbance spectrum of the protein solution.
13. Irradiate of the protein solution using the light source to obtain photoreduction of the FAD prosthetic group, while keeping the cuvette at room temperature in vertical position at a distance of about 15 cm from the light source, separated from it by the glass container filled with water, in order to eliminate infrared radiation. Irradiate the solution for single periods of 1–3 min, and gently mix the solution after each treatment. Record the absorbance spectrum after each illumination, allowing about 3 min equilibration before the measure. Repeated the operation until complete FAD reduction is obtained, which usually requires up to 15 min total irradiation time (*see* **Note 5**). For fully reduced AIF, the extinction coefficient $\varepsilon_{452} = 1.77\ \text{mM}^{-1}\ \text{cm}^{-1}$ is considered.
14. Make ten successive additions of 10 μl NAD^{+} solution each. Gently mix the solution after each of them and record the spectrum after 3–10 min equilibration at 25 °C, as no further spectral changes are detectable (Fig. 2a). Figure 2a shows selected spectra recorded during a typical titration of AIF. At the end of the titration, open the upper turncock in and admit air into the cuvette. Mix the solution, and allow FAD to gain full reoxidation, in order to verify that the original spectrum is recovered, indicating that no protein denaturation and/or FAD release had occurred. In the case of wild-type AIF, complete CT complex disappearance requires more than 1 h.

3.3 Data Analysis

Usually ligand binding to flavoproteins induces changes in their absorption spectrum, which can be exploited to calculate complex concentration. In the case of AIF, the absorbance of the solution at 750 nm represents an effective measurement of the protein–ligand CT complex concentration, since other species present in the sample provide no contribution to A_{750}. Thus, we have:

$$A_{750} = \varepsilon_{750}^{PL} \times [PL], \qquad (4)$$

where $\varepsilon_{750}{}^{PL}$ is the molar extinction coefficient of the CT complex and [PL] its concentration.

To take into consideration sample dilution due to titrant addition, we must consider that

$$[\mathrm{L}]_\mathrm{T} = \frac{V_a}{V_i + V_a}[\mathrm{L}]_\mathrm{S}\ \text{and} \tag{5}$$

$$[\mathrm{P}]_\mathrm{T} = \frac{V_\mathrm{i}}{V_\mathrm{i} + V_\mathrm{a}}[\mathrm{P}]_\mathrm{i}, \tag{6}$$

where $[\mathrm{L}]_\mathrm{T}$ and $[\mathrm{P}]_\mathrm{T}$ are total NAD^+ and total AIF concentration, respectively, $[\mathrm{L}]_\mathrm{s}$ the concentration of the NAD^+ titrant stock solution, $[\mathrm{P}]_\mathrm{i}$ the concentration of the target protein in the reaction mixture before the first addition of the titrant, and V_i and V_a the initial volume of the reaction mixture and the total titrant volume added, respectively. Equations 3, 4, 5, and 6 are combined to obtain the parametric expression of A_{750} as a function of V_a, which is used to fit the titration data points. Nonlinear regression analysis is performed by the data analysis software GraFit 5.0. The best fitting equation yields the estimate for both ${\varepsilon_{750}}^{PL}$ and K_d (Eqs. 3 and 4) of the complex. In Fig. 2b a typical result of data analysis is shown. While this experimental approach allows only a gross estimate of the K_d of the CT complex involving wild-type AIF, which is too low for an accurate determination (<50 nM), it proved extremely valuable for the analysis of variants of the protein displaying lower ligand affinity [3].

4 Notes

1. The assumption that the total amounts of target protein and titrant are known must be satisfied for the validity of Eq. 3. In addition, spectrophotometric data are strongly biased by particulate or aggregates suspended in the solution under analysis. For these reasons, it is recommended to clarify protein solutions just before use by centrifugation or microfiltration. Moreover, protein storage media and titration conditions should be chosen in order to prevent protein denaturation and/or aggregation. Possible solution additives to be considered to these aims are glycerol and/or mild detergents. Physicochemical parameters, like pH and ionic strength, should be considered as well. A useful approach for the rapid screening of several conditions is the ThermoFAD technique [9].
2. For the maintenance of the vacuum glassware, we suggest to clean it immediately after use first by removing most part of the grease with paper wipes, next by thoroughly rinsing it with water, and finally by eliminating residual grease traces by extensive rinsing with an organic solvent, such as cyclohexane. Occasionally, for deep cleaning, overnight treatment with a concentrated nitric acid solution is very effective.
3. It is important that the composition of the reaction mixture is not altered during the titration. Thus, the composition of the

solvent of the titrant working stock solution should match that of the reaction mixture.

4. NAD^+ is relatively stable in solution at pH 6.0, but less stable under more basic conditions. If a series of titration experiments is planned, a single concentrated NAD^+ stock may be prepared in water adjusting its pH to about 6.0. Next, individual working solutions of the titrant in the appropriate buffer can be prepared for each experiment, just before use.
5. Some flavoproteins can turn out particularly hard to be reduced photochemically. In these cases, the inclusion of micromolar amounts of an additional electron-transfer mediator can be considered. We found effective the use of methyl- or benzyl-viologen, which are colorless in the oxidized state, and whose low redox potential avoid the accumulation of their visible light–absorbing reduced forms during the photoreduction process.

Acknowledgments

This work was supported by grants Linea 2, 2017 and 2018, from Università degli Studi di Milano to A.A.

References

1. Brosey CA, Ho C, Long WZ, Singh S, Burnett K, Hura GL, Nix JC, Bowman GR, Ellenberger T, Tainer JA (2016) Defining NADH-driven allostery regulating apoptosis-inducing factor. Structure 24:2067–2079
2. Sevrioukova IF (2011) Apoptosis-inducing factor: structure, function, and redox regulation. Antioxid Redox Signal 14:2545–2579
3. Sorrentino L, Calogero AM, Pandini V, Vanoni MA, Sevrioukova IF, Aliverti A (2015) Key role of the adenylate moiety and integrity of the adenylate-binding site for the NAD(+)/H binding to mitochondrial apoptosis-inducing factor. Biochemistry 54:6996–7009
4. Sevrioukova IF (2016) Structure/function relations in AIFM1 variants associated with neurodegenerative disorders. J Mol Biol 428:3650–3665
5. Hangen E, Feraud O, Lachkar S, Mou H, Doti N, Fimia GM, Lam NV, Zhu C, Godin I, Muller K et al (2015) Interaction between AIF and CHCHD4 regulates respiratory chain biogenesis. Mol Cell 58:1001–1014
6. Modjtahedi N, Tokatlidis K, Dessen P, Kroemer G (2016) Mitochondrial proteins containing coiled-coil-helix-coiled-coil-helix (CHCH) domains in health and disease. Trends Biochem Sci 41:245–260
7. Massey V, Hemmerich P (1977) A photochemical procedure for reduction of oxidation-reduction proteins employing deazariboflavin as catalyst. J Biol Chem 252:5612–5614
8. Williams Jr CH, Arscott LD, Metthews RG, Thorpe C, Wilkinson KD (1979) Methodology employed for anaerobic spectrophotometric titrations and for assisted data analysis. Methods Enzymol 62:185–198
9. Forneris F, Orru R, Bonivento D, Chiarelli LR, Mattevi A (2009) ThermoFAD, a thermofluor-adapted flavin ad hoc detection system for protein folding and ligand binding. FEBS J 276:2833–2840

Chapter 12

Ligand Binding in Allosteric Flavoproteins: Part 2. Quantitative Analysis of the Redox-Dependent Interaction of the Apoptosis-Inducing Factor (AIF) with Its Protein Partner

Paolo Cocomazzi, Delia Tarantino, Eloise Mastrangelo, and Alessandro Aliverti

Abstract

To perform their action usually flavoproteins interact transiently with a variety of molecular partners, whose binding is reciprocally affected and often controlled by the redox state of the bound flavin cofactor. As a case study, here we describe an approach for the quantitative characterization of the redox-controlled interaction of the mammalian apoptosis inducing factor (AIF) with one of its known protein partners, namely, the mitochondrial coiled-coil-helix–coiled-coil-helix domain-containing protein 4 (CHCHD4). In particular, we report a protocol for the titration of the flavoprotein in both in its oxidized and reduced states with CHCHD4, using an implementation of the MicroScale Thermophoresis (MST) technique.

Key words Protein–ligand interaction, Protein–protein interaction, Redox-linked binding, Microscale thermophoresis

1 Introduction

Allostery, the conformational coupling of distant sites in a macromolecule, is a common feature of proteins, and flavoenzymes are no exception. In particular, the redox state of the bound flavin cofactor can induce long-range conformational changes in a flavoprotein, which, in turn, can affect the affinity of the binding sites for its ligands. The apoptosis-inducing factor (AIF), a mitochondrial intermembrane-space FAD-containing protein, is paradigmatic in such behavior [1]. AIF possesses a three-domain organization, comprising a FAD-binding, an NAD-binding, and a C-terminal domain [2]. FAD reduction dramatically increases affinity for NAD^+, whose binding yield a very tight $FADH^-$-NAD^+ charge-transfer (CT) complex, extremely resistant to oxidation by O_2. In addition to lock the protein in the reduced state, CT formation

Maria Barile (ed.), *Flavins and Flavoproteins: Methods and Protocols*, Methods in Molecular Biology, vol. 2280, https://doi.org/10.1007/978-1-0716-1286-6_12,

triggers a general rearrangement of the protein tertiary structure, which induce a change also in its quaternary level of organization [1, 2]. Indeed, FAD reduction is strictly coupled to the transition from the monomeric to the dimeric state of AIF [3]. The ability of AIF to adopt a CT/dimeric state is crucial for its functions and, in particular, for its role in assisting the biogenesis and maintenance of respiratory complexes. This is testified by the observation that pathogenic variants of human AIF, hampered in undergoing such transition, are the cause of mitochondriopathies with often severe neurological and muscular symptoms [4]. Recently, it has been discovered that, in its CT/dimeric state, AIF forms a firm complex with another mitochondrial intermembrane-space protein, namely the mitochondrial intermembrane space import and assembly protein 40 (MIA40), also known as the coiled-coil-helix–coiled-coil-helix domain-containing protein 4 (CHCHD4) [5]. CHCHD4/MIA40 acts as a chaperone, assisting the import and folding of nuclear-encoded mitochondrial-targeted proteins as well as their maturation by the formation of disulfide bonds [6].

Thus, the detailed study of the interaction of AIF in different redox states with its protein ligand is of paramount importance to investigate its action at the molecular level and to interpret the pathological effects of some of its allelic variants. Here, we describe a protocol for the functional characterization on a quantitative basis of complex formation between both the oxidized/monomeric and CT/dimeric forms of AIF and its protein partner CHCHD4. This experimental procedure, with the appropriate modifications, could find applications in the study of other flavoproteins. In Chapter 11, we report a titration procedure to determine the dissociation constants of the complex between AIF in reduced state and its dinucleotide ligand NAD^+.

2 Materials

NADH is purchased from Sigma-Aldrich. All other chemicals are analytical grade. Water used for the preparation of all solutions is purified using Elix and Milli-Q systems connected in series to attain a final conductivity of about 18 MΩ × cm at 25 °C. All solutions are filtered through low-extractable 0.22 μm filters.

2.1 Instrumentation and Other Devices

1. Metal-chelate chromatography cartridge: 5 ml HisTrap HP cartridge (GE Healthcare).
2. Desalting cartridge: 5 ml HiTrap HP Cartridge (GE Healthcare).
3. Microfuge: Any refrigerated bench centrifuge suitable to accommodate Eppendorf-type tubes.

4. Ultrafiltration device: any centrifugation-based ultrafiltration device with the appropriate molecular cutoff.
5. MicroScale Thermophoresis (MST) apparatus: Monolith NT.115 (NanoTemper Technologies GmbH, Munich, Germany).
6. Data analysis software: MO.Affinity Analysis software (NanoTemper Technologies GmbH).
7. Capillaries for MST: Monolith NT.115 standard treated capillaries (Nano Temper Technologies GmbH, Munich, Germany) (*see* **Note 1**).

2.2 Solutions

1. Target flavoprotein and ligand stock solutions: in our case, recombinant mouse AIFΔ1–101 (hereafter indicated as AIF) and its protein ligand CHCHD4 are both overproduced in *Escherichia coli* as protein fusion linked to a C-terminal His-tag, removable by thrombin proteolysis, using plasmids based on pKK223-3 and pET21a vectors, respectively. For AIF, expression and purification procedures are reported elsewhere [3]. The protein is typically obtained at a final concentration of about 0.5 mM in 50 mM Tris-HCl, pH 7.4 (at 25 °C), 10% glycerol. CHCHD4 is expressed in SHuffle T7 strain and isolated by nickel-chelate chromatography followed by gel filtration. Concentrations of AIF and CHCHD4 are determined spectrophotometrically on the basis of $\varepsilon_{452} = 12.84\ \text{mM}^{-1}\ \text{cm}^{-1}$ and $\varepsilon_{280} = 13.325\ \text{mM}^{-1}\ \text{cm}^{-1}$, respectively (*see* **Note 2**).
2. Protein labeling solutions: Labeling Kit RAD-tris-NTA (NanoTemper Technologies GmbH). The kit includes 50× concentrated dye solution and 10× concentrated PBS-T buffer (*see* **Note 3**).
3. NADH stock solution: 10 mM NADH in 10 mM Tris-HCl, pH 8.0. (*see* **Note 4**).
4. Thrombin stock solution: thrombin is available from different suppliers. We purchase it from Sigma-Aldrich. The lyophilized powder, which includes salt and Tris-HCl, pH 7.0, is dissolved in water before use (*see* **Note 5**).
5. PMSF stock solution: 200 mM PMSF in methanol (*see* **Note 6**).
6. Protein digestion medium: the composition of the thrombin digestion medium is a compromise between the conditions required for protease activity and those compatible with the stability of the fusion protein. For CHCHD4, 50 mM Tris-HCl, pH 8.0, 200 mM NaCl, 2 mM DTT is used.
7. Metal-chelate chromatography mobile phase: for CHCHD4, 50 mM Tris-HCl, pH 8.0, 200 mM NaCl, 10 mM imidazole is used.
8. Titration medium: the composition depends on the system under study. In the case of the AIF/CHCHD5 couple, 50 mM Tris-HCl, pH 8.0, 200 mM NaCl, 4 mM DTT (*see* **Note 2**).

3 Methods

3.1 Principle of the Method

The technique is based on the measure of the concentrations of ligand-free and ligand bound forms of the target protein in solution through the record of the fluorescence emission intensity of sample mixtures. To obtain such data, the target protein must be intrinsically fluorescent or labeled with an appropriate dye. MST measurements are performed on proteins and their ligands free in solution, using virtually any solvent, and allow the determination of dissociation constants over a very wide range of values. Strengths of the technique are the low concentration required for the target protein, the very small sample volumes—which are lower than 4 μl per titration point—and the confinement of samples within tiny glass capillaries that protect them from the contact with air. In brief, the instrument generates a locally restricted, highly precise temperature change in the sample by infrared laser light and monitors the resulting change in fluorescence [7, 8].

Fluorescence change under irradiation is expected to follow an exponential decay time-course because of the progressive temperature increase of the monitored solution portion. Such change is due the combination of two different physical effects: temperature-related intensity change and thermophoresis, both of which are potentially affected by ligand binding. Thermophoresis is the movement of a chemical species along a temperature gradient, resulting in a change of its concentration as a function of temperature. To analyze the effect of protein–ligand interaction on the temperature dependent emission change, the normalized fluorescence (F_{norm}) is considered, that is, the ratio between the emission at a definite time after laser activation (F_1) and that recorded at the baseline, before the irradiation start (F_0):

$$F_{\text{norm}} = \frac{F_1}{F_0} \tag{1}$$

The ligand-free protein and the protein–ligand complex contribute to F_{norm} at different extents, so that:

$$F_{\text{norm}} = f^{\text{P}}_{\text{norm}} \times (1 - [\text{PL}]) + f^{\text{PL}}_{\text{norm}} \times [\text{PL}], \tag{2}$$

where [PL] is the protein–ligand complex concentration while $f^{\text{P}}_{\text{norm}}$ and $f^{\text{PL}}_{\text{norm}}$ are the specific normalized emission intensities of the free and bound protein, respectively.

Based on a 1:1 binding stoichiometry, the dependence of [PL] from the total ligand concentration $[\text{L}]_\text{T}$ is expected to obey to the following equation:

$$[\text{PL}] = \frac{[\text{P}]_\text{T} + [\text{L}]_\text{T} + K_d - \sqrt{([\text{P}]_\text{T} + [\text{L}]_\text{T} + K_d)^2 - 4[\text{P}]_\text{T}[\text{L}]_\text{T}}}{2}, \tag{3}$$

where [PL] represents the concentration of the protein–ligand complex, $[P]_T$ that of total target protein—both free and ligand bound—and $[L]_T$ that of total ligand—both free and protein bound.

In our case, purified recombinant AIF and CHCHD4 need to be chemically modified in order to be suitable as the target protein and the ligand, respectively, in MST titrations. AIF is labeled with a fluorescent dye, taking advantage from its His-tag (*see* **Note 3**), while the tag of CHCHD4 is removed to prevent possible interference in subsequent analyses due to sequestration of the label.

3.2 AIF Labeling

The His-tagged form of AIF is labeled with the red fluorescent dye NT-647 using the Monolith NT His-Tag Labeling Kit RAD-tris-NTA according to the manufacturer directions. AIF is labeled under both its oxidized/monomeric and reduced CT/dimeric states.

1. Prepare the working dye solution by diluting 1:50 in 1× PBS-T the concentrated dye solution.
2. Prepare two 100 μl incubation mixtures in 1× PBS-T, each containing 200 nM AIF and either no nucleotide or 500 μM NADH, respectively.
3. Incubate the two AIF mixtures 1 h on ice to allow completion of the reaction yielding the CT/dimeric form of AIF, where NADH is present.
4. Add 100 μl working label solution to each AIF incubation mixture.
5. Incubate the two labeling mixtures for 30 min at room temperature in the dark.
6. Clarify the two labeled protein solutions by centrifugation at 15,000 × *g* for 10 min at 4 °C in a microfuge.

3.3 Preparation of Tag-Free CHCHD4

1. For tag removal, incubate purified CHCHD4 for 2 h at 37 °C with thrombin at a ratio of 10 units protease per milligram substrate in protein digestion medium (*see* **Note 5**).
2. Stop the proteolytic reaction by adding PMSF stock solution to obtain 1 mM final inhibitor concentration.
3. In order to remove residual tagged CHCHD4, separate the reaction mixture on a HisTrap HP cartridge using the metal-chelate chromatography mobile phase.
4. Change the buffer of the untagged protein solution by gel filtration on the HiTrap HP cartridge equilibrated in titration medium (*see* **Note 7**).
5. Concentrate the purified protein ligand by ultrafiltration to about 200 μM protein.
6. Clarify the ligand solution by centrifugation at 15,000 × *g* for 10 min at 4 °C in a microfuge, just before use.

3.4 Binding Assays

1. For each titration experiment, prepare 16 microfuge tubes, numbered from 1 to 16.
2. In each tube, with the exception of tube 1, place a volume of 10 μl titration medium.
3. Add 20 μ untagged titrant CHCHD4 solution to tube no. 1.
4. Transfer 10 μl of the content of tube no. 1 to tube no. 2 and carefully mix its content by pipetting.
5. Repeat the operation by transferring 10 μl of the content of tube no. 2 to tube no. 3, and so on, until tube no. 16 is reached.
6. Discard 10 μl of the solution of tube no. 16 is in order to have 10 μl 1:1 progressive serial dilutions of the titrant CHCHD4 in all test tubes.
7. To complete the samples, add 10 μl labeled target flavoprotein AIF—in either oxidized or CT complex state—to each of the 16 tubes of the titration series.
8. After extensive mixing by pipetting, transfer the content of each of the 16 tubes to a set of 16 Monolith NT.115 standard treated capillaries (*see* **Note 1**).

In the following two paragraphs, the typical results of two titration experiments on the AIF/CHCHD4 couple are reported, aimed to show expected possible outcomes of the application of the method. In the first case, the thermophoresis effect is used to quantify ligand binding, while in the second the observed ligand-induced fluorescence enhancement is exploited as the binding signal to cast the titration plot.

3.4.1 CHCHD4 Titration of AIF in Its Oxidized Monomeric State

MST measurements are performed at the baseline temperature of 24 °C with a fixed concentration of 50 nM AIF, and concentrations of CHCHD4 varied in the 3.85 nM to 126 μM range. Laser excitation power and MST power were set to 20% and 40%, respectively (*see* **Note 8**).

The capillary scan clearly shows that protein–protein interaction provides no fluorescence quenching or enhancing (Fig. 1a), while a clear ligand dose-dependent thermophoresis effect is evident from the recorded MST traces (Fig. 1b), where the normalized fluorescence change ($F_{norm} = F_1/F_0$) is reported as a function of irradiation time. To cast the titration plot, baseline emission (F_0) and measured emission selected over a specific irradiation time interval (F_1), selected in order to maximize signal-to-noise ratio (Fig. 1b), are considered. The effect of ligand binding generates a 14.4‰ maximal variation of the thermophoretic effect (Fig. 1c). Interpolation of data points with the theoretical equation for a 1:1 binding stoichiometry (Fig. 1c), obtained by combining Eqs. 2 and 3, yields a K_d value of 10.7 ± 2.4 μM for the AIF-CHCHD4 complex.

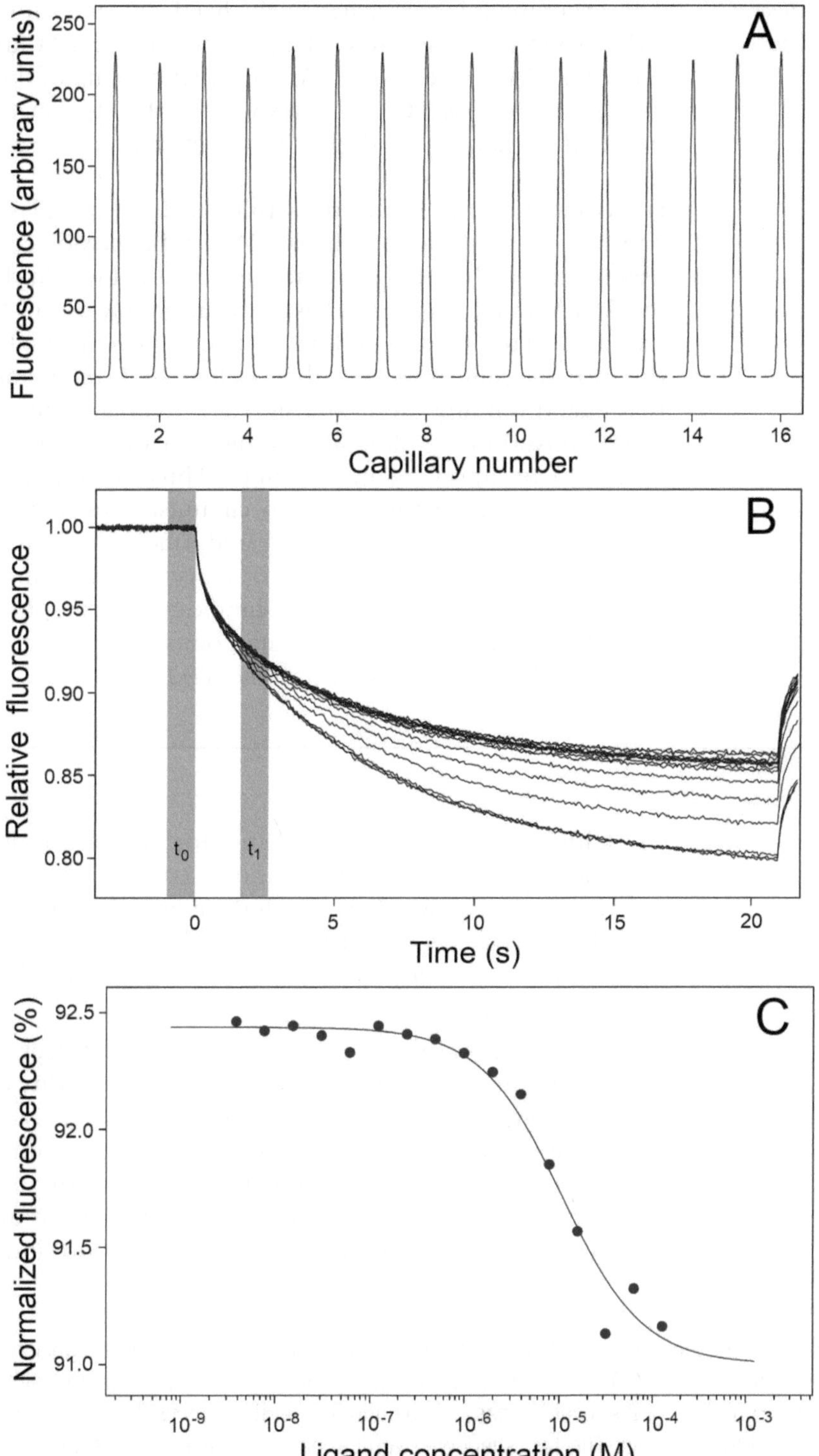

Fig. 1 Titration of AIF in its oxidized monomeric state with CHCHD4 using the MicroScale Monolith procedure. (**a**) Lateral fluorescence emission scan of the capillary set recorded at baseline temperature, before laser

3.4.2 CHCHD4 Titration of AIF in the Reduced Dimeric State

This experiment is performed at the fixed concentration of 50 mM AIF, also in this case. Here, the presence of 125 μM final concentration of NADH keeps the FAD prosthetic group of the protein in its reduced state as a CT complex with NAD^+. Under these conditions, AIF is known to adopt a homodimeric quaternary structure. The low reactivity of the AIF CT complex toward O_2 and the embedment of the solutions within small-caliber glass capillaries prevents protein-bound reduced FAD to be reoxidized during the experiment. CHCHD4 concentration is varied within the 3.85 nM to 126 μM range, as before. Laser excitation power and MST power are adjusted to 60% and 40%, respectively (*see* **Note 8**). However, in this case, the capillary scan at the baseline 24 °C temperature reveals that ligand binding induced a significant enhancement in fluorescence emission (Fig. 2a), suggesting a change in the fluorophore environment upon ligand binding. Thus, as suggested by the instrument manufacturer in such circumstances, the raw fluorescence change recorded before infrared irradiation is used to monitor protein–protein interaction. Anyway, also in our case, CHCHD4 binding induced a significant thermophoresis effect (not shown). Interpolation of data points with the same equation as before (Fig. 2b) yields a K_d value of 92 ± 16 nM.

4 Notes

1. Different types of glass capillaries for MST are available, either uncoated or modified with various coatings. The choice should be oriented to avoid protein and/or ligand interaction with the glass surface. A capillary selection kit is available from Nano Temper Technologies GmbH, comprising three different kinds of capillaries.
2. The assumption that the total amounts of target protein and titrant are known must be satisfied for the validity of Eq. 3. In addition, spectrophotometric data are strongly biased by particulate or aggregates suspended in the solution under analysis. For these reasons, it is recommended to clarify protein solutions just before use by centrifugation or microfiltration. Moreover, protein storage media and titration conditions should be

Fig. 1 (continued) irradiation. The number identifying the position of each of the 16 capillaries is reported below the horizontal axis. (**b**) MST traces of the 16 samples, showing the time-course of AIF concentration decrease as a consequence of the local increase in solution temperature, as well as the effect of the ligand concentration on thermophoresis. The two shadowed vertical bands represent the regions from which raw fluorescence emission data (F_0 and F_1, respectively) were obtained. (**c**) Titration plot reporting normalized fluorescence change ($F_{norm} = F_1/F_0$) determined by the thermophoretic effect as a function of CHCHD4 concentration. Dots represent data point while the solid line is the best-fitting curve corresponding to the interpolating equation. Note the logarithmic scale of the horizontal axis

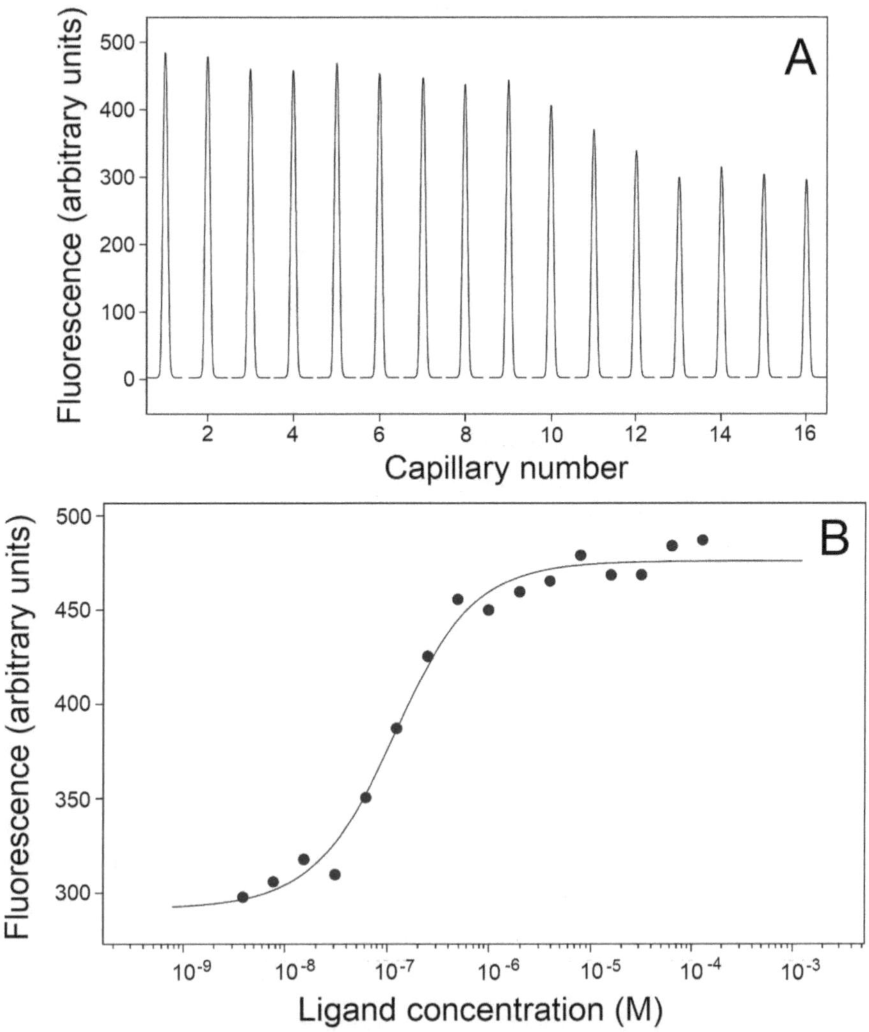

Fig. 2 Titration of AIF in its CT-complex dimeric state with CHCHD4 using the MicroScale Monolith procedure. (**a**) Lateral fluorescence emission scan of the capillary set recorded at baseline temperature, before laser irradiation. The number identifying the position of each of the 16 capillaries is reported below the horizontal axis. Note the fluorescence emission enhancement induced by the presence of CHCHD4 in a concentration-dependent fashion. (**b**) Titration plot reporting fluorescence change determined by CHCHD4-induced emission enhancement, as a function of CHCHD4 concentration. Dots represent data point, while the solid line is the best-fitting curve corresponding to the interpolating equation. Note the logarithmic scale of the horizontal axis

chosen in order to prevent protein denaturation and/or aggregation. Possible solution additives to be considered to these aims are glycerol and/or mild detergents. Physicochemical parameters, like pH and ionic strength, should be considered as well. A useful approach for the rapid screening of several conditions is the ThermoFAD technique [9].

3. In the case the target protein does not display a suitable intrinsic fluorescence, it must be labeled. A large variety of strategies

are available. The choice of the best solution should be guided by the feature of the protein itself—such as the presence of an affinity tag or reactive amino acid side chains on the protein surface, whose modification is not expected to interfere with ligand binding. Nano Temper Technologies GmbH offers different labeling kits covering a variety of modification chemistries.

4. NADH solutions are relatively stable at pH 8.0. We routinely prepare NADH stock solutions in 10 mM Tris-HCl, pH 8.0.
5. Digestion conditions should be experimentally found for each individual fusion protein.
6. PMSF solutions in absolute methanol or ethanol are stable at 4 °C or below, and can be safely stored for weeks.
7. It is important that the composition of the medium of the ligand working solution matches that of the titration medium in order to guarantee that the composition of titration mixtures is constant as the concertation of the titrant is varied.
8. Instrument settings should be empirically determined for each individual protein.

Acknowledgments

This work was supported by grants Linea 2, 2017 and 2018, from Università degli Studi di Milano to A.A. This chapter is dedicated to the memory of Delia Tarantino.

References

1. Brosey CA, Ho C, Long WZ, Singh S, Burnett K, Hura GL, Nix JC, Bowman GR, Ellenberger T, Tainer JA (2016) Defining NADH-driven allostery regulating apoptosis-inducing factor. Structure 24:2067–2079
2. Sevrioukova IF (2011) Apoptosis-inducing factor: structure, function, and redox regulation. Antioxid Redox Signal 14:2545–2579
3. Sorrentino L, Calogero AM, Pandini V, Vanoni MA, Sevrioukova IF, Aliverti A (2015) Key role of the adenylate moiety and integrity of the adenylate-binding site for the NAD(+)/H binding to mitochondrial apoptosis-inducing factor. Biochemistry 54:6996–7009
4. Sevrioukova IF (2016) Structure/function relations in AIFM1 variants associated with neurodegenerative disorders. J Mol Biol 428:3650–3665
5. Hangen E, Feraud O, Lachkar S, Mou H, Doti N, Fimia GM, Lam NV, Zhu C, Godin I, Muller K et al (2015) Interaction between AIF and CHCHD4 regulates respiratory chain biogenesis. Mol Cell 58:1001–1014
6. Modjtahedi N, Tokatlidis K, Dessen P, Kroemer G (2016) Mitochondrial proteins containing coiled-coil-helix-coiled-coil-helix (CHCH) domains in health and disease. Trends Biochem Sci 41:245–260
7. Duhr S, Braun D (2006) Why molecules move along a temperature gradient. Proc Natl Acad Sci U S A 103:19678–19682
8. Wienken CJ, Baaske P, Rothbauer U, Braun D, Duhr S (2010) Protein-binding assays in biological liquids using microscale thermophoresis. Nat Commun 1:100
9. Forneris F, Orru R, Bonivento D, Chiarelli LR, Mattevi A (2009) ThermoFAD, a Thermofluor-adapted flavin ad hoc detection system for protein folding and ligand binding. FEBS J 276:2833–2840

Chapter 13

Using D- and L-Amino Acid Oxidases to Generate the Imino Acid Substrate to Measure the Activity of the Novel Rid (Enamine/Imine Deaminase) Class of Enzymes

Stefania Digiovanni, Genny Degani, Laura Popolo, and Maria Antonietta Vanoni

Abstract

This chapter describes a method to assay the activity of reactive intermediate deaminases (Rid), a large family of conserved soluble enzymes, which have been proposed to prevent damages from metabolic intermediates such as the highly reactive and unstable compounds enamines/imines. In this method, the flavin adenine dinucleotide-dependent L- or D-amino acid oxidases generate an imino acid starting from a L- or D- amino acid, respectively. This reaction is coupled to the hydrolysis of the imino acid to the corresponding α-keto acid and ammonium ion catalyzed by a Rid enzyme. The spectrophotometric assay consists of measuring the decrease of the initial rate of formation of the semicarbazone, derived from the spontaneous reaction of the imino acid and semicarbazide, caused by the presence of the Rid enzyme. The set-up and testing of this method imply a preliminary characterization of the ability of the amino acid oxidase to release the imino acid required for the subsequent reactions. To this purpose, the activity of the L- or D-amino acid oxidases with different amino acids can be measured as production of hydrogen peroxide or formation of semicarbazone in parallel assays. The advantages and limitations of this assay of Rid activity are discussed.

Key words Rid family, Metabolic damage, 2-Aminoacrylate, Activity assay, Semicarbazone, L-Amino acid oxidase, D-Amino acid oxidase, Flavoprotein oxidase, Oxidoreductase, Imino acids

1 Introduction

The metabolism is a complex network of highly coordinated enzyme-catalyzed reactions that transform molecules through a series of metabolic intermediates. Among hundreds of metabolites produced from catabolic and anabolic pathways, small molecules endowed with high reactivity generate a phenomenon called metabolite damage. Cells protect their components by adopting

Stefania Digiovanni and Genny Degani contributed equally with all other contributors.

Maria Barile (ed.), *Flavins and Flavoproteins: Methods and Protocols*, Methods in Molecular Biology, vol. 2280, https://doi.org/10.1007/978-1-0716-1286-6_13, © Springer Science+Business Media, LLC, part of Springer Nature 2021 strategies primarily aimed to: (a) repair the anomalous metabolite and reconvert it to the original form (metabolite repair), (b) preempt damage by transforming the harmful compound into a nontoxic one, and/or (c) divert excess flux out of pathways (directed over-flow) [1–3]. Scavenging of reactive oxygen species by enzymes like catalase or superoxide dismutase is most likely the best-known example of how cells protect themselves from potentially harmful metabolites. However, recent biochemical studies on metabolite repair enzymes or preemptive mechanisms, driven by comparative genomics, demonstrated the existence of networks of conserved enzymes that deal with other types of metabolite damage across all living organisms [3–5]. The knowledge acquired with these studies has the potential to contribute to understand the molecular bases of rare human metabolic diseases, and it may also provide clues for novel biotechnological applications.

Recently, members of the ubiquitous YjgF/YER057c/UK114 protein family have been shown to play a role in a novel damage preempting mechanism [5–7]. Indeed, members of this family have been reported to degrade enamines/imines, which are highly reactive, although short-lived, intermediates produced by several enzymes central to amino acid metabolism. For this reason, the YjgF protein from *Salmonella enterica*, the archetypal member of the family, was recently renamed RidA (from *R*eactive *i*ntermediate *d*eaminase *A*) [6, 7] and the acronym is being applied to other members of the family [5–7].

1.1 RidA and 2-Aminoacrylate (2AA) Stress

To date, the best-characterized function of RidA is the degradation of 2-aminoacrylate (2AA), which is the product of the reactions of pyridoxal-5′-phosphate (PLP)-dependent serine/threonine dehydratases, with serine as the substrate (*see* Fig. 1), and of cysteine desulfhydrases with L-cysteine (not shown in the figure). As shown in Fig. 1, the free 2AA is believed to readily convert into the 2-iminopyruvate tautomer and RidA accelerates the rate of hydrolysis of 2-iminopyruvate to pyruvate and ammonium ion, preempting metabolite damage [5, 8]. The forced accumulation of 2AA in the cell, caused by the lack of RidA activity, was shown to reduce cell fitness and to cause pleiotropic phenotypes [7, 9, 10]. 2AA or 2-iminopyruvate likely form adducts with PLP or with active site residues of several PLP-dependent enzymes (e.g.: branched-chain aminotransferases (BCAT), transaminase B (IlvE), alanine racemase (Alr/DadX), and serine hydroxymethyltransferase (GlyA) [7, 11–13]) or of yet unknown targets that are irreversibly inactivated. Later, it was shown that, in vitro, RidA proteins also hydrolyze the 2-imino acids produced upon oxidation of a variety of L-α-amino acids by L-amino acid oxidase [14, 15].

Rid enzymes are soluble small homotrimeric proteins (128 amino acids, subunit mass 13,576 for *S. enterica* YjgF/RidA). On the basis of the crystal structure of several members of

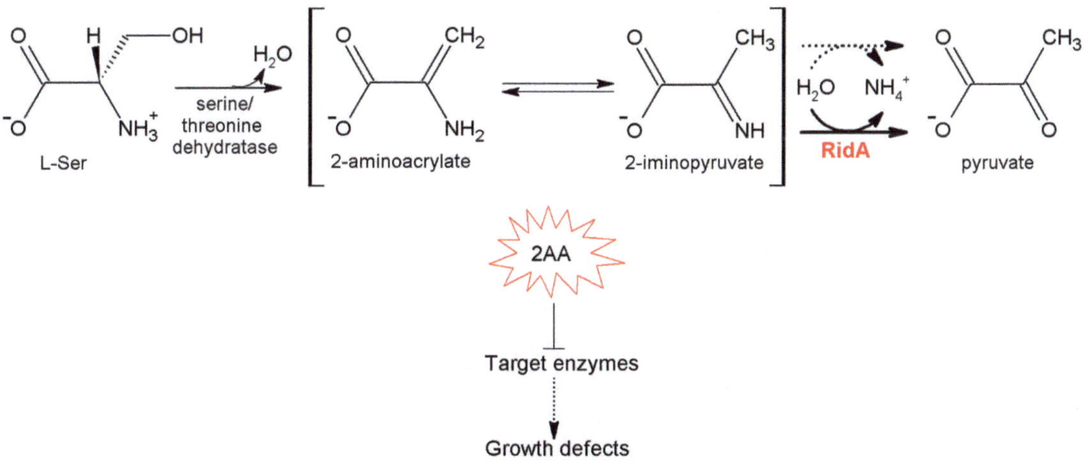

Fig. 1 Role of RidA in preempting accumulation of enamine/imine produced by imino acid metabolism. In the example, 2-aminoacrylate (2AA) released by serine dehydratase tautomerizes to 2-iminopyruvate, which spontaneously hydrolyzes to pyruvate and ammonium ion (dashed arrows). RidA catalyzes 2-iminopyruvate hydrolysis (thick arrow). The absence of RidA leads to accumulation of 2AA that can covalently modify and inactivate a number of PLP-dependent enzymes or other target enzymes and consequently reduce the fitness of the cell [7, 9, 10]

the RidA family, alone or in complex with α-keto acids (*E. coli* TdcF: PDB codes 2UYK (free) and 2UYN (in complex with 2-ketobutyrate), [16] or *Arabidopsis thaliana* RidA: 5HP7 (free) or 5HP8 (in complex with pyruvate), [13]), each homotrimer appears to harbor three equivalent active sites in a cleft at the interface between monomers. The precise mechanism of imino acid hydrolysis is not known. However, no metal cofactor was found in any of the RidA that have been characterized so far [5]. Recent studies demonstrated that UK114, a component of a perchloric acid-resistant fraction of goat (*Capra hircus*) liver, is indeed able to catalyze the hydrolysis of 2-iminopyruvate [8]. Whether the high antigenicity of goat UK114 (which is here renamed goat RidA), which results in cytotoxic antibodies [17–19], is related to its structural stability or to its catalytic activity [8] still needs to be clarified.

A key and essential step toward the elucidation of the biological role of RidA proteins is the characterization of their catalytic properties. In particular, determining their substrate specificity may help to shed light on their possible role in protecting cells from damage by toxic metabolites other than 2AA.

Initially, RidA activity was quantified in a coupled assay in which serine dehydratase or cysteine desulfhydrase generate 2AA from L-serine or L-cysteine, respectively [6, 7, 20, 21]. The initial velocity of pyruvate formation was measured by coupling the production of pyruvate from 2AA/2-iminopyruvate with the reaction of lactate dehydrogenase in the presence of NADH [22]. RidA was found to accelerate the observed rate of NADH oxidation paralleling the RidA-dependent increase of the rate of hydrolysis of the 2-iminopyruvate [22].

In order to explore the reaction specificity of RidA from different sources, Downs and coworkers cleverly adapted an early method of detection of the imino acid produced by the reaction of L- and D-amino acid oxidases [14, 15]. Such method was initially implemented to generate the RidA 2-imino acid substrate from a series of L-amino acids with L-amino acid oxidase during the study of *S. enterica* RidA [14] and several engineered variants of *Arabidopsis thaliana* RidA [13]. Later, the array of 2-imino acid substrates that could be produced was expanded using *Pseudomonas aeruginosa* D-arginine dehydrogenase (DauA) and D-arginine [15]. During the study of goat RidA, it was also shown that pig kidney D-amino acid oxidase and some of its D-amino acid substrates can be effectively used for the in situ generation of RidA substrates, independently confirming that RidA enzymes actually catalyze the hydrolysis of the imino acid rather than the corresponding enamine form [8].

1.2 RidA Deiminase Activity Assays Using L-Amino Acid and D-Amino Acid Oxidases

The method to measure the imine hydrolase activity of RidA (which will be here referred as the deiminase activity) is based on: (a) the ability of L- and D-amino acid oxidases to release the imino acid upon oxidation of their amino acid substrate, with concomitant reduction of molecular oxygen to H_2O_2, and (b) the high reactivity of imines with semicarbazide yielding the corresponding semicarbazone [23]. Therefore, in an assay containing the amino acid oxidase, the amino acid substrate and semicarbazide, the time-course of imino acid production can be readily detected as an increase of absorbance at 248 nm due to formation of the semicarbazone (ε_{248} for semicarbazone 10.3 mM^{-1} cm^{-1} [23]). In the presence of RidA, which accelerates the conversion of the imino acid into the corresponding α-keto acid and ammonium ion, the initial rate of formation of semicarbazone will decrease in a RidA-dependent fashion (*see* Fig. 2). In this method to assay RidA activity, the rapid spontaneous hydrolysis of imino acids requires that the imino acid substrate of RidA is enzymatically produced in situ by the FAD-dependent L- or D-amino acid oxidases from L- or D-amino acid substrates and oxygen. In order to minimize the rate of spontaneous hydrolysis of the imino acid, the reaction is carried out at pH 8.7. Indeed, Hafner and Wellner have shown that the first-order rate constant for the hydrolysis of the imino acid decreases from 6 min^{-1} at pH 7.7 to approximately 1.8 min^{-1} at pH 8.5 or higher [23]. Catalase needs to be included in the assays in order to avoid side reactions caused by H_2O_2, the second product of the amino acid oxidases. Semicarbazide concentration has been set at a value that allows to trap approximately half of the released imino acid [23] while it does not inhibit the amino acid

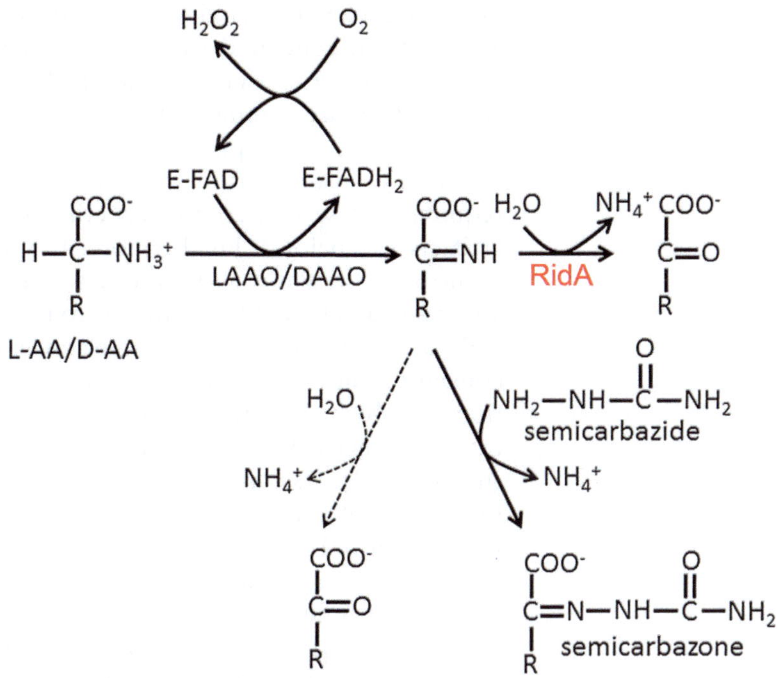

Fig. 2 Scheme of the RidA activity assay. The assay is based on the use of FAD-dependent LAAO and DAAO to generate the imino acid starting from a L- or D-amino acid, respectively. This reaction is followed by the hydrolysis of the imino acid to the corresponding α-keto acid and ammonium ion catalyzed by RidA enzyme. The spectrophotometric assay consists of monitoring at 248 nm the formation of the semicarbazone from the imino acid and semicarbazide. The presence of a Rid enzyme causes a decrease of the initial rate of formation of the semicarbazone (*see* Fig. 3). The dashed line indicates the spontaneous hydrolysis of the imino acid to the α-keto acid and ammonium ion

oxidase reaction [8]. In the final assay, three competing reactions are present, namely: (a) spontaneous hydrolysis of the imino acid to yield the α-keto acid and ammonium ion; (b) RidA-catalyzed imino acid hydrolysis, and (c) addition of semicarbazide to the imino acid with formation of the semicarbazone (*see* Fig. 2).

Although complex, this assay method offers several advantages. First, a large number of widely distributed FAD-dependent oxidases able to catalyze the oxidation of D- and L-amino acids exists beyond the prototypical snake venom L-amino acid oxidase and mammalian D-amino acid oxidase (for a recent review *see* ref. 24). Notable examples of microbial enzymes are the L-amino acid oxidase from *Rhodococcus opacus*, which has a natural broad substrate specificity [25], *P. aeruginosa* FAD-dependent D-arginine dehydrogenase [15], *Trichoderma viride* L-lysine oxidase [26] or *Streptomyces* sp. X-119-6 L-glutamate oxidase [27], offering the possibility to explore the specificity of RidA enzymes toward a broad range of imino acids. Second, engineered forms of amino acid oxidases have already been generated to expand their substrate specificity with respect to L- or D-α-amino acids [28–30]. Finally, several amino acid oxidases are commercially available.

However, the implementation of the assay to the study of a given RidA using various amino acid oxidase–amino acid pairs, requires that several control experiments are carried out to limit artifacts (*see* below). Furthermore, the intrinsic instability of the imino acid does not allow the determination of the steady-state kinetic parameters of RidA, that is, to measure the initial velocity of RidA at varying (known) substrate concentrations covering a broad concentration range. Rather, assays are carried out under conditions in which the initial velocity of the RidA reaction reflects the catalytic efficiency of the RidA enzyme with a given imino acid substrate [8]. The pH-dependence of the rate of spontaneous hydrolysis of the 2-imino acid, which, as mentioned above, increases at pH values below 8.5 [23], also hampers studies of the RidA activity at physiological pH.

On the other hand, thanks to their deiminating activity, RidA enzymes can be a valuable molecular tool to dissect the catalytic mechanism of oxidases since it can be easily established whether the imino acid or the α-keto acid is released in solution from the oxidase active site [31]. Furthermore, in several of the reported and potential applications of natural and engineered variants of various amino acid oxidases (e.g., in bioconversions [28, 29, 32]), the deiminase activity of RidA may be of help to avoid undesired side reactions of the (reactive) imino acid intermediate. RidA would be beneficial, in particular, when reactions need to be carried out at pH above neutrality, where the half-life of the imino acid is higher [23]. The high yield of production of goat RidA (200 mg from 9 g of starting *E. coli* wet paste [8]) coupled to its remarkable conformational stability would make it an excellent candidate for such applications [8].

2 Materials

All solutions are prepared with ultrapure water and analytical grade reagents. Enzymes were from Sigma-Aldrich or prepared in the laboratory.

2.1 Buffers and Reagents

1. *Buffer A*: 50 mM tetrasodium diphosphate decahydrate (sodium pyrophosphate, NaPP), pH 8.7. For 250 ml, 5.58 g NaPP are dissolved in approximately 240 ml water under magnetic stirring. The pH is adjusted to 8.7 with 10 M HCl and the volume is brought to 250 ml in a cylinder. The buffer is stored in aliquots in Pyrex bottles at 4 °C (*see* **Note 1**).

2. *Semicarbazide solution*: 100 mM semicarbazide hydrochloride neutralized. To prepare 1.5 ml, 16.7 mg semicarbazide hydrochloride) are added to a 2 ml-Eppendorf tube containing 1335 μl of water and 150 μl buffer A. After vortexing to dissolve, 15 μl of 10 M NaOH are added and mixed by vortexing. The final pH, which should be approximately 7, is checked with pH-indicator paper. The solution is freshly prepared, wrapped in aluminum foil to protect it from light and stored on ice except for small (100 μl) working aliquots, which are equilibrated at room temperature (typically 23–25 °C).
3. *Amino acid stock solutions*: 50 mM solutions (typically 10 ml) L- or D-amino acids are prepared in Buffer A. The pH is checked by spotting 2–3 μl aliquots on pH-indicator paper, and there is usually no need to adjust to pH 8.7. The amino acid solutions are stored in aliquots at −20 °C.
4. *4X-PBS:* 40 mM sodium phosphate buffer, pH 7.4, 552 mM NaCl, 10.8 mM KCl. To prepare 100 ml, $NaH_2PO_4 \cdot H_2O$ (552 mg), NaCl (3.23 g) and KCl (81 mg) are dissolved in ultrapure water, in a beaker under magnetic stirring; the pH is adjusted to 7.4 with 1 M NaOH and the volume is brought to 100 ml. 1–10 ml aliquots are stored at −20 °C. For the working PBS solution, a fourfold dilution in water is prepared and stored in the cold room or frozen at −20 °C.
5. *Saline solution*: 0.9% (w/v) NaCl, stored at 4 °C or −20 °C.

2.2 Enzymes

1. *Catalase*: 1 mg/ml bovine liver catalase) is prepared in PBS and stored in aliquots (50–100 μl) at −20 °C (*see* **Note 2**).
2. *L-Amino acid oxidase (LAAO)*: the commercial enzyme from *Crotalus adamanteus,* Type IV, ≥3.0 units/mg protein (aqueous suspension) is used and stored at 4 °C. To prepare the working solution, after careful mixing, the suspension is diluted tenfold in PBS and used within 3 days with storage on ice (during the experiment) or at 4 °C. Care must be taken to shield the enzyme solution from light with aluminum foil.
3. *D-amino acid oxidase* (*DAAO*): native porcine kidney DAAO was purified according to a previously published protocol [33]. The enzyme is also commercially available.
4. *RidA enzyme*: in the example reported in this chapter, we used a purified preparation of recombinant goat RidA (monomer mass, 14,579; UniProt Accession number P80601; 3 mg/ml in saline solution [8]). The goat RidA solution was stored frozen in small aliquots at −20 °C. For the activity assays, the stock enzyme solution was thawed on the bench, gently mixed by hand and briefly spun in the cold to collect it at the bottom of the test tube. The enzyme solution was diluted 10- to 100-fold in saline solution in order to explore various

concentration ranges in the assays while adding 1–10 μl aliquots to minimize pipetting errors. The working solutions were stored on ice during the experiment.

2.3 Equipment

1. HP 8452A diode-array UV-visible spectrophotometer (Agilent Technologies, Santa Clara, CA, U.S.A), which records absorption spectra between 190 and 1100 nm. The jacketed sample cell holder is connected to a cryothermostat set at 25 °C. Data acquisition and analysis is done with the HP Chemstation (*see* **Note 3**).
2. Black quartz microcuvettes (Hellma Analytics, 1 cm light path, 100 μl cell volume) (*see* **Note 4**).
3. Grafit or other suitable software can be used for data fitting.
4. Microplate reader Ensight (PerkinElmer, Waltham, MA, U.S. A.)
5. UV-star 96-well microplates (Greiner Bio-One International GmbH).
6. Multichannel pipettes.

3 Methods

3.1 RidA Activity Assay: The Principle

The RidA activity assay is based on the use of FAD-dependent LAAO or DAAO to generate in situ an imino acid starting from a L- or D-amino acid, respectively. The imino acid readily reacts nonenzymatically with semicarbazide to yield the corresponding semicarbazone resulting in an increase of absorbance in the UV (*see* Fig. 2). In the assay, Rid enzymes catalyze the hydrolysis of the imino acid to the corresponding α-keto acid and ammonium ion. This activity competes with both the spontaneous reaction of the imino acid with semicarbazide and its spontaneous hydrolysis (*see* Fig. 2). The formation of the semicarbazone is monitored spectrophotometrically at 248 nm in a continuous assay. Therefore, Rid activity is revealed by the decrease of the initial rate of formation of semicarbazone in the assay. To avoid interference from the H_2O_2 product of LAAO and DAAO, catalase is included [23]. To reduce the rate of spontaneous hydrolysis to α-keto acid and ammonium ion, the assay is carried out at pH 8.7 that increases the half-life of the imino acid derived from amino acid oxidation by LAAO or DAAO [8, 23, 34]. The semicarbazide concentration in the assay is close to that which traps 50% of the imino acid at this pH [8]. Higher concentrations may interfere with the enzyme's activities.

3.1.1 Assay Protocol

A 150 μl-reaction mix is setup by sequential addition of the components to a quartz microcuvette in the following order:

1. Buffer A (to reach a final volume of 150 μl);
2. Semicarbazide solution (15 μl);
3. Catalase (1 μl);
4. LAAO or DAAO (1–10 μl);
5. RidA (if present, 1–20 μl).

After the addition of each component and mixing, absorption spectra are recorded (*see* **Note 5**).

The reaction is initiated by the addition of the amino acid solution (15 μl) to the assay mixture that had been equilibrated at 25 °C in the spectrophotometer thermostated cell for at least 3 min (*see* **Note 6**). Spectra are recorded in "Kinetic mode" every 10–20 s for the first 300 s of reaction with an increase of 10% of the time interval between spectra between 300 and 1500 s (*see* Fig. 3a). The absorbance values at 248 nm are used to determine the initial velocity (typically in the first 30–120 s of reaction) expressed as ΔA_{248}/min. An example of the time-course of spectral and absorbance changes at 248 nm observed in the absence and presence of goat RidA is shown in Fig. 3a, b.

The initial velocity measured in the absence (v_o) and in the presence of a given concentration of RidA (v_{RidA}) can also be expressed as increase of semicarbazone concentration in the unit time (μM/min) by using an extinction coefficient of 10.3 mM^{-1} cm^{-1} [23] (*see* Fig. 3c). The initial part of the curve, describing the dependence of the initial velocity of the reaction upon RidA concentration, is linear (*see* Fig. 3c). Here, the slope $[(v_o - v_{RidA})/v_o]/[RidA]$ is a measure of the specific activity of RidA with the given imino acid under the chosen experimental conditions (μmol imino acid hydrolyzed per min per μmol RidA).

3.1.2 Assay Optimization and Controls

The amount of LAAO or DAAO in the assay with a given amino acid is established in a series of preliminary assays carried out in the absence of RidA in order to obtain an initial velocity of semicarbazone formation from the imino acid (IA) corresponding to 0.2–0.3 ΔA_{248}/min (*see* Fig. 4, right axes). With an extinction coefficient of 10.3 mM^{-1} cm^{-1} this value corresponds to 19–29 μM/min. Under these conditions, the increase of absorbance at 248 nm in the absence of RidA is linear for at least 3 min and one can reliably measure the initial reaction velocity when RidA causes up to a 90–95% decrease (*see* Fig. 3b). Whether and to what extent the amino acid oxidase actually releases the imino acid in solution under the given experimental conditions also need to be checked. As an example, the comparison of the initial velocity of H_2O_2 production of the amino acid oxidase (v_{H2O2}) with that of

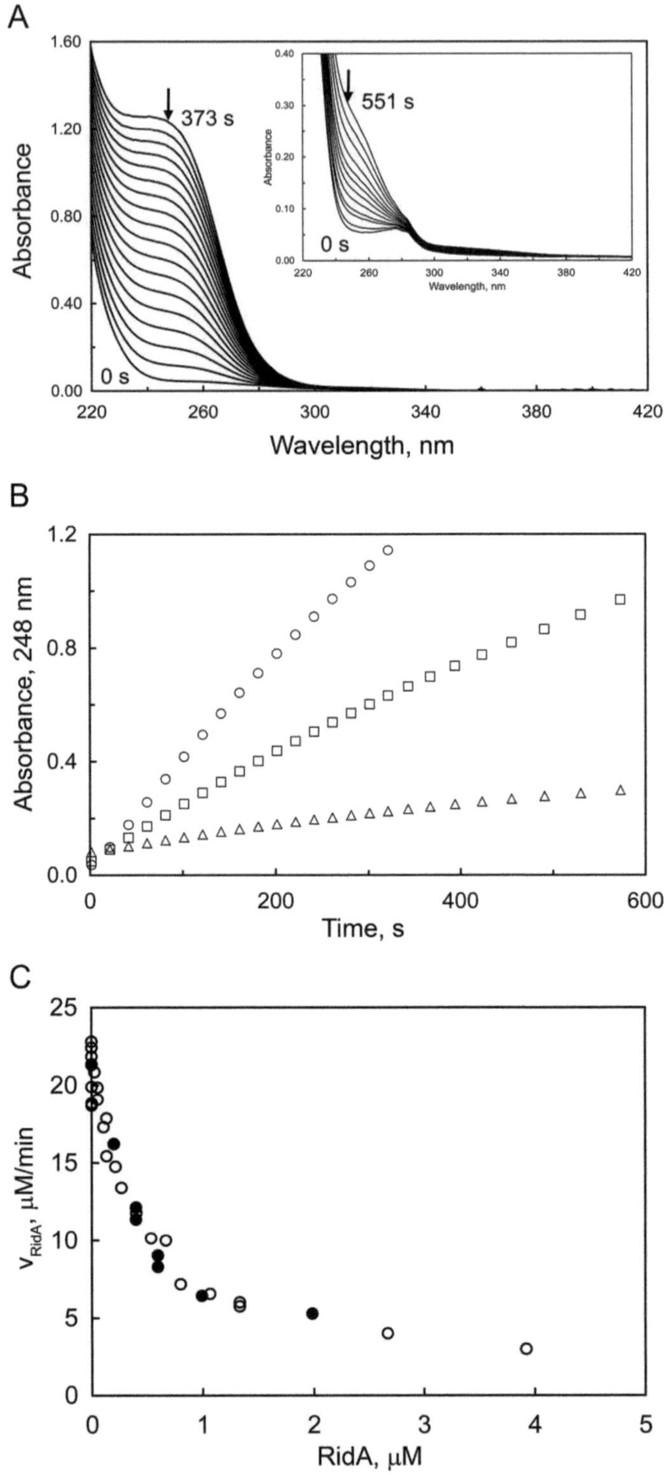

Fig. 3 Measurement of the deiminase activity of goat RidA. (**a**) Selected absorption spectra recorded at different times from the addition of L-Leu (5 mM) to a 150 μl reaction mixture containing semicarbazide (10 mM),

semicarbazone formation from the imino acid (v_{IA}) in parallel assays, at varying amino acid oxidase concentration, demonstrated that indeed the semicarbazide concentration in the assay traps approximately 50% of the imino acid released in solution by LAAO and DAAO with L-Leu and D-Ala, respectively (*see* Fig.4). To monitor H_2O_2 production, the assay contains 5 mM L- or D-amino acid, horse radish peroxidase (HRP, 10 μg/ml), *o*-dianisidine (0.4 mM final concentration from a 8 mM stock solution in 20% Triton X-100) and varying amounts of the amino acid oxidase which is added last to initiate the assay [35, 36]. The initial velocity of H_2O_2 production (v_{H2O2}) is measured from the absorbance changes at 436 nm using an extinction coefficient of 11.6 mM^{-1} cm^{-1} for oxidation of *o*-dianisidine [35].

On the other hand, the activity of LAAO with L-serine proved to be too low to be used to generate the 2-imino 3-hydroxypyruvate product (not shown). Although D-serine is a DAAO substrate, we could not use DAAO and D-serine to generate the corresponding imino acid since no semicarbazone formation was observed under enzyme and D-Ser concentrations that demonstrated H_2O_2 formation (not shown). Thus, it appears that hydrolysis of the imino acid resulting from D-serine oxidation takes place prior to release of the imino acid from DAAO active site, while the imino acid derived from D-Ala is quantitatively released as such into the medium. Therefore, a similar experimental scheme can be used to establish if other amino acid oxidases release the imino acid in the context of their mechanistic characterization or during the characterization of Rid proteins. Furthermore, monitoring semicarbazone formation may provide a convenient spectrophotometric assay for amino acid oxidases, as an alternative to polarographic methods when the amino acid reacts with the components of the HRP-coupled assay of H_2O_2. For example, tyrosine oxidation by HRP makes it impossible to measure the amino acid oxidase activity

Fig. 3 (continued) catalase (1 μg), and LAAO (~1.1 μg) in the absence of RidA (main panel) and in the presence of 3.9 μM RidA (inset). The arrow indicates the wavelength selected to measure semicarbazone formation (248 nm). Note that in the inset, the absorbance peak centered at 280 nm is due to the presence of RidA. (**b**) Time course of the absorbance changes at 248 nm in the absence (circles) or presence of goat RidA (squares, 0.4 μM; triangles, 3.9 μM) extracted from the spectra recorded with the diode array spectrophotometer. (**c**) Scatter graph of initial velocity values of semicarbazone formation measured in the absence or presence of RidA (v_{RidA}). Increasing amounts of RidA decrease the rate of production of semicarbazone from the imino acid formed from L-Leu and LAAO (open circles) or D-Leu and DAAO (closed circles). Assay conditions were L-Leu or D-Leu (5 mM), semicarbazide (10 mM), LAAO (1.1 μg) or DAAO (2.4 μg), catalase (1 μg) in 50 mM sodium pyrophosphate, pH 8.7 at 25 °C. Assay volume: 150 μl

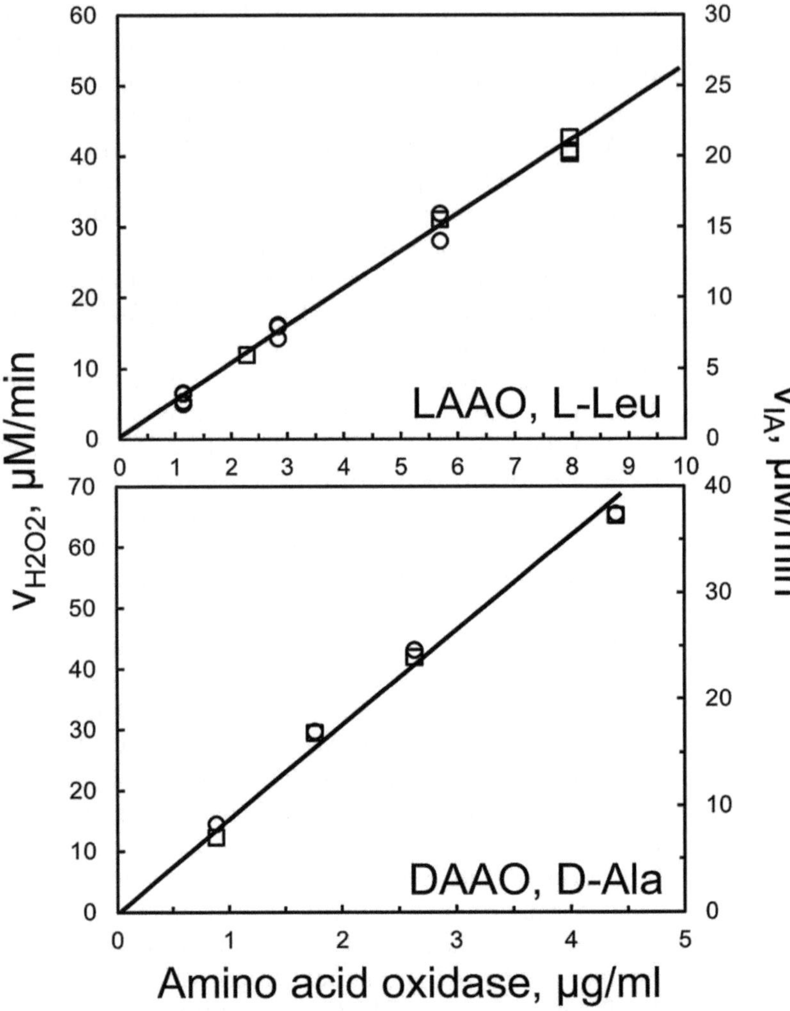

Fig. 4 Establishing the amount of amino acid oxidase to be used in the assay and the extent of release of imino acid. The initial velocity of production of H_2O_2 (v_{H2O2}, circles, *left axis*) catalyzed by varying amounts of LAAO and L-Leu (*Top panel*) or DAAO in the presence of D-Ala (*Lower panel*) is compared directly with that of imino acid production ($v_{IA,}$ squares, *right axis*). The resulting v_{H2O2} and v_{IA} values have been plotted on two different scales to show both the linearity of the dependence of the initial velocities upon the amino acid oxidase concentrations and the ratio of approximately 2 between v_{H2O2} and v_{IA} [8]

by monitoring H_2O_2 production in HRP-coupled assays [37]. If needed, the two types of assays (H_2O_2 production coupled to HRP and imino acid release coupled to semicarbazone formation) can be used to determine the steady-state kinetic parameters of the reaction of the amino acid oxidase with the given amino acid (*see* Fig. 5).

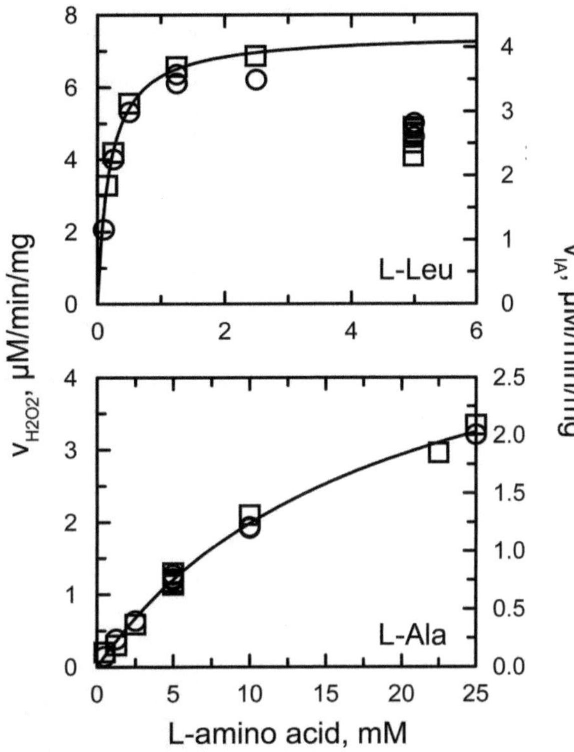

Fig. 5 Determination of the steady-state kinetic parameters of the LAAO reaction with L-Leu and L-Ala as substrates by monitoring H_2O_2 or imino acid production. The HRP-coupled assay of H_2O_2 production ($v_{H2O2,}$ circles, *left axis*) or the assay of semicarbazone formation from the imino acid produced ($v_{IA,}$ squares, *right axis*) can be effectively used to determine the steady-state kinetic parameters of the oxidase reaction, further confirming that the amount of imino acid trapped by semicarbazide is approximately 50% the amount of H_2O_2 produced by the enzyme. The curves show the fit of the data obtained by monitoring semicarbazone formation with the Michaelis–Menten equation [$v = (V_{max}{*}S)/(K_m + S)$] with v, initial reaction velocity, V_{max}, apparent maximum velocity; K_m, apparent Michaelis constant for the amino acid substrate. The data obtained with L-Leu above 3 mM, where substrate inhibition is observed, were not taken into account for the fitting. The resulting values were V_{max}, 4.2 ± 0.06 μM/min/mg and K_m, 0.19 ± 0.01 mM for L-Leu; V_{max}, 3.5 ± 0.25 μM/min/mg and K_m, 18.4 ± 2.4 mM for L-Ala

3.1.3 Data Analysis and Determination of RidA Specificity

For data analysis, the initial velocity measured in the presence of a given concentration of RidA (v_{RidA}) is divided by the initial velocity measured in the absence of RidA (v_o) (*see* **Note 7**). The v_{RidA}/v_o ratio is multiplied by 100 to express the reaction velocity in percent (v, %). Using the logarithmic scale to represent RidA concentration provides a clear visualization of the preference of RidA toward the imino acids derived from the different L- and D-amino acids and LAAO or DAAO, respectively. In the example of Fig. 6a, goat RidA shows a preference for 2-imino acids derived from Ala > L-

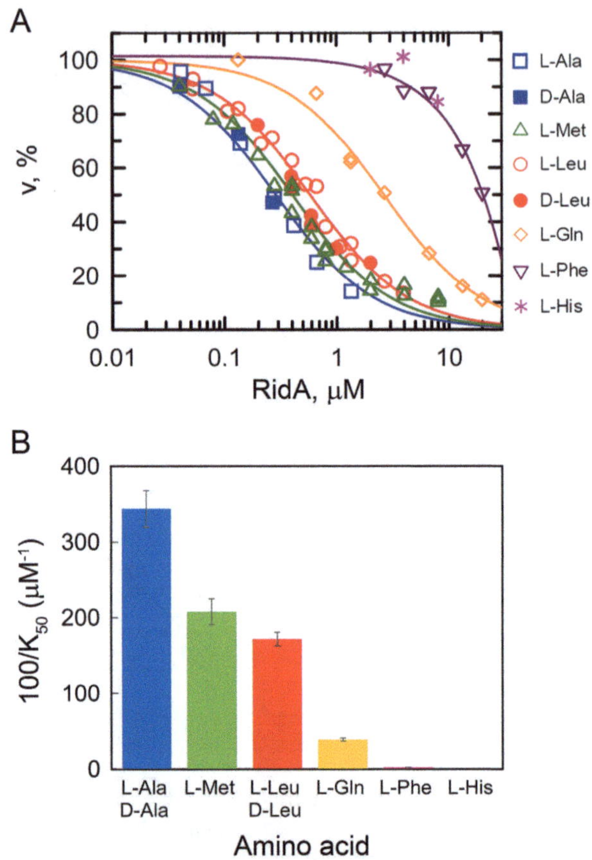

Fig. 6 Determination of substrate specificity of RidA enzymes. (**a**) The indicated amino acids were incubated with 10 mM semicarbazide in 50 mM sodium pyrophosphate buffer, pH 8.7, in the presence of catalase (10 μg/ml), LAAO or DAAO and the indicated concentrations of goat RidA at 25 °C. The amount of LAAO (5–800 μg/ml) and DAAO (2–20 μg/ml) was established in preliminary experiments (*see* Figs. 3c and 4) in order to obtain for each amino acid a similar value of initial velocity of semicarbazone formation in the absence of RidA (0.2–0.3 ΔA_{248}/min). The observed relative velocity (*v*, %) is plotted as a function of RidA concentration. The curves are the best fit of the data to Eq. 1. The data obtained with DAAO or LAAO and D- or L-alanine and leucine overlapped demonstrating that the RidA substrate is the common imino acid product of the two reactions and that the excess L- or D-amino acid present does not seem to interfere with RidA activity. Therefore, the data obtained from D- and L- Ala or D- and L-Leu were fitted together to Eq. 1. (**b**) shows the calculated values of $100/K_{50}$, which is an estimate of the catalytic efficiency of RidA with the specific imino acid and provides a convenient way to express the substrate preference of RidA. For L-Phe, which appears to yield a poor substrate of RidA, the *v* % values data were fitted to a straight line, the slope of which corresponds to $100/K_{50}$. Little or no activity was observed with L-His and therefore the data were not fitted

Met ~ Leu > L-Gln whereas little or no activity was detected for L-Phe and L-His [8]. The specificity is similar to that reported previously for *S. enterica* RidA [14, 15].

The values of the observed relative reaction velocity (v, %) as a function of RidA concentration can be fitted with the Grafit program (or equivalent) to Eq. 1, whose derivation is described in detail in [8].

$$v = \frac{100}{1 + \frac{[\text{RidA}]}{K_{50}}} \tag{1}$$

In the equation, [RidA] is the concentration of RidA in the assay; v is the percent residual activity measured at a given concentration of RidA [$v = (v_{\text{RidA}}/v_{\text{o}}) \times 100$]; K_{50} is the RidA concentration that halves the observed initial velocity of semicarbazone formation. As discussed in [8], under the assay conditions, the dependency of the observed reaction velocity (expressed as %, v) upon RidA concentration is unaltered when the initial velocity of semicarbazone formation in the absence of RidA is between 0.1 and 0.4 ΔA_{248}/min. Therefore, K_{50} is a parameter that is inversely correlated with the catalytic efficiency of a RidA with a specific imino acid. Indeed, the slope of the initial part of the curves, which is a measure of the specific activity of RidA for a given imino acid, can be mathematically derived from Eq. 1 as $100/K_{50}$. This value results from the first derivative of Eq. 1 with RidA concentration approaching zero. The $100/K_{50}$ value is the convenient parameter to express the catalytic efficiency of RidA with the different imino acids in a tabular form [8] or bar graph (*see* Fig. 6b).

3.2 Set up of the RidA Activity Assay in a Microplate Format

The assay described above (*see* Subheading 3.1) can also be carried out in a 96-well UV-transparent flat-bottomed microplate in order to explore several conditions at once. We here describe, as a proof-of-principle, the method applied to measure goat RidA activity in the presence of LAAO and L-Leu.

3.2.1 Protocol

The reaction mixtures, containing all the reagents except the substrate, are prepared in microtubes in a 500 μl-volume suitable to obtain a triplicate for each reaction. Incubate in a thermomixer at 25 °C for 2 min (*see* **Note 8**).

1. Turn on the plate-reader and set the chamber incubator at 25 °C. Set the reader to carry out a "single read" with the measurement sequence of 120 repeats every 10 s (total time, 20 min), with the following parameters: λ, 248 nm; number of flashes, 30; measurement height, 7.5 mm.

2. Fill the wells of a standard 96-well plate with the amino acid solution (in a volume greater than that required) and place it in the plate reader to equilibrate at 25 °C for 2–3 min.
3. For each condition, add to three wells of the UV-transparent plate, 135 μl of each reaction mixture lacking the amino acid, or 150 μl of buffer.
4. Start the reaction by adding 15 μl of the prewarmed L-leucine solution with a multichannel pipette to the sample wells and mix by gently pipetting up and down three times.
5. Place immediately the plate in the microplate reader and start the protocol.
6. Once finished, export the data in Excel format.
7. The absorbance values from each well are plotted in a graph as a function of time (in seconds) and visually inspected for: (a) consistency of the replicates; (b) absence of significant time-dependent absorbance changes in the control samples, and (c) detection of a common interval in which absorbance changes are linear with time (in our case 40–180 s, *see* Fig. 7a).
8. For each assay, calculate the slope ($\Delta A_{248}/s$) of the linear part of the curve. For simplicity convert the value to $\Delta A_{248}/min$ and into initial velocity of the reaction v_0 or v_{RidA} in μM/min, by taking into account the extinction coefficient for semicarbazone formation and the experimentally determined light path (*see* **Note 9**).

3.2.2 Assay Optimization and Estimate of RidA Activity in a Microplate Format

In preliminary experiments the amount of LAAO was varied, in the absence of RidA, in order to establish (a) the linearity of the dependence of the initial velocity of the reaction as a function of LAAO concentration in the assays (*see* Fig.7b); (b) the match of the activity measured in the plate format and that measured in the microcuvette (compare Fig. 7b with Fig. 4, *Top*); (c) the amount of LAAO to use at a fixed level to test RidA activity (1.3 μg in 150 μl) in order to obtain an initial $\Delta A_{248}/min$ of 0.12–0.13 (corresponding to 28–30 μM/min). Under these conditions, with the LAAO/L-Leu couple, the calculated K_{50} was 0.65 ± 0.03 μM (*see* Fig. 7c), which is similar to that determined in the microcuvette format (0.58 ± 0.03 μM, *see* Fig. 6a).

While the microplate format provides a convenient way to carry out parallel assays under a variety of conditions, it should be noted that the number of such assays may be limited by the ability to start the reaction simultaneously (or at known time intervals) in a large number of wells in order to ensure that the linear part of the A_{248} traces is actually recorded. In the example reported here, we set up a maximum of nine parallel assays in order to start the reaction simultaneously with the automatic pipette.

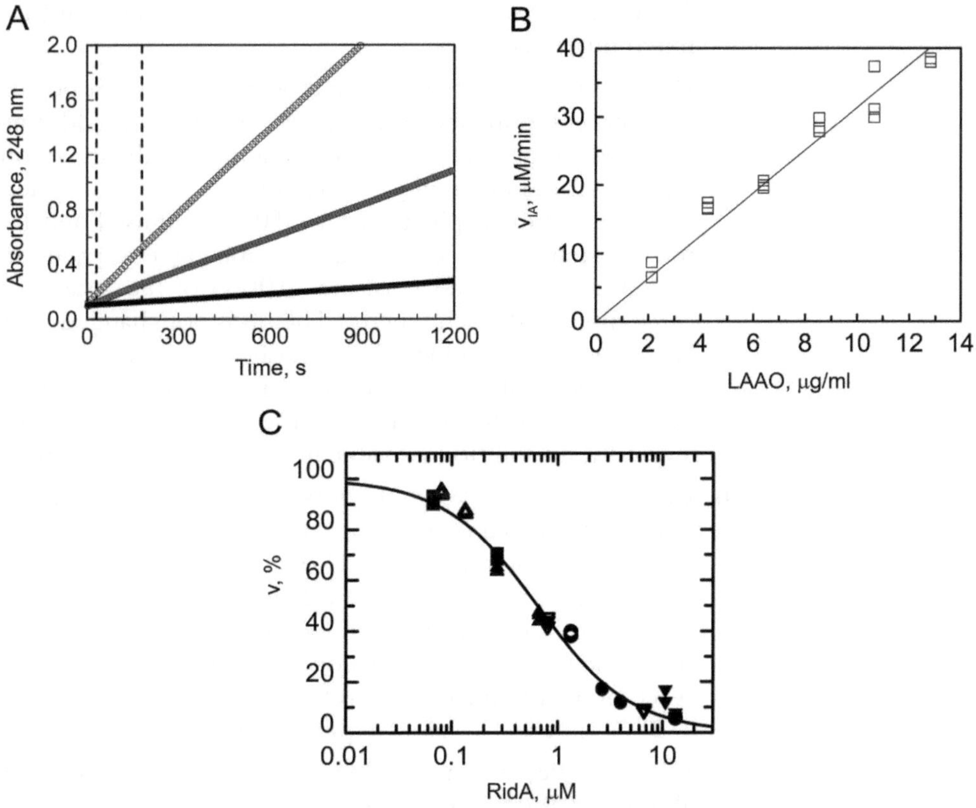

Fig. 7 Assay of the RidA activity in a microplate format. (**a**) Time-course of the absorbance changes at 248 nm measured in a well of a microplate containing 5 mM L-Leu, 10 mM semicarbazide, 1 μg catalase, and 1.3 μg LAAO in the absence of RidA (empty circles) and in the presence of two different concentrations of RidA (1.3 μM, grey circles; 13.3 μM, black circles) in a final assay volume of 150 μl. (**b**) Linear dependence of initial velocity of imino acid production as a function of LAAO concentration. (**c**) Determination of the K_{50} of RidA for the product of L-Leu oxidation by LAAO in the microplate format. Different symbols correspond to different series of assays carried out under the same experimental condition. The curve is the best fit of the data to Eq. 1, which yielded a K_{50} of 0.65 ± 0.03 μM

4 Notes

1. Since the pH of the NaPP buffer is particularly sensitive to dilution, the pH must be adjusted when the solution volume is close to the final one.
2. If convenient, a 10 mg/ml solution can be prepared and stored at −20 °C to be used as a stock to obtain the working 1 mg/ml solution. Both solutions are stable frozen for several years.
3. The assays can be carried out in a dual beam spectrophotometer set at 248 nm. However, the diode array spectrophotometer offers the advantage to monitor the entire absorption spectrum of the reaction over time. Thus, it makes possible to easily detect artifacts (e.g., mixing problems, development of turbidity) or any other anomalous behavior.

4. Assays can be scaled up to 0.6–1 ml and carried out in semi-micro quartz cuvettes.
5. Due to the geometry of the microcuvettes it is not possible to mix the components by inversion or by using adder-mixers [38]. Therefore, each reagent is added with an automatic pipette. At the beginning of each assay, a fresh tip is placed on a 200 μl-automatic pipette set at ~100 μl and used only for the mixing steps. After each addition, mixing is carried out by rapidly taking up and releasing ~100 μl of the assay solution three times with such pipette. The tip is not changed until the last component is added. In this way, the reagents are kept at the desired final concentration with the 150 μl assay solution being either in the cuvette or in the tip of the mixing pipette. Typically, no solution is visible in the mixing pipette tip.
6. Care is taken to keep aliquots of all reagents (except the enzymes) at room temperature (typically 23–25 °C) to accelerate the equilibration at 25 °C of the reaction mixture in the spectrophotometer's jacketed cell holder. As an alternative, the aliquots of the reagents can be equilibrated at 25 °C in a thermostated bath or a thermomixer. The subsequent addition of components into the cuvette is done at time intervals of 1–2 min in order to allow equilibration at 25 °C.
7. It is advisable, within the same day, to repeat the reaction in the absence of RidA in order to monitor the stability of the assay components. The average of initial velocity measured in the absence of RidA during the day is used as v_0. Replicates of RidA assays are obviously also required.
8. During assay optimization, the following controls are also set up in triplicate directly in the wells:

Control	NaPP	Semicarbazide	Catalase	RidA[a]	LAAO[a]	L-Leu
No RidA	117 μl	15 μl	1 μl	0	2 μl	15 μl
No LAAO	117 μl	15 μl	1 μl	2 μl	0	15 μl
No L-Leu	130 μl	15 μl	1 μl	2 μl	2 μl	0

[a]volumes/final concentrations depend on the specific series of assays

However, no significant time-dependent absorbance changes at 248 nm were observed with these controls, so that the data obtained in the complete assay did not need to be corrected.

9. In order to compare the activities measured with the two methods, the light path of the sample in the microplate should be determined. Once defined the final volume of the reaction mix (150 μl), the absorbance of solutions of different concentrations of a molecule absorbing in the same UV region as the

semicarbazone (in our case L-tyrosine, 0.25–1 mM) is measured with the HP diode array spectrophotometer in the microcuvette (1 cm light path) and with the plate reader (with 150 μl solution in each well). The light path of the well is the slope of the line relating the absorbance at 280 nm measured in the well with that measured in the microcuvette. With the microplate used in this protocol, a total assay volume of 150 μl corresponds to a light path of 0.418 ± 0.005 cm.

Acknowledgments

The authors are grateful to Prof. Alberto Bartorelli Cusani and Dr. Francesco Baggi Sisini for the generous financial support to this research. We wish to thank Mr. Alessandro Lucini Paioni for the assistance in the enzymatic assays. S.D. is a recipient of a fellowship financed by Alalia S.r.l (Turin, Italy). G.D. is a recipient of a Post-Doctoral fellowship from the University of Milan (Italy). Stefania Digiovanni and Genny Degani contributed equally to this work.

References

1. de Crecy-Lagard V, Haas D, Hanson AD (2018) Newly-discovered enzymes that function in metabolite damage-control. Curr Opin Chem Biol 47:101–108
2. Frelin O, Huang L, Hasnain G et al (2015) A directed-overflow and damage-control *N*-glycosidase in riboflavin biosynthesis. Biochem J 466(1):137–145
3. Linster CL, Van Schaftingen E, Hanson AD (2013) Metabolite damage and its repair or pre-emption. Nat Chem Biol 9(2):72–80
4. Peracchi A, Veiga-da-Cunha M, Kuhara T et al (2017) Nit1 is a metabolite repair enzyme that hydrolyzes deaminated glutathione. Proc Natl Acad Sci U S A 114(16):E3233–E3242
5. Borchert AJ, Ernst DC, Downs DM (2019) Reactive enamines and imines in vivo: lessons from the RidA paradigm. Trends Biochem Sci. https://doi.org/10.1016/j.tibs.2019.04.011
6. Lambrecht JA, Flynn JM, Downs DM (2012) Conserved YjgF protein family deaminates reactive enamine/imine intermediates of pyridoxal 5′-phosphate (PLP)-dependent enzyme reactions. J Biol Chem 287(5):3454–3461
7. Lambrecht JA, Schmitz GE, Downs DM (2013) RidA proteins prevent metabolic damage inflicted by PLP-dependent dehydratases in all domains of life. MBio 4(1):e00033–e00013
8. Degani G, Barbiroli A, Regazzoni L et al (2018) Imine deaminase activity and conformational stability of UK114, the mammalian member of the Rid protein family active in amino acid metabolism. Int J Mol Sci 19 (4):945–963
9. Borchert AJ, Downs DM (2017) The response to 2-aminoacrylate differs in *Escherichia coli* and *Salmonella enterica*, despite shared metabolic components. J Bacteriol 199(14): e00140–e00117
10. Borchert AJ, Downs DM (2017) Endogenously generated 2-aminoacrylate inhibits motility in *Salmonella enterica*. Sci Rep 7 (1):12971
11. Flynn JM, Downs DM (2013) In the absence of RidA, endogenous 2-aminoacrylate inactivates alanine racemases by modifying the pyridoxal 5′-phosphate cofactor. J Bacteriol 195 (16):3603–3609
12. Flynn JM, Christopherson MR, Downs DM (2013) Decreased coenzyme A levels in RidA mutant strains of *Salmonella enterica* result from inactivated serine hydroxymethyltransferase. Mol Microbiol 89(4):751–759
13. Liu X, Zeng J, Chen X et al (2016) Crystal structures of RidA, an important enzyme for the prevention of toxic side products. Sci Rep 6:30494

14. Niehaus TD, Gerdes S, Hodge-Hanson K et al (2015) Genomic and experimental evidence for multiple metabolic functions in the RidA/YjgF/YER057c/UK114 (Rid) protein family. BMC Genomics 16:382
15. Hodge-Hanson KM, Downs DM (2017) Members of the Rid protein family have broad imine deaminase activity and can accelerate the *Pseudomonas aeruginosa* D-arginine dehydrogenase (DauA) reaction in vitro. PLoS One 12(9):e0185544
16. Burman JD, Stevenson CE, Sawers RG et al (2007) The crystal structure of *Escherichia coli* TdcF, a member of the highly conserved YjgF/YER057c/UK114 family. BMC Struct Biol 7:30
17. Bartorelli A, Bussolati B, Millesimo M et al (1996) Antibody-dependent cytotoxic activity on human cancer cells expressing UK 114 tumor membrane antigen. Int J Oncol 8 (3):543–548
18. Bartorelli A, Biancardi C, Cavalca V et al (1996) Purification and partial characterization of proteins present in a perchloric acid extract of goat liver (UK101). J Tumor Marker Oncol 11(1):57–61
19. Bussolati G, Geuna M, Bussolati B et al (1997) Cytolytic and tumor inhibitory antibodies against UK114 protein in the sera of cancer patients. Int J Oncol 10:779–785
20. Lambrecht JA, Browne BA, Downs DM (2010) Members of the YjgF/YER057c/UK114 family of proteins inhibit phosphoribosylamine synthesis in vitro. J Biol Chem 285 (45):34401–34407
21. Ernst DC, Lambrecht JA, Schomer RA et al (2014) Endogenous synthesis of 2-aminoacrylate contributes to cysteine sensitivity in *Salmonella enterica*. J Bacteriol 196 (18):3335–3342
22. Niehaus TD, Nguyen TN, Gidda SK et al (2014) Arabidopsis and maize RidA proteins preempt reactive enamine/imine damage to branched-chain amino acid biosynthesis in plastids. Plant Cell 26(7):3010–3022
23. Hafner EW, Wellner D (1979) Reactivity of the imino acids formed in the amino acid oxidase reaction. Biochemistry 18(3):411–417
24. Campillo-Brocal JC, Lucas-Elio P, Sanchez-Amat A (2015) Distribution in different organisms of amino acid oxidases with FAD or a quinone as cofactor and their role as antimicrobial proteins in marine bacteria. Mar Drugs 13 (12):7403–7418
25. Geueke B, Hummel W (2003) Heterologous expression of *Rhodococcus opacus* L-amino acid oxidase in *Streptomyces lividans*. Protein Expr Purif 28(2):303–309
26. Amano M, Mizuguchi H, Sano T et al (2015) Recombinant expression, molecular characterization and crystal structure of antitumor enzyme, L-lysine α-oxidase from *Trichoderma viride*. J Biochem 157(6):549–559
27. Arima J, Tamura T, Kusakabe H et al (2003) Recombinant expression, biochemical characterization and stabilization through proteolysis of an L-glutamate oxidase from *Streptomyces sp. X-119-6*. J Biochem 134(6):805–812
28. Hossain GS, Li J, Shin HD et al (2014) L-Amino acid oxidases from microbial sources: types, properties, functions, and applications. Appl Microbiol Biotechnol 98(4):1507–1515
29. Pollegioni L, Molla G (2011) New biotech applications from evolved D-amino acid oxidases. Trends Biotechnol 29(6):276–283
30. Subramanian K, Gora A, Spruijt R et al (2018) Modulating D-amino acid oxidase (DAAO) substrate specificity through facilitated solvent access. PLoS One 13(6):e0198990
31. Xu XL, Grant GA (2016) Mutagenic and chemical analyses provide new insight into enzyme activation and mechanism of the type 2 iron-sulfur L-serine dehydratase from *Legionella pneumophila*. Arch Biochem Biophys 596:108–117
32. Takahashi S, Abe K, Kera Y (2015) Bacterial D-amino acid oxidases: recent findings and future perspectives. Bioengineered 6 (4):237–241
33. Curti B, Ronchi S, Branzoli U et al (1973) Improved purification, amino acid analysis and molecular weight of homogenous D-amino acid oxidase from pig kidney. Biochim Biophys Acta 327(2):266–273
34. Hillebrand GG, Dye JL, Suelter CH (1979) Formation of an intermediate and its rate of conversion to pyruvate during the tryptophanase-catalyzed degradation of *S*-o-nitrophenyl-L-cysteine. Biochemistry 18 (9):1751–1755
35. Vanoni MA, Curti B (2007) D-amino acid oxidase activity assays. In: Konno R (ed) D-amino acids: a new frontier in amino acid and protein research. Hauppauge, Nova Science Publishers, pp 467–476
36. Rosini E, Caldinelli L, Piubelli L (2017) Assays of D-amino acid oxidase activity. Front Mol Biosci 4:102
37. Bayse GS, Michaels AW, Morrison M (1972) The peroxidase-catalyzed oxidation of tyrosine. Biochim Biophys Acta 284(1):34–42
38. Cook PF, Cleland WW (2007) Enzyme kinetics and mechanism. Chapter 3 Enzyme assays. Garland Science, New York

Chapter 14

The In Vitro Production of prFMN for Reconstitution of UbiD Enzymes

Stephen A. Marshall, Karl Fisher, and David Leys

Abstract

Prenylated flavin (prFMN) is a modified FMN cofactor, the isoalloxazine is extended by an additional six membered nonaromatic ring. The modification confers azomethine ylide characteristics on the oxidised prFMN, allowing it to support the reversible nonoxidative decarboxylation of unsaturated acids by the UbiD family of decarboxylases. In absence of a chemical synthesis route for prFMN, enzymatic production by the flavin prenyltransferase, UbiX, is required for in vitro reconstitution of prFMN-dependent enzymes. Here we provide an overview of the methods for producing prFMN in vitro using the flavin prenyltransferase UbiX, and the subsequent reconstitution and activation of UbiD enzymes.

Key words Flavin, Prenylation, Decarboxylation, Carboxylation, Cofactor reconstitution, UbiX, UbiD

1 Introduction

The UbiD-UbiX decarboxylase system is involved in a variety of reversible nonoxidative (de)carboxylation processes [1, 2]. The UbiX-UbiD catalyzed enzymatic reaction is essential in the bacterial production of quinones involved in aerobic growth [3], in the anaerobic catabolism of aromatic hydrocarbons [4–6], and in the production of chemicals involved in the spoilage of food and beverages [7, 8]. Applications have involved the recent use of the UbiD-UbiX decarboxylase system to fix CO_2 in the valorisation of terminal alkenes and (hetero)aromatic compounds in industrial biotechnology [9].

The study and application of UbiD enzymes has garnered interest in recent years since the discovery that UbiD requires the unusual prenylated flavin (prFMN) cofactor [10]. Indeed, the literature was previously confused as to which protein, UbiX or UbiD (or both), was responsible for (de)carboxylase activity [11–13]. It is now understood that the cellular (de)carboxylation activity phenotype relies on the UbiD decarboxylases, that are in turn dependent upon the activity of UbiX to produce the required prFMN cofactor [14] (Fig. 1).

Maria Barile (ed.), *Flavins and Flavoproteins: Methods and Protocols*, Methods in Molecular Biology, vol. 2280,
https://doi.org/10.1007/978-1-0716-1286-6_14,

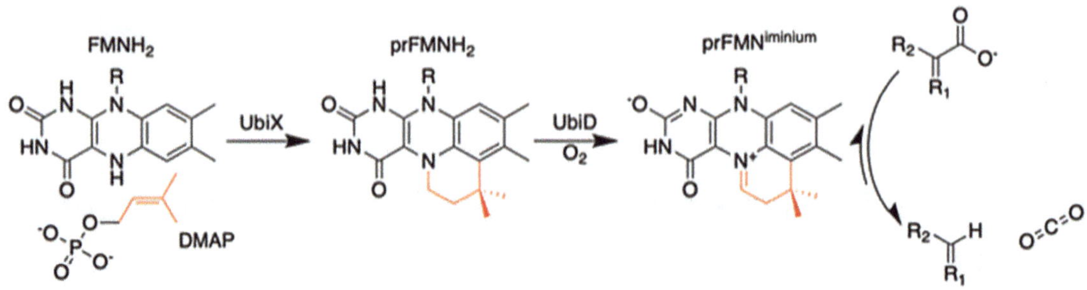

Fig. 1 UbiX and UbiD activities. UbiX is a flavin prenyltransferase which adds a fourth nonaromatic ring to $FMNH_2$ using a prenyl donor (DMAP). UbiD proteins bind the resulting $prFMNH_2$, and upon exposure to an oxidant (often O_2), convert it to $prFMN^{iminium}$. The latter is capable of catalyzing the reversible decarboxylation of a variety of α-β unsaturated acids and aromatic compounds

The native UbiX expression levels appear sufficient to support in vivo activity of associated UbiD enzymes. However, when over-expressing and purifying UbiD enzymes at higher levels for in vitro biochemical or biophysical studies, prFMN incorporation efficiency is insufficient when relying upon the native UbiX expression levels. Hence, UbiX also has to be overexpressed to achieve a high degree of *holo*-UbiD formation [9, 14]. Unfortunately, in some cases UbiD enzymes do not readily purify in the prFMN-bound *holo* state (despite UbiX co-expression), and therefore require in vitro reconstitution [15, 16].

The efficient in vitro production of prFMN requires the pre-nylation of reduced FMN by UbiX, using a dimethylallyl (pyro)-phosphate prenyl donor (DMA(P)P) under anoxic conditions [10, 17, 18]. The reaction depends on reduced FMN ($FMNH_2$) and ultimately forms reduced prFMN, both of which are readily oxidised by molecular O_2. Recently, reports have been published detailing the aerobic production of prFMN and in vitro activation of UbiD enzymes using a multienzyme-one pot synthesis of prFMN starting from flavin and prenol [19]. This system is reliant on the continued reduction of FMN by Fre reductase with the concomitant oxidation of NADH (recycled by formate dehydrogenase), and the phosphorylation of prenol by a promiscuous kinase (ThiM) with the consumption of ATP. This method appears to be suitable for UbiD reconstitution on a modest scale, and thus for example for assessing of the (de)carboxylase activity of a range of UbiD homologs. However, large-scale production of pure *holo*-UbiD at the multiple mg level, as required for detailed mechanistic, biophysical, and crystallographic studies, requires a more robust route to prFMN. As such, this chapter will concentrate on the widely applicable anaerobic production of prFMN using UbiX from Pseudomonas aeruginosa (*Pa*UbiX), which has been extensively used to aid the study of multiple UbiD enzymes to great effect [9, 14–16, 20]. The specific UbiX used for the prenylation reaction impacts upon the reaction, due to the differing substrate specificities of distinct UbiX enzymes [18]. In addition, the reconstitution of UbiD enzymes is discussed, including methods for prFMN oxidation to the catalytically relevant iminium form.

2 Materials

The production of prFMN should be performed in a buffer optimized for the stability of the UbiD enzyme being reconstituted. Reagents can be dissolved in water to a high concentration and stored at −20 °C. We recommend the use of flavin reductase (Fre) and NADH as reductant, to avoid issues associated with sodium dithionite decomposition [15] (*see* **Note 1**). The volumes given in the method relate to a typical 500 μL reaction with 1 mM FMN (*see* **Note 2**).

2.1 prFMN Production

1. ≈500 μM purified *Pa*UbiX (*see* **Note 3**).
2. ≈250 μM purified E. coli Fre reductase (*see* **Note 4**).
3. 100 mM NADH.
4. 100 mM DMAP (*see* **Notes 5** and **6**).
5. 25 mM FMN.
6. Buffer appropriate for optimum UbiD activity/stability (=buffer A).
7. Microfuge concentrators (10 kDa cutoff).

2.2 UbiD Reconstitution

1. >200 μM purified *apo*-UbiD.
2. Buffer appropriate for optimum UbiD activity/stability (=buffer A).
3. Desalting column (e.g., GE Healthcare PD Minitrap Sephadex G-25).
4. 1 M $MnCl_2$.
5. 1 M KCl.
6. Sodium dithionite.
7. Sodium ferricyanide (if using chemical oxidant method).

2.3 Essential Equipment

1. Anaerobic chamber (ideally <5 ppm O_2).
2. Microfuge (in anoxic chamber).
3. UV-Vis spectrophotometer.

3 Methods

3.1 prFMN Production

1. Purge the buffer A with oxygen free N_2 for at least 1 h (for approximately 200 mL of buffer) and seal with a Suba Seal (or similar) to transport into anaerobic chamber. When

introducing large volumes into the chamber, it is ideal to leave exposed to the anaerobic environment overnight prior to use to equilibrate to the nitrogen atmosphere.

2. Add 50 μL UbiX, 5 μL Fre and 20 μL FMN to a microfuge tube. In separate tubes, place sufficient NADH powder and dissolved DMAP for the desired reaction and place all tubes in the airlock port to enter the chamber. Fully purge the airlock port and allow the UbiX mixture to equilibrate against the anaerobic atmosphere (for a small volume 10 min should be sufficient).

Carry out all subsequent procedures in an anaerobic chamber with an oxygen concentration of $\leq$5 ppm.

3. Add 375 μL buffer A to the UbiX/Fre/FMN mixture (to make it up to the desired volume less the volume of DMAP and NADH still to be added).
4. Add buffer A to NADH to a give a final concentration of 100 mM, ensure it is dissolved prior to adding 25 μL NADH to the UbiX/Fre/FMN mixture to reduce. Mix and wait for reduction as evidenced by gradual disappearance of the yellow color. Any oxygen present will be reduced at this stage.
5. Add 25 μL DMAP to the reaction and mix. The final concentrations should be: 1 mM FMN, 5 mM DMAP, 5 mM NADH, $\approx$50 μM UbiX, $\approx$2.5 μM Fre.
6. Allow at least an hour for the reaction to proceed to completion.
7. Place the reaction mixture in a 500 μL microfugal spin concentrator, and spin at a speed appropriate for the concentrator used. Allow enough time for all liquid to pass through the concentrator, typically 20 min.
8. The prFMN is in the filtrate along with residual NAD(H) while UbiX and Fre are retained on the filter (*see* **Note 7**).

3.2 UbiD Reconstitution

1. Take *apo*-UbiD and a desalting column into the anaerobic chamber, allowing sufficient time for the airlock port to purge.
2. Wash the desalting column with 1 CV buffer A containing $\approx$50 mM sodium dithionite to remove any residual traces of oxygen.
3. Wash the column with >4 CV anaerobic buffer A to remove all the sodium dithionite.
4. Desalt *apo*-UbiD as per the resin manufacturer's instructions to render the protein anaerobic. Then reequilibrate the column ready for desalting protein with prFMN bound. Alternatively use two columns.

5. Measure the concentration of *apo*-UbiD after the desalting step, an estimate using a calculated extinction coefficient at 280 nm is suitable.
6. Add $MnCl_2$ and KCl to the filtrate to a final concentration of 1 mM (*see* **Note 8**).
7. Add the prFMN filtrate to *apo*-UbiD in at least a 2:1 prFMN: UbiD ratio, assuming complete conversion of FMN to prFMN. Incubate for a minimum of 5 min to ensure complex formation.
8. Desalt the UbiD-prFMN mixture into buffer A. This step separates *holo*-UbiD from free prFMN, NAD(H), Mn^{2+}, and K^{+}.
9. A UV-vis spectrum should be taken at this point to estimate the final concentration of *holo*-UbiD.
10. Oxidise $UbiD^{prFMN}$ by exposure to air (*see* **Note 9**). A UV-Vis spectrum should be measured to assess the changes in the cofactor. Appearance of a peak in the 550 nm region is indicative of a radical species, this species reminiscent of $prFMN^{radical}$ seen in UbiX (Fig. 2). If this is present in high concentrations then a purple color will be apparent. In most instances oxidation by air exposure is sufficient to convert $prFMNH_2$ to $prFMN^{iminium}$ (i.e., the active form of the protein). This has spectral features in the 300–500 nm range, although these are

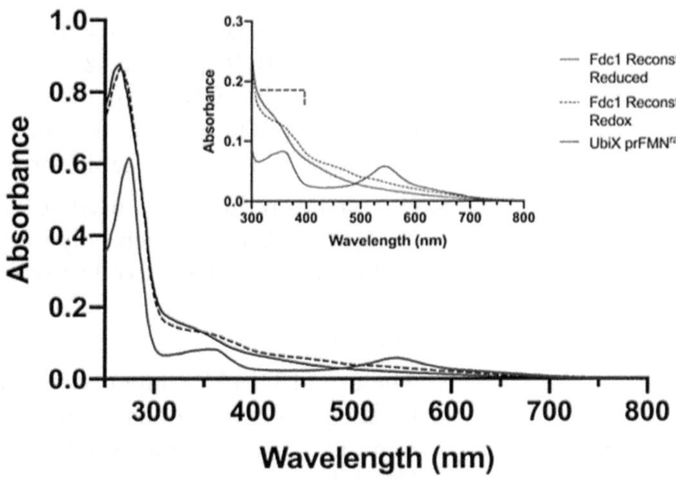

Fig. 2 UV-vis spectra of prFMN species. Spectra of the model UbiD, *An*Fdc1, reconstituted with prFMN in the reduced state and oxidized states. Red line highlights the shoulder which corresponds to $prFMN^{iminium}$. The $prFMNH_2$ has few distinguishable peaks and is reminiscent of the spectrum of $FMNH_2$. $prFMN^{radical}$ in UbiX (purple line) has a characteristic peak centred at 550 nm and can be seen in UbiD enzymes, often in combination with other prFMN species

not distinct peaks. A shoulder on the 280 nm peak, centred around 350 nm is most indicative of $prFMN^{iminium}$ (Fig. 2). Alternatively oxidize $UbiD^{prFMN}$ by the stoichiometric addition of potassium ferricyanide in a 2:1 molar ratio to UbiD [21]. Activity assays can be performed in the presence of the oxidant; however, the protein can be further desalted to remove contaminants prior to spectroscopic analysis of the UbiD enzyme.

4 Notes

1. Sodium dithionite can be used as a reductant when titrated into FMN. It is best to follow this process using a spectrophotometer to ensure full reduction of FMN in absence of excess dithionite. In cases where excess dithionite has been used, we have observed the formation of sulfite adducts to prFMN [15].
2. The UbiX reaction is scalable; the volume of the reaction can be increased when a large volume of UbiD is being reconstituted. Similarly, if a high concentration of UbiD is required (e.g., for spectroscopic studies), the concentrations of the components can be increased to avoid excessive dilution of UbiD during the reconstitution. In our hands, a reaction consisting of 5 mM FMN, 20 mM DMAP, and 20 mM NADH, with 500 μM UbiX and 100 μM Fre was successfully used to reconstitute UbiD (>2 mM) [15].
3. The UbiX suggested for use here is from Pseudomonas aeruginosa (PA4019) [10, 18, 22]. The expression, solubility, and stability of this UbiX homolog is higher than that of others tested [18]. The protein is easily purified from the heterologous host and remains stable for several months at 4 °C.
4. E. coli Fre reductase is used to efficiently reduce FMN with NADH [23], and as such only low concentrations are required in the reaction. The use of Fre avoids complications that can be encountered with the use of chemical reductants such as dithionite.
5. The use of DMAP (dimethylallyl monophosphate) is required for use with P. aeruginosa UbiX [10]. If using a different UbiX homolog, the substrate specificity of UbiX (i.e., DMAPP or DMAP) needs to be determined [18]. A comparison of simple single turnover reactions of UbiX (bound with FMN) reduced with minimal amounts of sodium dithionite and a small excess of DMAP or DMAPP should be able to determine the substrate specificity due to the appearance of the 550 nm UV-vis peak and purple coloration of the sample upon oxidation.

6. DMAP can be purchased, although it can be prohibitively expensive. Alternatively, DMAP can be synthesized following established protocols [21], or can by produced in situ through the phosphorylation of prenol by a promiscuous kinase (such as ThiM) with the concomitant consumption of ATP [19].
7. The concentrator can be removed from the anaerobic chamber at this point. Exposure to oxygen should result in a strong purple color, confirming successful prFMN production. If at this point the filter appears yellow, the reaction has been unsuccessful and should be repeated.
8. The addition of Mn^{2+} has been found essential for the binding of prFMN in many UbiD enzymes. This may not be essential for all UbiD enzymes; some reports include binding of Fe^{2+} instead of Mn^{2+} [24]. Divalent cation screening may be required for some homologs of UbiD to identify the cation that is required.
9. prFMN^{iminium} can be obtained by the exposure of *holo*-UbiD to oxygen. In some cases, prFMN^{iminium} is sensitive to oxygen upon prolonged exposure causing rapid decay in activity [16]. This can be mitigated by returning *holo*-UbiD to the anaerobic environment to stabilize activity. Oxygen sensitivity appears to be dependent upon the UbiD homolog under study, and UbiD features contributing to this are, as yet, undiscovered. In Fdc1, continued exposure to air has no impact on activity, however it has been observed that exposure to light can impact on activity by irreversible isomerization of prFMN^{iminium} to prFMN^{ketimine} [25].

Acknowledgments

This project has received funding from the European Research Council (ERC) under the European Union's Horizon 2020 research and innovation program (Grant agreement No. 695013).

References

1. Marshall SA, Payne KA, Leys D (2017) The UbiX-UbiD system: The biosynthesis and use of prenylated flavin (prFMN). Arch Biochem Biophys 632:209–221
2. Leys D (2018) Flavin metamorphosis: cofactor transformation through prenylation. Curr Opin Chem Biol 47:117–125. https://doi.org/10.1016/j.cbpa.2018.09.024
3. Aussel L, Pierrel F, Loiseau L, Lombard M, Fontecave M, Barras F (2014) Biosynthesis and physiology of coenzyme Q in bacteria. Biochim Biophys Acta 1837(7):1004–1011. https://doi.org/10.1016/j.bbabio.2014.01.015
4. Luo F, Gitiafroz R, Devine CE, Gong Y, Hug LA, Raskin L, Edwards EA (2014) Metatranscriptome of an anaerobic benzene-degrading, nitrate-reducing enrichment culture reveals involvement of carboxylation in benzene ring activation. Appl Environ Microbiol 80 (14):4095–4107
5. Ebenau-Jehle C, Mergelsberg M, Fischer S, Brüls T, Jehmlich N, von Bergen M, Boll M (2017) An unusual strategy for the anoxic

biodegradation of phthalate. ISME J 11 (1):224–236

6. Meckenstock RU, Boll M, Mouttaki H, Koelschbach JS, Cunha Tarouco P, Weyrauch P, Dong X, Himmelberg AM (2016) Anaerobic degradation of benzene and polycyclic aromatic hydrocarbons. J Mol Microbiol Biotechnol 26(1–3):92–118
7. Stratford M, Plumridge A, Archer DB (2007) Decarboxylation of sorbic acid by spoilage yeasts is associated with the PAD1 gene. Appl Environ Microbiol 73(20):6534–6542. https://doi.org/10.1128/AEM.01246-07
8. Plumridge A, Melin P, Stratford M, Novodvorska M, Shunburne L, Dyer PS, Roubos JA, Menke H, Stark J, Stam H, Archer DB (2010) The decarboxylation of the weak-acid preservative, sorbic acid, is encoded by linked genes in Aspergillus spp. Fungal Genet Biol 47 (8):683–692. https://doi.org/10.1016/j.fgb.2010.04.011
9. Payne KAP, Marshall SA, Fisher K, Cliff MJ, Cannas DM, Yan C, Heyes DJ, Parker DA, Larrosa I, Leys D (2019) Enzymatic carboxylation of 2-furoic acid yields 2,5-furandicarboxylic acid (FDCA). ACS Catal 9(4):2854–2865. https://doi.org/10.1021/acscatal.8b04862
10. White MD, Payne KA, Fisher K, Marshall SA, Parker D, Rattray NJ, Trivedi DK, Goodacre R, Rigby SE, Scrutton NS, Hay S, Leys D (2015) UbiX is a flavin prenyltransferase required for bacterial ubiquinone biosynthesis. Nature 522 (7557):502–506. https://doi.org/10.1038/nature14559
11. Zhang H, Javor G (2003) Regulation of the isofunctional genes ubiD and ubiX of the ubiquinone biosynthetic pathway of Escherichia coli. FEMS Microbiol Lett 223(1):67–72
12. Meganathan R (2001) Ubiquinone biosynthesis in microorganisms. FEMS Microbiol Lett 203(2):131–139
13. Mukai N, Masaki K, Fujii T, Kawamukai M, Iefuji H (2010) PAD1 and FDC1 are essential for the decarboxylation of phenylacrylic acids in Saccharomyces cerevisiae. J Biosci Bioeng 109(6):564–569. https://doi.org/10.1016/j.jbiosc.2009.11.011
14. Payne KA, White MD, Fisher K, Khara B, Bailey SS, Parker D, Rattray NJ, Trivedi DK, Goodacre R, Beveridge R, Barran P, Rigby SE, Scrutton NS, Hay S, Leys D (2015) New cofactor supports alpha,beta-unsaturated acid decarboxylation via 1,3-dipolar cycloaddition. Nature 522(7557):497–501. https://doi.org/10.1038/nature14560
15. Marshall SA, Fisher K, Cheallaigh AN, White MD, Payne KA, Parker D, Rigby SE, Leys D (2017) Oxidative maturation and structural characterization of prenylated-FMN binding by UbiD, a decarboxylase involved in bacterial ubiquinone biosynthesis. J Biol Chem 292 (11):4623–4637
16. Payer SE, Marshall SA, Bärland N, Sheng X, Reiter T, Dordic A, Steinkellner G, Wuensch C, Kaltwasser S, Fisher K, Rigby SEJ, Macheroux P, Vonck J, Gruber K, Faber K, Himo F, Leys D, Pavkov-Keller T, Glueck SM (2017) Regioselective para-carboxylation of catechols with a prenylated flavin dependent decarboxylase. Angew Chem Int Ed 56 (44):13893–13897. https://doi.org/10.1002/anie.201708091
17. Arunrattanamook N, Marsh ENG (2018) Kinetic characterization of prenyl-flavin synthase from Saccharomyces cerevisiae. Biochemistry 57(5):696–700. https://doi.org/10.1021/acs.biochem.7b01131
18. Marshall SA, Payne KAP, Fisher K, White MD, Ní Cheallaigh A, Balaikaite A, Rigby SEJ, Leys D (2019) The UbiX flavin prenyltransferase reaction mechanism resembles class I terpene cyclase chemistry. Nat Commun 10(1):2357. https://doi.org/10.1038/s41467-019-10220-1
19. Batyrova KA, Khusnutdinova AN, Wang P-H, Di Leo R, Flick R, Edwards EA, Savchenko A, Yakunin AF (2020) Biocatalytic in vitro and in vivo FMN prenylation and (de) carboxylase activation. ACS Chem Biol 15(7):1874–1882
20. Bailey SS, Payne KAP, Saaret A, Marshall SA, Gostimskaya I, Kosov I, Fisher K, Hay S, Leys D (2019) Enzymatic control of cycloadduct conformation ensures reversible 1,3-dipolar cycloaddition in a prFMN-dependent decarboxylase. Nat Chem. https://doi.org/10.1038/s41557-019-0324-8
21. Marshall SA, Payne KA, Fisher K, Gahloth D, Bailey SS, Balaikaite A, Saaret A, Gostimskaya I, Aleku G, Huang H, Rigby SE, Procter D, Leys D (2019) Heterologous production, reconstitution and EPR spectroscopic analysis of prFMN dependent enzymes. Methods Enzymol 620:489–508
22. Kopec J, Schnell R, Schneider G (2011) Structure of PA4019, a putative aromatic acid decarboxylase from Pseudomonas aeruginosa. Acta Crystallogr Sect F Struct Biol Cryst Commun 67:1184–1188. https://doi.org/10.1107/s174430911102923x
23. Spyrou G, Haggård-Ljungquist E, Krook M, Jörnvall H, Nilsson E, Reichard P (1991)

Characterization of the flavin reductase gene (fre) of Escherichia coli and construction of a plasmid for overproduction of the enzyme. J Bacteriol 173(12):3673. https://doi.org/10.1128/jb.173.12.3673-3679.1991

24. Mergelsberg M, Willistein M, Meyer H, Stärk HJ, Bechtel DF, Pierik AJ, Boll M (2017) Phthaloyl-coenzyme A decarboxylase from Thauera chlorobenzoica: the prenylated flavin-, K+-and Fe2+−dependent key enzyme of anaerobic phthalate degradation. Environ Microbiol 19(9):3734–3744

25. Bailey SS, Payne KA, Fisher K, Marshall SA, Cliff MJ, Spiess R, Parker DA, Rigby SE, Leys D (2018) The role of conserved residues in Fdc decarboxylase in prenylated flavin mononucleotide oxidative maturation, cofactor isomerization, and catalysis. J Biol Chem 293(7):2272–2287

Part IV

Bioanalytical Applications of Flavoenzymes

Chapter 15

Alcohol Oxidase from the Methylotrophic Yeast *Ogataea polymorpha*: Isolation, Purification, and Bioanalytical Application

Halyna M. Klepach, Andriy E. Zakalskiy, Oksana M. Zakalska, Galina Z. Gayda, Oleh V. Smutok, and Mykhailo V. Gonchar

Abstract

Alcohol oxidase (EC 1.1.3.13; AOX) is a flavoprotein that catalyzes the oxidation of primary short-chain alcohols to corresponding carbonyl compounds with a concomitant release of hydrogen peroxide. It is a key enzyme of methanol metabolism in methylotrophic yeasts, catalyzing the first step of methanol oxidation to formaldehyde.

Here we describe the isolation and purification of AOX from the thermotolerant methylotrophic yeast *Ogataea* (*Hansenula*) *polymorpha*, and using this enzyme in enzymatic assay of ethanol, simultaneous analysis of methanol and formaldehyde, and in construction of amperometric biosensors selective to primary alcohols and formaldehyde.

Key words Alcohol oxidase, Isolation and purification, Methylotrophic yeast, Primary alcohols and formaldehyde, Enzymatic assay, Amperometric biosensor

1 Introduction

Alcohol oxidase (AOX, EC 1.1.3.13) is a peroxisomal enzyme detected in several genera of methylotrophic yeasts, such as *Candida*, *Pichia*, and *Ogataea* (*Hansenula*), that utilize methanol as a sole carbon and energy source [1]. The enzyme catalyzes the oxidation of methanol with oxygen to formaldehyde and hydrogen peroxide. AOX is a flavoprotein that contains FAD as a prosthetic group linked to the apoenzyme by noncovalent bonds. The native protein is an octamer consisting of eight identical FAD-containing subunits (664 amino acid residues) with a total molecular weight of about 600 kDa. The protein has a rather low isoelectric point: for example, the pI AOX from *Candida pastoris* is 5.7. Possibly, due to the high content of AOX in peroxisomes and the lower pH in these organelles (5.8–6.0), the enzyme easily transforms in vivo into an

Maria Barile (ed.), *Flavins and Flavoproteins: Methods and Protocols*, Methods in Molecular Biology, vol. 2280, https://doi.org/10.1007/978-1-0716-1286-6_15,

crystalloid state, fixed by electron microscopy. The structure of the corresponding genes and proteins are known for AOX of different origin, but a full X-ray analysis of the enzyme has been performed only for AOX from *Pichia pastoris* [2].

AOX is isolated in a highly purified state from several types of methylotrophic yeast. As a rule, a multistage procedure is used for purification, which includes mechanical destruction of cells, fractional precipitation with ammonium sulfate, ion exchange chromatography on DEAE-cellulose and gel filtration. The specific activity of the purified enzyme varies widely—from 5 to 30 and even 50 μmol/(min·mg), depending on the type of the yeast and purification scheme.

AOX is not a highly specific enzyme and, in addition to methanol, is able to oxidize in vitro other short aliphatic primary alcohols (from ethanol to hexanol), with virtually no effect on their secondary and tertiary isomers. The catalytic activity of the enzyme decreases with increasing length of the carbon chain of the substrate. Formaldehyde, which in its hydrated form is methylene glycol, is also oxidized by AOX with an efficiency of 13% when compared to methanol (Table 1).

AOX is able to oxidize allyl alcohol (relative efficiency compared to methanol—from 30 to 89%) with the formation of a very toxic compound, acrolein. This reaction is used for the positive selection of strains with reduced AOX activity (in particular, with impaired peroxisome biogenesis) by their resistance to allyl alcohol [5].

The concentration of dissolved oxygen as the second substrate of AOX significantly affects the catalytic activity of the enzyme. K_M for oxygen is 0.8 mM [6], which is significantly higher than the value of 0.23 mM, the saturated oxygen concentration at 30 °C in aqueous solutions at an air pressure of 1 atm. Moreover, oxygen concentration affects the kinetic parameters of AOX relative to alcohols. For example, the K_M for methanol, determined in oxygen saturated solutions, is three fold higher than the value obtained under air saturation.

Table 1
K_M (mM) values for AOX from various sources in reactions with alcohols [3] and formaldehyde [4]

	Substrate				
Yeast	**Methanol**	**Ethanol**	**Propanol-1**	***n*-Butanol-1**	**Formaldehyde**
O. polymorpha	0.712	2.7	27.3	54.6	2.6
P. pastoris	0.845	3.68	20.26	18.06	
C. boidinii	0.417	1.78	6.06	10.56	

Besides of to molecular oxygen, other electronic acceptors do not support the alcohol oxidase reaction, in particular, NAD^+, potassium ferricyanide, 2,6-dichlorophenolindophenol, methyl violet, methylene blue, and cytochrome *c*.

The temperature optimum of AOX of different origin differs slightly and is most shifted to higher temperatures for the enzyme from *O. polymorpha* (up to 45 °C), reflecting, perhaps, the relative thermophilicity of the strain and strongly contrasting with the behavior of the enzyme from *P. pastoris* (25 °C [3], although data with a value of 40 °C were presented [7]). AOX from *O. polymorpha* differs significantly from similar enzymes of other methylotrophic yeasts also in pH-optimum value: this enzyme has a wide range of optimum action—from pH 7 to 11, while for other AOXs there is a peak of maximum activity at pH of about 8. In addition, AOX from *O. polymorpha* is usually resistant to the inhibitory effect of chloride, whereas an enzyme from *P. pastoris* is very sensitive to the action of this anion.

The product of the alcohol oxidase reaction—hydrogen peroxide—adversely affects the enzymatic properties of the AOX, causing its inactivation. AOXs of most yeast species lose half their activity at a H_2O_2 concentration of 1–10 mM, while the enzyme from *P. pastoris* is markedly more resistant to peroxide inactivation, maintaining activity at 1 M H_2O_2 for 20 min at 30 °C [7]. The second oxidation product of methanol, formaldehyde, although being a substrate of AOX, is an inhibitor of the enzyme, especially at high concentrations. This is probably due to the ability of CH_2O to react with the functional groups of the protein, in particular, lysine residues. The functional role of covalent reversible adducts of AOX with formaldehyde in vivo is unknown, although it is suggested a specific buffer role of AOX in peroxisomes in maintaining a lower level of free formaldehyde in these organelles [7]. In addition to the binding of excess formaldehyde, this process can lead to a decrease in AOX activity and inhibition of further accumulation of this toxic metabolite in the cells.

AOX is completely inactivated by reagents for the sulfhydryl group (1 mM *p*-chloromercuribenzoate and 10 mM iodoacetamide), heavy metal ions (1 mM Cd^{2+}, Cu^{2+}, Hg^{2+}, and Ag^+), and sodium azide NaN_3. No metal activates AOX, which may explain the lack of enzyme inhibition by chelating agents, in particular, ethylenediaminetetraacetic acid (EDTA).

AOX is synthesized in cells of methylotrophically grown yeasts in large quantities—up to 37–40% of the total cellular protein, and with complete localization of the enzyme in peroxisomes, although the synthesis of the monomer chains occurs in the cytoplasm on membrane-bound endoplasmic reticulum. The need for protein synthesis is associated with a low AOX affinity for oxygen and methanol. The mechanism by which AOX enters the peroxisome and assembles into the crystal lattice has been extensively studied in recent years. AOX monomer (Mr = 74 kDa) is known to be synthesized in the cytosol in the mature form, but how it binds to the FAD, translocates into the peroxisome, assembles into the octamer and is incorporated into the crystalloid matrix has not been studied in detail. AOX oligomerization has been shown to occur only after protein transfer to peroxisomes, after the prebinding of the monomers to the FAD [8].

A mutant strain C-105 (*gcr1 catX*) of the thermotolerant methylotrophic yeast *O. polymorpha* capable to overproduce AOX in glucose medium and avoiding catalase was selected [9, 10]. The simple scheme for the isolation and chromatographic purification of AOX was proposed and highly purified enzyme preparations were obtained [11, 12]. Partually purified enzyme was successfully used as a cheap catalyst of the bioreactor for removing formaldehyde from environmental media [13–15] and in enzymatic assay of primary alcohols and formaldehyde. Highly purified AOX was used as the biorecognition element of biosensors sensitive to primary alcohols and formaldehyde [16–22]. The fabrication and properties of a reagentless bienzyme amperometric biosensor based on alcohol oxidase/peroxidase in combination with an Os-complex modified electrodeposition paint was described [17].

There are published several reviews on AOXs from different yeast sources, including old papers [3, 7, 8, 23, 24] and more recent ones [25–27]. Undying scientific interest in AOXs is caused by complicity of the enzyme, being a model for structural and functional theoretical studies of flavoproteins. Due to ability of AOXs for irreversible and relatively selective oxidation of alcohols, easy availability, and no requirement in external cofactors, these enzymes are regarding as promising catalytic tools for modern technologies [25]: analytical biotechnology (enzymatic kits and biosensors for alcohol assay), biotransformation (synthesis of flavor and fragrance compounds, organic synthesis of optically pure compounds), and bioremediation (removing toxic methanol and formaldehyde contaminations).

The current review is devoted mainly to methodological aspects of the work with AOX from the yeast *O. polymorpha*: its isolation and purification and protocols of analytical procedures (enzymatic assay of alcohols, simultaneous analysis of methanol and formaldehyde, biosensor construction and their evaluation on real samples of food technology), for fields where authors of this review have many years' experience.

2 Materials and Methods

2.1 Strain, Medium and Cultivation Conditions

1. As a producer of AOX, the mutant strain of the thermotolerant methylotrophic yeast *O. polymorpha* C-105 (*gcr1 catX*), selected in the Institute of Cell Biology, National Academy of Sciences of Ukraine, is used. It has impairment in glucose catabolite repression of AOX synthesis, is catalase-defective, and has the ability to over-produce AOX in glucose medium [10, 16].
2. Cells of the mutant *O. polymorpha* C-105 are cultivated in flasks on a shaker (200 rpm) up to the late-logarithmic growth phase (~36 h) at 30 °C in medium containing (per 1 L of deionized water) 10 g glucose, 3.5 g $(NH_4)_2SO_4$, 1.0 g KH_2PO_4, 0.5 g $MgSO_4 \cdot 7H_2O$, 0.1 g $CaCl_2$, and 2.0 g yeast extract. The pH of the medium is adjusted to 5.5 with KOH.
3. The cells are harvested by centrifugation (5000 × *g*, 10 min) and washed twice with water and once with 0.02 M phosphate buffer (PB), pH 7.5. The cells can be stored at −60 °C until use. The freezing process did not reduce the AOX activity of the cells.

2.2 Determination of Cells Concentration

The biomass of *O. polymorpha* cells (mg dry weight cells per 1 mL suspension) is determined by measuring optical density (D) of the suspension at 600 nm using a spectrophotometer (1 cm cuvette). The calculation is performed according to the equation:

$$C = D_{600} \cdot N/1.66,$$

where C—cells concentration; mg/mL; D_{600}—optical density at 600 nm; N—dilution of the initial suspension; 1.66—calibration coefficient.

2.3 Permeabilization of Yeast Cells

Freshly grown cells of the yeast *O. polymorpha* are washed from the medium and resuspended in 0.05 M PB, pH 7.5, containing 2 mM EDTA, to the cell concentration 20–50 mg dry weight cells in 1 mL buffer. The equal volume of aqueous solution of cetyltrimethylammonium bromide (2 mg/mL) is added to the suspension and mixture is incubated at 30 °C with careful shaking during 15 min. After permeabilization, the cells are washed three times with 0.02 M PB, pH 7.5, containing 2 mM EDTA.

2.4 Assay of AOX Activity

1. The AOX activity is measured in a reaction mixture with a final volume of 3.0 mL, containing 50 mM PB, pH 7.5, 0.25 mM *o*-dianisidine, horseradish peroxidase (0.07 mg/mL), 10 mM methanol, and appropriate amount of enzyme or permeabilized cells. Incubation is carried out for 5–10 min at 30 °C. The enzymatic reaction is stopped by the addition of 0.8 mL 12 M HCl.

2. The generated colored product is determined spectrophotometrically at 525 nm [28]. The millimolar extinction coefficient of the formed dye in acidic solution (2.5 M HCl) is shown to be 13.38 mM^{-1} cm^{-1}. One unit (U) of AOX activity is defined as the amount of enzyme releasing 1 μmol H_2O_2 per 1 min at 30 °C under standard assay conditions.
3. Protein concentration is determined by the Lowry method.

2.5 Preparation of Cell-Free Extracts

1. After washing, freshly grown cells (about 15 g wet weight cells from 1 L of culture) are resuspended in 0.1 M PB, pH 7.5, containing 2 mM EDTA and 0.4 mM phenylmethylsulfonyl fluoride (PMSF) to the final cell concentration 100–130 mg dry weight cell in 1 mL of buffer (approx. 1.0 g wet weight cells per 1 mL).
2. The cells are disrupted with glass beads ($d = 0.45$–0.50 mm) (2/3 of suspension volume) in Bead-Beater on ice for 6 min. After removal of cell debris by a low speed centrifugation (5000 × *g*, 10 min, 4 °C), the supernatant is centrifuged at 40,000 × *g* for 45 min at 4 °C. The supernatant is used as the cell-free extract for isolation of AOX.

2.6 Ammonium Sulfate Fractionation

AOX is isolated from cell-free extract of the mutant *O. polymorpha* C-105 by a two-step fractionation with ammonium sulfate (30–70% of saturation) in the presence of 2 mM EDTA and 0.4 mM PMSF to inhibit proteases.

1. To remove ballast proteins, crystalline ammonium sulfate fine powder is added to cell-free extract to 30% saturation under constant stirring and keeping the mixture on ice. The pH of the mixture is maintained at 7.5 with 2 M KOH. After complete dissolving of the salt, the mixture is left for 3–5 h at 4 °C. The pellet, formed at 30% saturation of ammonium sulfate, is discarded by centrifugation at 10,000 × *g* for 30 min at 4 °C.
2. To the supernatant, ammonium sulfate is added to 70% saturation and the mixture is left for 5 h at 4 °C. The obtained AOX precipitate is collected by centrifugation at 10,000 × *g* for 30 min at 4 °C, and washed three times with 70% saturated ammonium sulfate in 0.05 M PB, pH 7.5, containing 2 mM EDTA.

2.7 Dialysis

The AOX preparation obtained at 70% saturation ammonium sulfate is dialyzed against three changes (3 L each) of 0.01 M PB, pH 7.0, containing 1 mM EDTA, at 4 °C for 12 h under constant stirring. The insoluble pellet is removed by centrifugation at 10,000 × *g* for 10 min at 4 °C and discarded.

Table 2
Isolation and purification of AOX from the cells of mutant strain *O. polymorpha* C-105 (*gcr1 catX*)

Purification step	Total protein, mg	Total activity, U	Specific activity, U/mg	Yield[a], %	Purification, fold
Crude extract	340.0	1020.0	3.0	100	1.0
Protein precipitation with $(NH_4)_2SO_4$ (30–70% of saturation)	146.0	730.0	5.0	71.6	1.6
DEAE-Toyopearl 650 M chromatography	14.6	336.6	23.0	33.0	7.7

[a]Calculations of yields were carried out per 1 L batch culture that corresponds to 15 g wet weight cells

2.8 Ion-Exchange Liquid Chromatography

After dialysis, the protein solution is applied to a DEAE-Toyopearl 650 M column (1.6 × 23 cm), which is previously equilibrated with 0.05 M PB, pH 7.0. After removal of unspecific bound proteins by washing with the same buffer, enzyme is eluted from the column using a linear gradient of PB, pH 7.0, with the concentration range 0.05–0.5 M. The AOX appeared as a peak at 0.3 M concentration of the buffer. Obtained purified enzyme preparation exhibited specific activity at least of 23.0 U/mg. For AOX long-term storage, the fractions possessing the highest AOX activity are pooled together and enzyme is precipitated at 70% saturation of ammonium sulfate using 100% saturated solution of ammonium sulfate in 0.05 M PB, pH 7.5, containing 2 mM EDTA and 0.1 mM PMSF. A summary of a typical purification procedure is presented in Table 2 [11] and purity of the AOX preparation is illustrated in Fig. 1. AOX preparations can be stored at −10 °C without significant loss of activity for at least 6 months as a suspension in 0.05 M PB, pH 7.5 at 70% ammonium sulfate saturation.

2.9 Lyophilization of AOX

To obtain lyophilized powder of AOX, enzyme preparation after ammonium sulfate fractionation is mixed in a ratio 2:1 (v/v) with 24% trehalose solution or 1.5% gelatin and is subjected to freeze-drying at −55 °C. Trehalose is found to be the most effective stabilizing agent for AOX during lyophilization. It ensured maintenance of the enzyme's activity up to 85% during 12 months of storage of the dried powder at 4 °C, while the lyophilization of AOX without the addition of stabilizer leads to the loss of more than 60% of the original activity.

2.10 Analysis of AOX Preparations by Electrophoresis

1. Electrophoresis of proteins is performed under native (according to Ornstein and Davis) or denaturing (according to Laemmli) conditions, using 7.5% PAG or 10% PAG plus 0.1% SDS, respectively, and Tris–glycine buffer, pH 8.3.

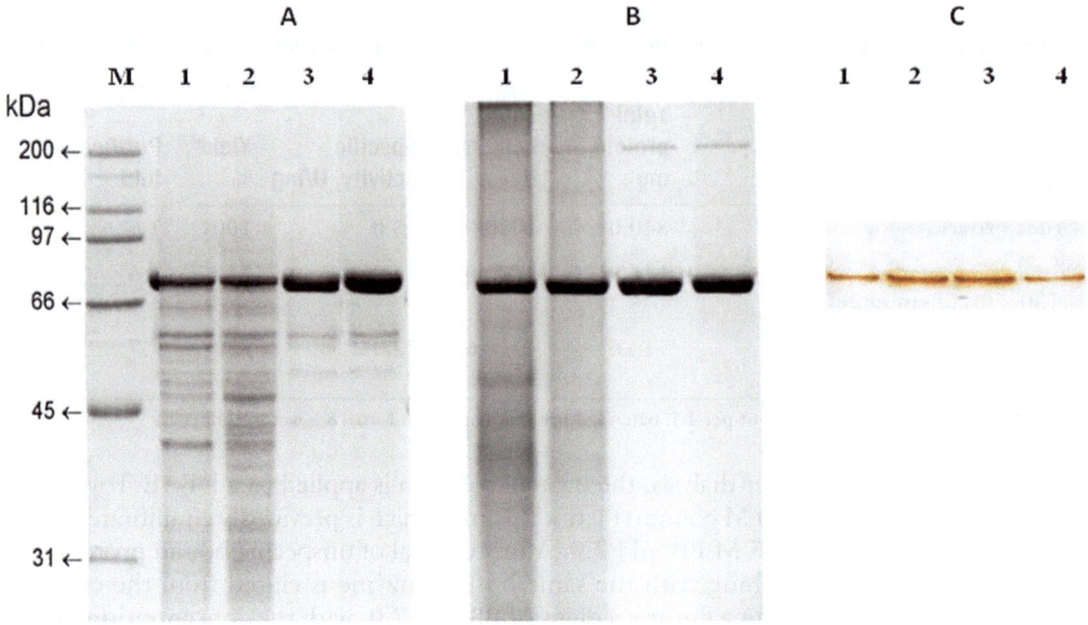

Fig. 1 Electrophoretic patterns of AOX preparations after SDS (**a**) and native (**b**, **c**) electrophoreses in 10% and 7.5% PAG, respectively; M—protein standards; lane 1—cell-free extract, 30 μg protein; lane 2—crude enzyme after second step of precipitation with ammonium sulfate (at 30–70% of saturation), 20 μg protein; lane 3—eluate from DEAE-Toyopearl (23 U/mg protein), 5 μg protein; lane 4—commercial preparation of AOX from *Pichia pastoris*, 5 μg protein. The visualization of AOX activity (**c**)

2. The protein zones are stained with Coomassie brilliant blue: R-250 for SDS-PAG (Fig. 1a) and G-250 for native PAG (Fig. 1b). Visualization of AOX for native PAGE in fractions of chromatographic eluates is presented on Fig. 1c. The enzyme is detected as a brown precipitate in a mixture of 0.01 M methanol, 0.05 mg/mL horseradish peroxidase, and 0.6 mM benzidine in 0.05 M PB, pH 7.0. After colored zones appearance, PAG are washed with water and with 10% acetic acid.

2.11 AOX-Based Enzymatic Assay of Ethanol, Methanol, and Formaldehyde

2.11.1 Assay of Ethanol Using Enzymatic Kit ALCOTEST

Alcohol oxidase of the yeast *O. polymorpha* together with commercial horseradish peroxidase are the enzymes exploited for quantitative determination of ethanol. As a chromogen, tetramethylbenzidine hydrochloride (TMB) can be used. Using these key components, the enzymatic kit *ALCOTEST* is developed for determination of ethanol in alcoholic beverages [29], fermented musts and wine products [30], human serum and blood [31].

The enzymatic kit *ALCOTEST* consists of three components:

1. "Chromogen," a mixture of chromogen and buffer components.
2. "Enzymes," a suspension of alcohol oxidase and peroxidase in ammonium sulfate solution.
3. "Standard," a calibrated ethanol solution, 10 g/L (stabilized).

1. Preparation of reagent and standard solutions: The content of bottle 1 ("Chromogen") is dissolved in 350 mL distilled water at 100 °C and then cooled to room temperature. The final concentration of TMB is 0.067 mM in 0.05 M PB, at pH 7.0. The solution is stable for 2 weeks at 20 °C, kept in the dark.
   (a) The content of bottle 2 ("Enzymes") containing 27 units of AOX and 2 mg horseradish peroxidase with RZ (Reinheitzahl) 3.0 in 0.5 mL 60% saturated ammonium sulfate was dissolved in cooled solution of chromogen. Fresh solution is prepared before analysis. If the number of samples to be analyzed are less than 100 (4 mL reaction mixture) corresponding aliquots of the well-stirred enzymes suspension and of the chromogen solution are mixed.
   (b) Ethanol solution from the bottle 3 ("Standard") is diluted 100-fold to the concentration of 0.1 g/L and used the same day.
2. Sample preparation: Samples of the tested alcoholic beverages are diluted with water to ethanol concentration of about 0.05–0.5 g/L. Usual dilution factor is 500–2000.
3. Analysis procedure:
   (a) In the separate test tubes, add 0.1 mL aliquots of the sample, standard or water (for blank).
   (b) To the test tubes, portions of 3.5 mL reagent solution are added in fixed time (sequentially in 30 s intervals).
   (c) The solutions are well mixed and left to stand at room temperature for 15–20 min.
   (d) After incubation during 10–15 min, the reaction is stopped by addition of 0.5 mL 0.8 M HCl.
   (e) Absorbance at 450 nm of the reaction mixtures is read against the blank. If the assay is performed using the "fixed time" regime, timing of the start and termination of the assay is very important. Therefore, the reaction is started (and also ended) by adding respective reagent sequentially in 30 s intervals.
4. Calculations: The calculation of ethanol content in alcoholic beverages is performed using the following formula:

$$C = A_{\text{sample}} \cdot 0.1 \cdot n / A_{\text{standard}}$$

or $C_{\text{vol\%}} = 0.13 \cdot C$ (with respect to 96.5% ethanol),

where C—ethanol mass concentration, expressed in g/L; A_{sample}—optical density of the sample solutions; A_{standard}—optical density of the standard solution; n—dilution factor for initial sample; $C_{\text{vol\%}}$—ethanol volumetric content in %.

2.11.2 Assay of Formaldehyde in Fish Food Products

AOX-peroxidase method using enzymatic kit ALCOTEST can be applied for determination of formaldehyde [32]. Formaldehyde (FA) is classified as a mutagen and a possible carcinogen [33]. High level of FA toxicity necessitates the control over the content of this substance in environment, industrial products, medical preparations, and even in some food products, in particular, fish food products [15, 34, 35]. The AOX-based method compared with routine chemical approaches has several advantages: high sensitivity, good linearity, and insensitivity to the interference effect of the sample's components and usage of nonaggressive reagents for the sample pretreatment and assay procedure [32]. The linearity of calibration curve for this method is kept even at high optical densities—up to 0.9 which corresponds to 15 μM FA in final reaction mixture (15 nmol/mL), and the threshold sensitivity of the method is about 0.8 nmol/mL.

It is demonstrated that some fish products (hake and cod) contain high levels of toxic FA, up to 100 mg/kg wet weight, while the content of FA in carp is negligible [32].

1. Preparation of reagent and standard solutions:
   (a) Sample of solution of AOX from *O. polymorpha* with specific activity 3–5 μmol/(min·mg) of protein (3–5 U/mg).
   (b) Sample of horseradish peroxidase solution with specific activity 900 U/mg and RZ = 3.0.
   (c) Mixture "Enzymes" a suspension of AOX oxidase (42 U/mL) and peroxidase (4 mg/mL) in 60% saturated ammonium sulfate solution.
   (d) Mixture "Chromogen" a mixture of the chromogen TMB and buffer components.
   (e) The final reagent is prepared by dissolution of 490 mg mixture "Chromogen" in 60 mL water. The mixture is heated until it is dissolved, cooled to room temperature and supplemented with 0.5 mL of the enzyme mixture. The final composition of the prepared reagent is as follows: 75 mM PB, pH 7.5; 0.08 mM TMB; 0.05 mg/mL horseradish peroxidase with RZ = 3.0; 0.35 U/mL AOX.
   (f) Formaldehyde calibration solution is prepared by hydrolysis of paraformaldehyde in water (1 M concentration) in a sealed ampoule at 105 °C for 12 h and by dilution of hydrolysate to the necessary concentration.
   (g) FA solution ("Standard") is freshly prepared by 5000-fold dilution of the initial 1 M FA to the final concentration of 0.2 mM (or 6 mg/L) before analysis.
2. Sample preparation: For assay of FA in fish meat (muscle tissues of frozen hake and cod, as well as of freshly killed carp), the

following deproteinization procedure is used: 2.5 g cut muscle tissue is ground in a ceramic pot. After adding 10 mL water, the mixture is ground again, mixed with 2 mL of Carrez reagent I (15%, m/v, aqueous solution of $K_3Fe(CN)_6 \cdot 3H_2O$), 2 mL Carrez reagent II (30%, m/v, aqueous solution of $ZnSO_4 \cdot 7H_2O$), and 34 mL water (to obtain 50 mL volume) and stirred. Precipitated proteins are removed by filtering through folded filter. The filtrate is stored at 4 °C before the analysis.

3. Analysis procedure:
   (a) In separate test tubes, add 0.2 mL aliquots of the analyzed extract (5–50 nmol CH_2O), standards, and dilute with water to the volume of 0.5; or to the blank sample, add 0.5 mL water.
   (b) To the test tubes, add portions of 3 mL of freshly prepared reagent in fixed time (sequentially in 30 s intervals).
   (c) The samples are incubated at room temperature for 25 min.
   (d) The reaction mixture is stopped by adding 0.5 mL 0.8 M HCl to each sample.
   (e) The optical density of solutions is measured at 450 nm against blank sample which contains water instead of the analyzed extract.
4. Calculations: The calculation of FA content in fish products (in mg per 1 kg of wet weight of the sample) is performed using the following formula:

$$C_{mg/kg} = 600 \cdot 0.2 \cdot A_e/A_c = 120 \cdot A_e/A_c,$$

where A_e and A_c—values of optical density of the experimental and calibration samples, respectively; 0.2—millimolar (mM) concentration of FA in the calibration solution; 600—coefficient which arises from extraction ratio (2.5 g of fish tissue per 50 mL of the extract); the volume of extract taken for analysis (1 mL) and molecular weight of FA (30), that is, $600 = (50/2.5) \cdot 30$.

2.11.3 Simultaneous Assay of Methanol and Formaldehyde in Wastewater

An enzymochemical method using AOX and 3-methyl-2-benzothiazolinone hydrazone (MBTH) is assigned for the simultaneous analysis of methanol and formaldehyde in mixtures, including industrial wastewaters. AOX from the yeast *O. polymorpha* is used as a methanol oxidizing agent and MBTH as a reagent for colorimetric determination of generated formaldehyde [36].

1. Preparation of reagent and standard solutions:

(a) 25 mM MOPS/KOH buffer, pH 7.0. 5.23 g 3-(*N*-morpholino)propanesulfonic acid (MOPS) is dissolved in 1 L water. Mix 250 mL MOPS solution (5.23 g/L) and 87 mL 25 mM KOH (14.03 g/L). Dilute to 0.5 L with water. Store at 4 °C.

(b) Crude preparation of *O. polymorpha* AOX with specific activity 4 μmol/(min·mg) of protein (4 U/mg).

(c) Reagent **1** (0.05% MBTH-hydrochloride solution in 25 mM MOPS/KOH buffer, pH 7.0): 0.05 g MBTH hydrochloride is dissolved in 100 mL 25 mM MOPS/KOH buffer, pH 7.0.

(d) Reagent **2** (reaction mixture containing AOX): 0.1 mL crude preparation of *O. polymorpha* AOX (with activity of 50 U/mL) is dissolved in 10 mL Reagent **1** (final activity 0.5 U/mL).

(e) Reagent **3** (0.1% $FeCl_3$ solution in 30 mM HCl): 122 mg $FeCl_3 \cdot 6H_2O$ is dissolved in 100 mL 30 mM HCl. It is stable for at least 10 days of storage at 20 °C.

(f) FA calibration solution is prepared by hydrolysis of paraformaldehyde in water (1 M concentration) in a sealed ampoule at 105 °C for 12 h and by dilution of hydrolysate to the necessary concentration.

(g) FA solution (Standard **1**) is freshly prepared by 5000-fold dilution of the initial 1 M FA to the final concentration of 0.2 mM (or 6 mg/L) before analysis.

(h) Methanol solution (Standard **2**) is freshly prepared by dilution of the initial methanol to the final concentration of 0.156 mM (or 5 mg/L) before analysis.

2. Sample preparation: The wastewater samples are frozen and stored at −25 °C before analysis.

   (a) Assay of FA and methanol are carried out in distillates.

   (b) The sample's distillation is carried out in accordance with recommendations of the Polish Standard PN-71C/04568 [37].

   (c) After distillation, samples are diluted (2–100-fold), immediately analyzed or frozen and stored at −25 °C before analysis.

3. Assay of Formaldehyde :

   (a) In separate test tubes, add 0.2 mL aliquots of the diluted samples or Standard **1**, or water (for blank).

   (b) To the test tubes, add 1.8 mL Reagent **1**.

   (c) Samples are incubated for 15 min at 30 °C and Reagent **3** is added.

(d) Mixtures are incubated for 20 min at 30 °C and their optical densities are measured at 670 nm against a blank containing all components except a tested sample.

(e) Calculations: The formaldehyde content in the test samples is calculated using a calibration curve [36]. Calibration curve is constructed using formaldehyde calibration solutions. For calibration, formaldehyde solution (12 mg/L) is used. The linearity of calibration curve for MBTH-method is kept even at high optical densities—up to 0.8 which corresponds to 18.30 μM FA in final reaction mixture (18.3 nmol/mL), and the threshold sensitivity of the method is about 0.25 nmol/mL.

The calculation of formaldehyde content in the tested samples is performed according to the following formula:

$$C = A_{\text{sample}} \cdot 0.2 \cdot n/A_{\text{standard}}$$

or $C_{\text{mg/L}} = 30 \cdot C$, where C—millimolar concentration of formaldehyde in mM; A_{sample}—optical density of the tested sample solutions; A_{standard}—optical density of the standard solution; n—dilution factor for initial sample; $C_{\text{mg/L}}$—mass concentration of formaldehyde in mg/L.

2.11.4 Assay of Methanol

(a) In the separate test tubes, add 0.2 mL aliquots of the diluted samples or Standard **2,** or water (for blank).

(b) To the test tubes, add 1.8 mL Reagent **2**.

(c) Samples are incubated for 15 min at 30 °C and Reagent **3** is added.

(d) The samples are incubated for 20 min at 30 °C.

(e) The optical density of solutions is measured at 670 nm against blank sample which contains all components except a tested sample.

(f) Calculations: Calculations of methanol content in tested samples are carried out using calibration curve [36]. Calibration curve is constructed using methanol calibration solutions. For calibration, methanol solution (5 mg/L) is used. The linearity of calibration curve for AOX-MBTH-method is kept even at high optical densities—up to 0.8 which corresponds to 17.8 μM methanol in final reaction mixture (17.80 nmol/mL), and the threshold sensitivity of the method is about 0.475 nmol/mL.

The calculation of methanol content in the tested samples is performed according to the following formulas:

$$C = \Delta D_{\text{sample}} \cdot 0.156 \cdot n/D_{\text{standard}}$$

or $C_{\text{mg/L}} = 32 \cdot C$, where C—millimolar concentration of methanol in mM; ΔD_{sample}—difference in optical densities of the samples tested for the content FA + methanol and samples tested for only FA; $D_{standard}$—optical density of the solution with methanol standard; *n*—dilution factor for initial sample; and $C_{mg/L}$—mass concentration of methanol in mg/L.

2.12 Application of AOX in Construction of Biosensors

The use of AOX in biosensorics is illustrated here by the fabrication of a reagentless bienzyme amperometric biosensor based on AOX (specific activity of 30 U/mg protein), commercial peroxidase (HRP, 30 U/mg protein) in combination with an Os-complex modified electrodeposition paint [17].

1. Electrodes and measurement:

   Constant-potential amperometry in a three-electrode configuration with Ag/AgCl/KCl (3 M) reference electrode and a Pt-wire counter electrode is used. As a working electrode, graphite rod (type RW001, 3.05 mm diameter) sealed in a glass tube using epoxy glue is exploited. Before sensor preparation, the graphite electrodes are polished on emery paper (P1200). Amperometric measurements are carried out using a bipotentiostat EP30 Biometra connected to a personal computer via a RS232 port for data acquisition. Measurements are performed in steady-state mode using a standard cell in volume 25 mL at 25 °C at continuous stirring. Between experiments, the enzyme electrodes are stored in 0.1 M PB, pH 7.6, at 4 °C.

2. Selection of electrodeposition paints: The paints are synthesized in the laboratory of Analytical Chemistry—Elektroanalytik & Sensorik, Ruhr-Universität Bochum, Germany, according to strategy of the radical copolymerization of acrylic acid (anodic paints) or (dimethylamino)ethylmethacrylate (cathodic paints) with suitable comonomers.

   A lot of osmium complex-integrated anodic paints are screened by their electron transferring activity and enzyme-inactivating/stabilizing property on the model of HRP-containing electrode. Cyclic voltammetry is used for these investigations. The most effective composition *AP59-Os* (Os-bis-*N*,*N'*-(2,2'-bipyridyl)-dichloro complex—an anodic paint 59 supplemented by the osmium complex) demonstrated the ability for direct electrochemical communication with immobilized HRP in the presence of hydrogen peroxide under the lowest working potential (−50 mV vs. SCE). *AP59-Os* composition is chosen for construction of bienzyme AOX-based biosensor. The scheme of electron transfer for this biosensor is shown in Fig. 2.

3. Entrapment of HRP in a *AP59-Os* layer: One microliter of a freshly prepared mixture of horseradish peroxidase (400 U/mL) solution in 0.02 M PB, pH 7.6 and of *AP59-Os*, ratio 1:1, is dropped onto the surface of the working electrode. In a

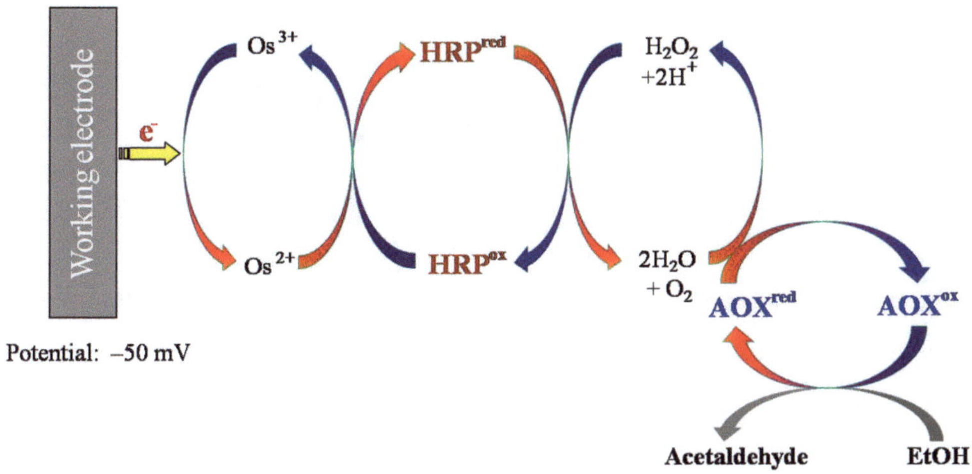

Fig. 2 Reaction scheme and electron-transfer pathway from ethanol *via* AOX and enzymatically generated H_2O_2 *via* HRP and polymer-bound Os-relays to the AOX-HRP-based electrode [17]

miniaturized electrochemical cell, the anodic paint is precipitated on the electrode surface using a potentiostatic pulse sequence of +2200 mV for 0.2 s and 0 mV for 5 s. During the electrochemical initiated water oxidation, the anodic paint is protonated concomitantly modulating the solubility of the carboxylate-containing polymer chains. The coprecipitated enzyme is thus entrapped in front of the electrode surface. The electrodes are rinsed with 0.02 M PB, pH 7.6.

4. Entrapment of AOX in a *CP9* layer: One microliter of AOX suspension (200 U/mL; pH 7.6) is dropped on the surface of the working electrode and allowed to dry. Forty microliters of the *CP9* suspension is added to the miniaturized electrochemical cell and the *CP9* is precipitated by means of a potentiostatic pulse sequence of −1200 mV for 0.2 s and 0 mV for 5 s. AOX is entrapped within the electrodeposited cathodic paint layer on the surface of the working electrode. The electrode is rinsed with 0.1 M PB, pH 7.6, before use. Finally, an optimal architecture of the developed alcohol-selective biosensor is proposed [17]: HRP- and *AP59-Os*-containing first layer followed by a second one consisting of AOX (or a mixture of AOX and HRP) entrapped in cathodic paint *CP9*.

The main bioanalytical properties of the constructed bienzyme sensor are presented in Fig. 3.

Acknowledgments

This work was financially supported, in part, by the Ministry of Education and Science of Ukraine (projects #0118U000297, #0119U100671, and Ukrainian-Lithuanian R&D project

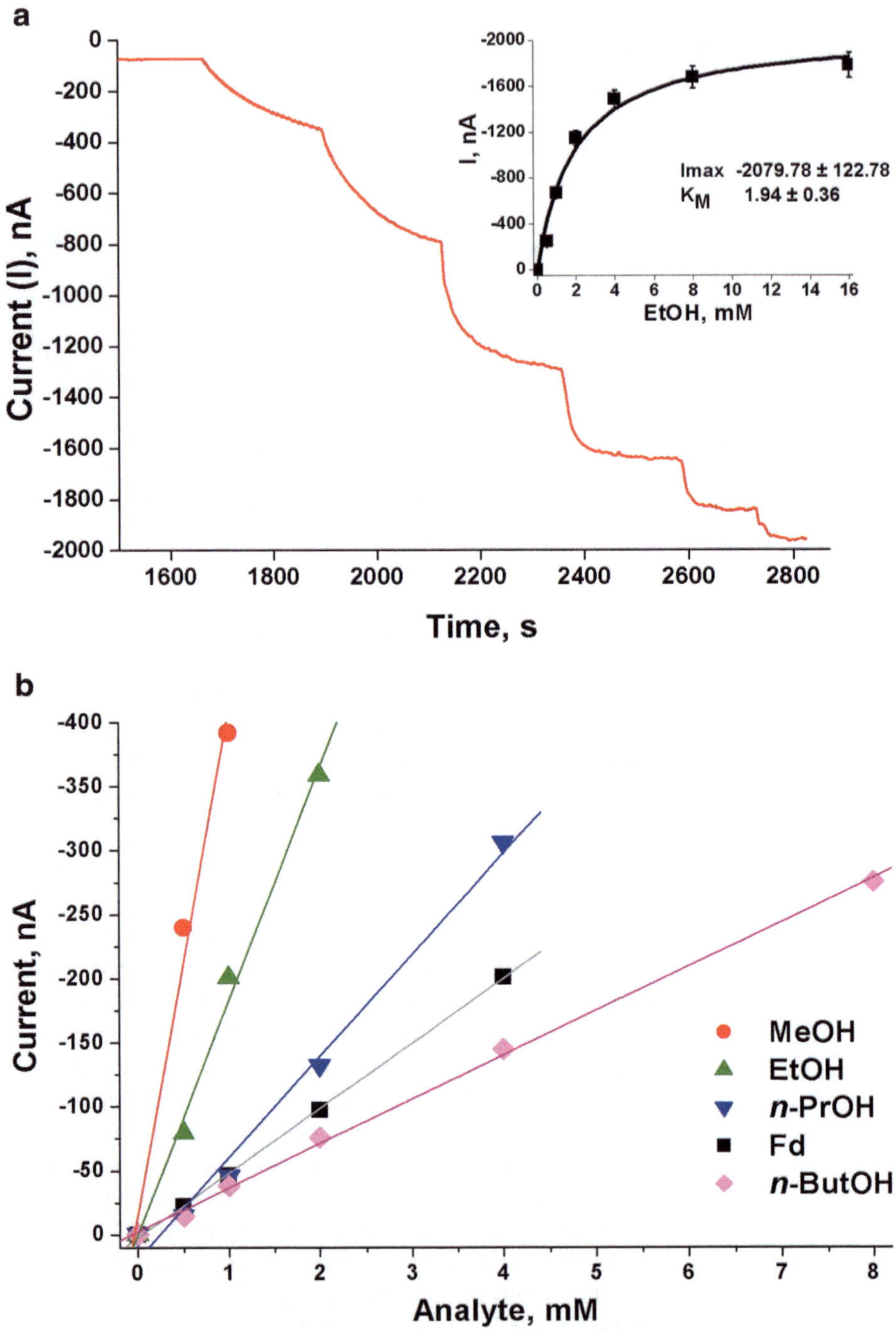

Fig. 3 The main bioanalytical properties of the constructed bienzyme sensor. (**a**) The typical dynamic response range of the sensor toward ethanol and corresponding calibration curve as dependence on added ethanol (inserted); (**b**) the linear dynamic range of sensor's response toward different analytes: MeOH, EtOH, n-PrOH, Fd, n-ButOH. 3.05 mm graphite electrode; −50 mV *vs.* Ag/AgCl

#0120U103398), National Academy of Sciences of Ukraine in the frame of the Scientific-Technical Program “Smart sensor devices of a new generation based on modern materials and technologies” (projects #13 and #10/3), National Research Foundation of Ukraine (project #48/02.2020).

References

1. Sahm H (1977) Metabolism of methanol by yeasts. Adv Biochem Eng 6:77–103
2. Koch C, Neumann P, Valerius O, Feussner I, Ficner R (2016) Crystal structure of alcohol oxidase from *Pichia pastoris*. PLoS One 11(2): e01498
3. Woodward J (1990) Biochemistry and applications of alcohol oxidase from methylotrophic yeasts. In: Codd GA, Dijkhuizen L, Tabita FR (eds) Autotrophic microbiology and one-carbon metabolism, vol 1. Springer, Dordrecht
4. Kato N, Omori G, Tani Y et al (1976) Alcohol oxidase of *Kloeckera sp.* and *Hansenula polymorpha*. Catalytic properties and subunit structure. Eur J Biochem 64:341–350
5. Johnson MA, Waterham HR, Ksheminska GP et al (1999) Positive selection of novel peroxisome biogenesis-defective mutants of the yeast *Pichia pastoris*. Genetics 151:1379–1391
6. Geissler J, Ghisla S, Kroneck P (1986) Flavin-dependent alcohol oxidase from yeast. Studies on the catalytic mechanism and inactivation during turnover. Eur J Biochem 160:93–100
7. Hopkins T, Muller F (1987) Biochemistry of alcohol oxidase. Microbial growth on C_1 compounds. In: Verseveld H, Duine J (eds) Proc 5th Int Symp. Springer, Dordrecht, pp 150–157
8. Ozimek P, van Dijk R, Latchev K et al (2003) Pyruvate carboxylase is an essential protein in the assembly of yeast peroxisomal oligomeric alcohol oxidase. Mol Biol Cell 14(2):786–797
9. Gonchar MV, Ksheminska GP, Hladarevska NM et al (1980) Genetics of respiratory enzymes in yeasts. In: Lachowicz TM (ed) Proc Int Conf. Wroclaw University Press, Wroclaw, pp 222–228
10. Gonchar M, Kostryk L, Sibirny A (1997) Cytochrome *c* peroxidase from a methylotrophic yeast: physiological role and isolation. Appl Microbiol Biotechnol 48:454–458
11. Shleev SV, Shumakovich GP, Nikitina OV et al (2006) Purification and characterization of alcohol oxidase from a genetically constructed over-producing strain of the methylotrophic yeast *Hansenula polymorpha*. Biochemistry (Mosc) 71:245–250
12. Gayda GZ, Demkiv OM, Klepach HM et al (2019) Effective technologies for isolating yeast oxido-reductases of analytical importance. In: Sibirny A (ed) Non-conventional yeasts: from basic research to application. Springer Nature, Cham
13. Sigawi S, Smutok O, Demkiv O et al (2011) Immobilized formaldehyde-metabolizing enzymes from *Hansenula polymorpha* for removal and control of airborne formaldehyde. J Biotechnol 153(3):138–144
14. Sibirny V, Demkiv O, Sigawi S et al (2011) Formaldehyde oxidizing enzymes and genetically modified yeast *Hansenula polymorpha* cells in monitoring and removal of formaldehyde. In: Einschlag FSG (ed) Waste water – evaluation and management. InTech, Rijeka
15. Gayda G, Demkiv O, Klepach H et al (2015) Formaldehyde: detection and biodegradation. In: Patton A (ed) Formaldehyde: synthesis, applications and potential health effects. In Nova Science Publishers Inc., New York, NY
16. Gonchar MV, Maidan MM, Moroz OM et al (1998) Microbial O_2 - and H_2O_2-electrode sensors for alcohol assays based on the use of permeabilized mutant yeast cells as the sensitive bioelements. Biosens Bioelectron 13 (9):945–952
17. Smutok O, Ngounou B, Pavlishko H et al (2006) A reagentless bienzyme amperometric biosensor based alcohol oxidase/peroxidase and an Os-complex modified electrodeposition paint. Sens Actuators B 113(2):590–598
18. Dmytruk KV, Smutok OV, Ryabova OB et al (2007) Isolation and characterization of mutated alcohol oxidases from the yeast *Hansenula polymorpha* with decreased affinity toward substrates and their use as selective elements of an amperometric biosensor. BMC Biotechnol 7:1–7
19. Smutok O, Gayda G, Dmytruk K et al (2011) Amperometric biosensors for lactate, alcohols, and glycerol assays in clinical diagnostics. In: Serra PA (ed) Biosensors - emerging materials and applications. InTech, Rijeka

20. Sigawi S, Smutok O, Demkiv O et al (2014) Detection of waterborne and airborne formaldehyde: from amperometric chemosensing to a visual biosensor based on alcohol oxidase. Materials (Basel) 7(2):1055–1068
21. Stasyuk N, Gayda G, Zakalskiy A et al (2019) Amperometric biosensors based on oxidases and PtRu nanoparticles as artificial peroxidase. Food Chem 285:213–220
22. Gayda GZ, Demkiv OM, Stasyuk NY et al (2019) Metallic nanoparticles obtained via "green" synthesis as a platform for biosensor construction. Appl Sci 9:720
23. Sahm H, Schütte H, Kula M (1982) Alcohol oxidase from *Candida boidinii*. Methods Enzymol 89:424–428
24. van der Klei IJ, Bystrykh L, Harder W et al (1990) Alcohol oxidase from *Hansenula polymorpha* CBS 4732. Methods Enzymol 188:420–427
25. Goswami P, Chinnadayyala SSR, Chakraborty M et al (2013) An overview on alcohol oxidases and their potential applications. Appl Microbiol Biotechnol 97:4259–4275
26. Romero E, Gadda G (2014) Alcohol oxidation by flavoenzymes. Biomol Concepts 5 (4):299–318
27. Gonchar M, Smutok O, Karkovska M et al (2017) Yeast-Based Biosensors for Clinical Diagnostics and Food Control. In: Sibirny A (ed) Biotechnology of Yeasts and Filamentous Fungi. Springer, Cham
28. Sibirnyi AA, Titorenko VI, Efremov BD et al (1987) Multiplicity of mechanisms of carbon catabolite repression involved in the synthesis of alcohol oxidase in the methylotrophic yeast *Pichia pinus*. Yeast 3:233–241
29. Gonchar MV, Maidan MM, Pavlishko HM et al (2001) A new oxidase-peroxidase kit for ethanol assays in alcoholic beverages. Food Technol Biotechnol 39(1):37–42
30. Pavlishko HM, Ryabinina OV, Zhilyakova TA et al (2005) Oxidase-peroxidase method of ethanol assay in fermented musts and wine products. Appl Biochem Microbiol 41 (6):604–609
31. Gonchar M, Maidan M, Pavlishko H et al (2002) Assay of ethanol in human serum and blood by the use of a new oxidase-peroxidase based kit. Visnyk of L'viv Univ Biol Ser 31:22–27
32. Sibirny V, Demkiv O, Klepach H et al (2011) Alcohol oxidase and formaldehyde dehydrogenase-based enzymatic methods for formaldehyde assay in fish food products. Food Chem 127(1):774–779
33. Feron VJ, Til HP, de Vrijer F et al (1991) Aldehydes: occurrence, carcinogenic potential, mechanism of action and risk assessment. Mutat Res 259:363–385
34. Gonchar MV, Grabek D, Oklejewich B et al (2005) A new enzymo-chemical method for simultaneous assay of methanol and formaldehyde. Ukr Biochem J 77(3):118–126
35. Demkiv OM, Klepach HM, Gayda GZ et al (2015) Methods of formaldehyde assay in vaccines. Acta Carpathica 24:209–219
36. Sibirny VA, Gonchar MV, Grabek-Lejko D et al (2008) Photometric assay of methanol and formaldehyde in industrial waste-waters using alcohol oxidase and 3-methyl-2-benzothiazolinone hydrazone. Int J Environ Anal Chem 88 (4):289–301
37. Polish Standard (Polska Norma) PN-71/C-04568 (1988) Water and waste water. Determination of methyl alcohol content, 5th edn. Polski Komitet Normalizacyjny, Warsaw

Chapter 16

Flavocytochrome b_2 of the Methylotrophic Yeast *Ogataea polymorpha*: Construction of Overproducers, Purification, and Bioanalytical Application

Oleh V. Smutok, Kostyantyn V. Dmytruk, Taras S. Kavetskyy, Andriy A. Sibirny, and Mykhailo V. Gonchar

Abstract

Flavocytochrome b_2 (EC 1.1.2.3; L-lactate cytochrome: *c* oxidoreductase, FC b_2) from the thermotolerant methylotrophic yeast *Ogataea polymorpha* is a thermostable enzyme—prospective for a highly selective L-lactate analysis in the medicine, nutrition sector, and quality control of commercial products. Here we describe the construction of FC b_2 producers by overexpression of the *CYB2* gene *O. polymorpha*, encoding FC b_2, under the control of a strong alcohol oxidase promoter in the frame of plasmid for multicopy integration with the next transformation of recipient strain *O. polymorpha* C-105 (*gcr1 catX*) impaired in the glucose repression and devoid of catalase activity. The selected recombinant strain *O. polymorpha* "tr1" (*gcr1 catX CYB2*), characterized by eightfold increased FC b_2 activity compared to the initial strain, was used as a source of the enzyme. For purification of FC b_2 a new method of affinity chromatography was developed and purified preparations of the enzyme were used for the construction of the highly selective enzymatic kits and amperometric biosensor for L-lactate analysis in human liquids and foods.

Key words Flavocytochrome b_2, *CYB2*, *Ogataea polymorpha*, Overproducers, Enzyme purification, L-lactate analysis, Enzymatic kits, Amperometric biosensor

1 Introduction

There was no information on the flavocytochrome b_2 (FC b_2) of methylotrophic yeast prior to our studies. However, the enzyme from *Saccharomyces cerevisiae* and non-conventional *Hansenula anomala* yeasts is well studied. The primary enzyme structure is known, the X-ray structural analysis was performed, the genes were cloned and recombinant forms of the protein were obtained [1–3]. For most nonconventional yeast species only fragmentary data focused on the purification and characterization of FC b_2 are available, and the cloning of the corresponding gene for *Kluyveromyces*

Maria Barile (ed.), *Flavins and Flavoproteins: Methods and Protocols*, Methods in Molecular Biology, vol. 2280, https://doi.org/10.1007/978-1-0716-1286-6_16,

lactis is described [4–6]. It was shown that different yeast species have a high FC b_2-gene homology.

FC b_2 is an integral mitochondrial protein of yeast cells providing dual function: electron transfer in the respiratory chain and L-lactate utilization. It is a homotetramer, each subunit of which consist of one polypeptide chain, containing one FMN and heme domains [7–10]. Owing to its unique catalytic properties (absence of the necessity of the exogenous cofactor, absolute selectivity in relation to L-lactate, and nonspecificity to the electron acceptor), FC b_2 is of utmost bioanalytical importance and can replace the unspecific mammalian NAD^+-dependent lactate dehydrogenase and high-cost bacterial lactate oxidase in bioanalytical application [11]. However, the use of FC b_2 from baker's yeast as well the other known microbial souces is limited due to the high lability of this enzyme and the complexity of its isolation, purification, and stabilization [12]. In our previous work, the screening of yeasts producing stable FC b_2 forms resulted in the selection of several *Ogataea polymorpha* strains as promising sources of a stable enzyme [13]. In this chapter, the combined results of the construction of the producer of the thermostable FC b_2 based on the preliminarily selected strain of the thermotolerant methylotrophic yeast *O. polymorpha* C-105 (*gcr1 catX*), enzyme purification, and its bioanalytical application are presented.

2 Materials and Methods

2.1 Yeast Strains

1. The following yeast strains from the collection of the Institute of Cell Biology, National Academy of Sciences of Ukraine are used: *Ogataea polymorpha* 356 line DL1; *O. polymorpha* CBS 4732; and *O. polymorpha* C-105 (*gcr1 catX*) [14, 15]. The initial strain C-105 is impaired in glucose catabolic repression due to a genetic defect of the glucose sensor [16], which provides the constitutive functioning of the alcohol oxidase (AOX) promoter, and the lack of catalase activity, which facilitates the isolation of the target enzyme [17].
2. For plasmid construction and amplification, the strain *Escherichia coli* DH5α is used [18].

2.2 Cloning the CYB2 Gene of O. polymorpha

1. The plasmid DNA isolation, restriction, ligation, electrophoresis in agarose gel, electrotransformation, and PCR are carried out by the standard methods [18].
2. The nucleotide sequence of the *O. polymorpha CYB2* gene is obtained from the Rhein Biotech database. The open reading frame (ORF) of the *O. polymorpha CYB2* gene along with the terminator sequence is isolated by PCR using the primers Sm1 (5′-CCC AAG CTT ATG TGG AGA ACC TCC TAT AG-3′)/

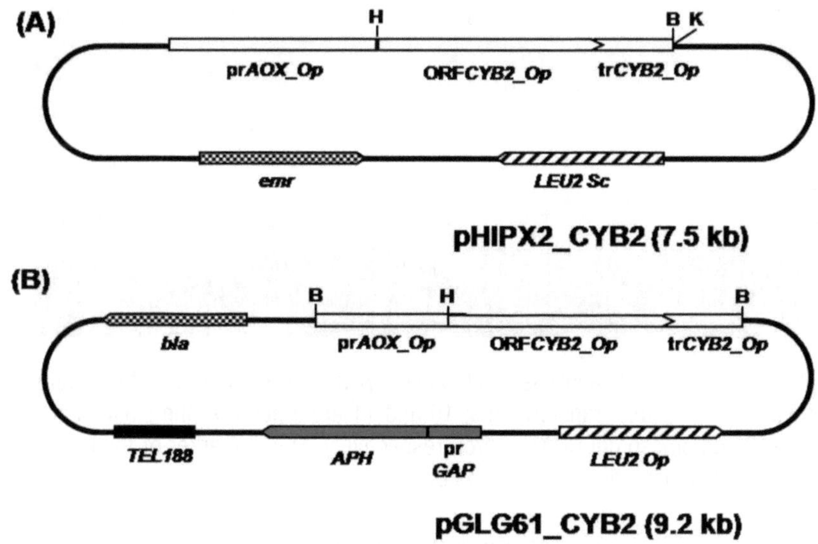

Fig. 1 Circular schemes of (**a**) the plasmid pHIPX2_CYB2 (7.5 kb) and (**b**) the plasmid pGLG61_CYB2 (9.2 kb). The expression cassette, consisting of the *AOX* gene promoter, the *CYB2* ORF, and the *O. polymorpha CYB2* terminator, is designated with the white line. *S. cerevisiae* or *O. polymorpha LEU2* genes are designated with hatched lines. Genes *emr* and *bla* conferring resistance to erythromycin and ampicillin are designated with cross-hatched lines. Gene *APH* conferring resistance to geneticin, linked to an impaired constitutive gene promoter, encoding glyceraldehyde phosphate dehydrogenase (*GAP*) is designated with the gray line. The *HARS36* ARS-element (*TEL188*) is designated with the black line. Restriction sites: *H* HindIII, *B* BamHI, *K* KpnI

Sm2 (5′-CCC GGT ACC GGA TCC CAA AAT AGA GCG CAA GAT TGC-3′) and the chromosomal DNA of *O. polymorpha* CBS 4732 as a template. The amplified fragment flanked by the HindIII and KpnI restriction sites is cloned before the *O. polymorpha* AOX gene promoter into the pHIPX2 plasmid [19] by substituting the *AOX* gene terminator. As a result, the plasmid is obtained, designated pHIPX2_-CYB2 (Fig. 1a).

3. The expression cassette consisting of the AOX gene promoter and the *CYB2* ORF with the terminator sequence is isolated by PCR using the primers Sm3 (5′-TGT GGA TCC TCG TTT AGA ACG TCC TG-3′)/Sm2 and the vector pHIPX2_CYB2 as a template. The PCR product preliminarily treated with BamHI restriction endonuclease was cloned to the vector pGLG61 [20] kindly provided by Dr. H.A. Kang. The constructed plasmid was named pGLG61_CYB2 (Fig. 1b).

2.3 Transformant Selection

1. A rich medium with increasing geneticin concentrations is inoculated with the transformants. The highest geneticin concentration at which the transformants grew is 1 mg/mL. The obtained transformants are stabilized by cultivation under nonselective conditions over the course of 10–12

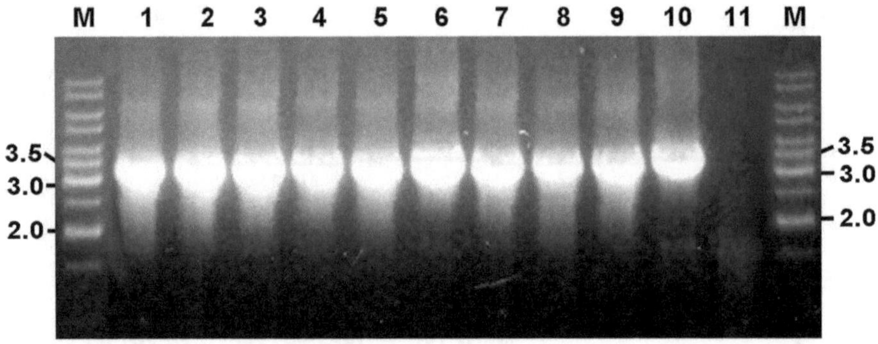

Fig. 2 Electrophoresis for the PCR analysis of *O. polymorpha* C-105 stable transformants with the plasmid pGLG61_CYB2. Lanes: 1–9 stable transformants; 10 and 11 are a positive (the plasmid pGLG61_CYB2) and negative (strain C-105 chromosomal DNA) control, respectively. M is a marker of the molecular mass of the fragments (the fragment values are expressed in kb)

generations with subsequent transfer onto selective medium with genecitin.

2. The presence of the expression cassette in the stable transformants is confirmed by PCR. Using the primers Sm3/Sm2 and the chromosomal DNA of stable transformants as a template, fragments of the predetermined size (~3.3 kb) are obtained (Fig. 2).

2.4 Assay of FC b_2 Activity

1. The activity of FC b_2 is determined spectrophotometrically by the level of potassium ferricyanide reduction monitored at $\lambda = 420$ nm [12]. The reaction mixture contained: 0.03 M phosphate buffer, pH 7.8; 0.03 M L-lactate; 1 mM EDTA; 0.08 mM $K_3Fe(CN)_6$; and 0.02 mL of the analyzed extract. The specific mass activity of the enzyme (SA, μmol/(min·mg) protein) is calculated by the formula (1), and the specific volume activity (VA, μmol/(min·mg) reaction mixture)—according to the formula (2):

$$\mathrm{SA} = \frac{\Delta E/\min \cdot V_{\mathrm{tot}} \cdot n}{\varepsilon_{\mathrm{mM}} \cdot C_{\mathrm{pr.}} \cdot V_{\mathrm{ex}}}; \quad (1)$$

$$\mathrm{VA} = \frac{\Delta E/\min \cdot V_{\mathrm{tot}} \cdot n}{\varepsilon_{\mathrm{mM}} \cdot V_{\mathrm{ex}}}. \quad (2)$$

where ΔE/min—change of the optical density at $\lambda = 420$ nm/min; V_{tot}—the volume of the reaction mixture, mL; n—dilution of the analyzed extract; ε_{mM}—the millimolar extinction coefficient of potassium ferricyanide which is equal to 1.04 mM^{-1} cm^{-1}; V_{ex}—volume of the analyzed extract, mL; $C_{pr.}$—protein concentration in the initial extract, mg/mL.

2. The specific activity of the enzyme is calculated by the formula, taking into account the difference between specific activity

(with L-lactate) and non-specific activity (without L-lactate): $SA = SA_{+Lact} - SA_{-Lact}$. 1 unit of enzyme activity (1 U) is taken the amount of enzyme that causes the oxidation of 1 μmol of L-lactate per 1 min under standard conditions.

2.5 Optimization of Conditions for FC b_2 Isolation

1. The optimization of the culture medium composition for the maximal production of FC b_2 by the cells of the selected transformants (1, 6, and 9, *see* Fig. 3) is carried out taking into account the optimal conditions of induction of the target gene in the *AOX* initial recipient strain *O. polymorpha* C-105 (*gcr1 catX*). This strain is impaired in catabolic repression; therefore, constitutive FC b_2 synthesis regulated by the *AOX* promoter occurs in the medium with glucose. In addition, it has been previously shown that the maximal *AOX* expression for this strain is on the synthetic medium containing increased (0.2%) content of yeast extract (YE) [15]. Considering this fact, the FC b_2 activities were determined in the transformants and the initial strain when the yeast cells grew in medium with 1% glucose (Glc) or a mixture of 1% Glc and 0.2% L-lactate (Lac) supplemented with 0.2% of the YE. Figure 3 shows the data on the FC b_2 activity in the cell-free extracts of the transformants and the initial strains in media of different carbon sources.
2. The FC b_2 activities in the transformants grown in medium with 1% Glc were 1.2–1.6 μmol/(min·mg) protein, thus

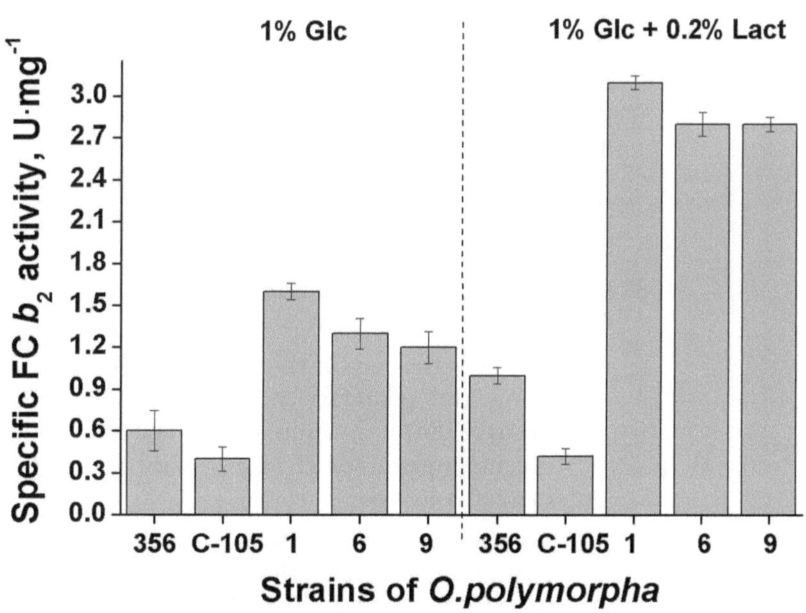

Fig. 3 Specific FC b_2 activity in cell-free extracts of the initial strains and the selected transformants grown in culture medium with 1% Glc or a mixture of 1% Glc and 0.2% Lac

exceeding three- to fourfold the enzyme activity of the initial strain C-105 (0.4 μmol/(min·mg) protein). When the yeast cells grew in a mixture of 1% Glc and 0.2% Lac, the enzyme activity in the transformants constituted 2.8–3.2 μmol/(min·mg) protein, whereas the activity of the initial strain C-105 was 0.4 μmol/(min·mg) protein. The difference in the enzyme activity was seven to eightfold. An increase in the FC b_2 activity in all the strains, when grown in medium with lactate, results from the inducing influence of the latter on the expression of the initial *CYB2* gene [6]. Thus, the medium containing 1% Glc and 0.2% Lac is the optimal one for the maximal production of an active form of FC b_2 by the transformants [21].

2.6 Affinity Chromatography of FC b_2 *from the Extracts of Recombinant Cells of* O. polymorha *"tr1" (*gcr1 catX CYB2*)*

1. For affinity purification of FC b_2, we synthesized an affinity sorbent based on aminopropyl silochrome, modified by commercial cytochrome *c* as a ligand. Cytochrome *c* from the bovine heart is covalently bound by glutaraldehyde-activated aminopropyl silochrome due to the Schiff base forming between protein lysine residues and aldehyde groups according to the following scheme: 50 mL 0.25 M phosphate buffer (PB), pH 8.0 and 5 mL 25% glutaraldehyde are added to 25 mL aminopropyl silochrome. The mixture is stirred overnight at a temperature of 28 °C with moderate shaking. The precipitate is washed with 250 mL 0.05 M PB, pH 8.0.
2. Modification of the activated sorbent is performed using cytochrome *c*. Solution of cytochrome *c* with a concentration of 2 mg/mL is added to the flask containing 10 mL of glutaraldehyde-activated aminopropyl silochrome in PB, pH 8.0. The sorbent is washed with distilled water under gentle stirring, followed by filtration through filter paper and air-dried [22]. It is shown that 5 mL of sorbent is able to bind about 9.5 mg of cytochrome *c*.
3. Cell-free extract of the strain *O. polymorha "tr1" (gcr1 catX CYB2)* is obtained by two stages:
   (a) The cells washed off the *culture* medium are disintegrated using the planetary homogenizer with 0.5-mm glass beads (8000 × *g*; 6 min; 4 °C). The obtained homogenate is centrifugated for 15 min at 9700 × *g*; 4 °C and cell-free extract collected.
   (b) To ensure the better yield of FC b_2, after destruction, the cell fragments are further treated with a mixture containing 10% butanol in 50 mM PB, pH 7.8 with the addition of 1 mM EDTA, 1 mM phenylmethylsulfonyl fluoride (PMSF), and 100 mM sodium lactate. The optimal conditions for the treatment of cell fragments are as follows:

the ratio of the cell fragments to the extracting mixture—2 g/10 mL; incubation time—20 h; 4 °C with a constant stirring using a magnetic stirrer. Cell-free extracts are separated from cell debris by centrifugation at 9700 × *g*; 15 min; 4 °C. The cell-free extracts obtained by two stages are combined and the total protein and the specific FC b_2 activity are determined.

4. The enzyme purification is performed as follows: 15 mL of cell-free extract with a total FC b_2 activity of 278 units and 359 mg of protein is applied to a column filled with immobilized cytochrome *c* on the aminopropyl silochrome. The column is washed with 25 mM PB, pH 7.8 and eluted bound proteins with increasing concentrations of the same buffer. Resulting, a purified preparation of FC b_2 with a specific activity of up to 10 U/mg is obtained. The yield of the enzyme by activity is 74%. To stabilize the enzyme, to the combined eluate fractions is added dry ammonium sulfate to 70% of saturation at 0 °C, maintaining a pH value of about 7.5.
5. The degree of purification of the target enzyme from ballast proteins is characterized by determining its activity and protein concentration in each eluate fraction and by PAG electrophoresis under denaturing conditions in the presence of SDS according to Laemmli [23] (Fig. 4).

2.7 Assay of L-lactate Using In-House FC b_2-Based Enzymatic Kit

1. Enzymo-photometric method/kit based on FC b_2-catalyzed enzymatic oxidation of L-lactate coupled with a nonenzymatic reaction between ferrocyanide $Fe(CN)_6^{4+}$ generated during enzymatic reaction and Fe^{3+} ions, resulting in the formation of intensively colored product of Prussian Blue. This approach is firstly developed by us for a very sensitive visualization of FC b_2 protein band in PAG after electrophoresis of cell-free extracts [24].
2. The L-lactate analysis by the proposed method includes the following operations: mixing 0.68 mL 25 mM PB, pH 7.7; 0.1 mL 30 mM $K_3Fe(CN)_6$, and 0.02 mL FC b_2 (with a volume activity of 1.1 U/mL). The reaction is started by adding 0.2 mL tested sample. The reaction is carried out in a dark place for 30 min at a temperature of 25 °C. The enzymatic reaction is stopped by the addition of 0.3 mL 200 mM $FeCl_3$ in 30 mM HCl, which at the same time leads to the formation of the final color product, Prussian Blue (Fig. 5). For its solubilization, 1.7 mL 0.9 M oxalic acid is added.
3. Measurement of the optical density of the reaction mixture is performed on a spectrophotometer in a 3 mL cell (1 cm optical path), at 680 nm against a blank sample (all components except the tested sample are added, and 0.2 mL distilled water is added

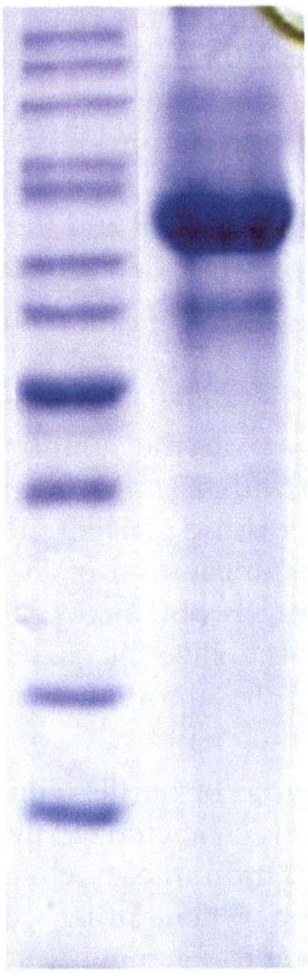

Fig. 4 Electrophoretic characterization of the purified preparation of FC b_2 from *O. polymorpha* "tr1" (*gcr1 catX CYB2*) obtained by affinity chromatography. The left column—molecular weight markers, right—purified FC b_2 preparation. SDS electrophoregram was made under denaturing conditions in 12% PAG, the proteins colored by Coomassie Brilliant Blue *R-250*

Fig. 5 The calibration picture for the FC b_2-based kit with an increasing L-lactate concentration (from left to right, 0–0.2 mM L-lactate)

instead) [25]. It should be noted that the optical density of the reaction mixture after the introduction of oxalic acid remained stable during 30–40 min.

4. At the optimal conditions for the analysis, the FC b_2-based kit provides linearity in the range of 0.004–0.27 mM L-lactate with the detection limit about 3.3 μM L-lactate.

5. The content of L-lactate in real food samples is determined by a calibration curve based on model L-lactate aqueous solutions. Prior to the analysis, the milk samples are mixed with 4% trichloroacetic acid (final concentration). The supernatants are neutralized by 1 M KOH and used for analysis. The ketchup samples are diluted with distilled water and filtrated through 0.2 μm pores microfilter. The wine samples are diluted 50-fold by distilled water. Additional treatment of the wine and juices samples is omitted [26].

2.8 Immobilization of FC b2 on the Magnetic Microparticles for Multiple Enzymatic Analysis of L-lactate

1. Immobilization of the FC b_2 on the surface of commercial silane coated iron oxide-based microparticles (1 mm particle size), functionalized with amino groups is performed due to electrostatic interactions of the protonated amino group of the microcarrier and the protein at pH 7.8.
2. The basic solution of magnetic microparticles (50 mg/mL at pH 7.0) is diluted to a final concentration of 1.5 mg/mL and mixed with an equal volume of purified enzyme preparation (with activity of 3.75 U/mL and a protein concentration of 1.5 mg/mL). The resulting suspension with a total volume of 400 μL is incubated for 1.5 h at 8 °C to immobilize the enzyme on a magnetic microcarrier. After incubation, the biofunctionalized magnetic microparticles are precipitated in a magnetic field and washed twice with 5.0 mM PB, pH 7.5. It is shown that 85% of the enzyme is bound with magnetic microcarriers; moreover, the specific FC b_2 activity increases 1.73-fold due to this process.
3. The obtained biofunctionalized magnetic microparticles with FC b_2 activity of 6.5 U/mL are used to test the possibility of reusing the enzyme in the enzymatic assay of L-lactate by the previously developed enzymatic kit (*see above*) [25]. It is proved that biofunctionalized magnetic microparticles can be reused for multiple (at least six times) enzymatic assay of L-lactate without decreasing the accuracy of the analysis.

2.9 Construction of FC b2-Based Mediatorless Amperometric Biosensor

1. Amperometric biosensor is evaluated using constant-potential amperometry in a three-electrode configuration using 4 mm diameter planar gold electrodes DRP-C220AT. Amperometric measurements are carried in a batch mode under continuous stirring in a standard 4 mL electrochemical cell at 25 °C with applied working potential −100 mV *vs.* Ag/AgCl reference electrode. An amperometric microcomputer-based analyzer system [27] and appropriate software for L-lactate analysis are developed in Institute of Cell Biology, National Academy of Science of Ukraine.
2. Modification of the surface of gold planar microelectrode by gold nanolayer is performed directly on the surface of planar

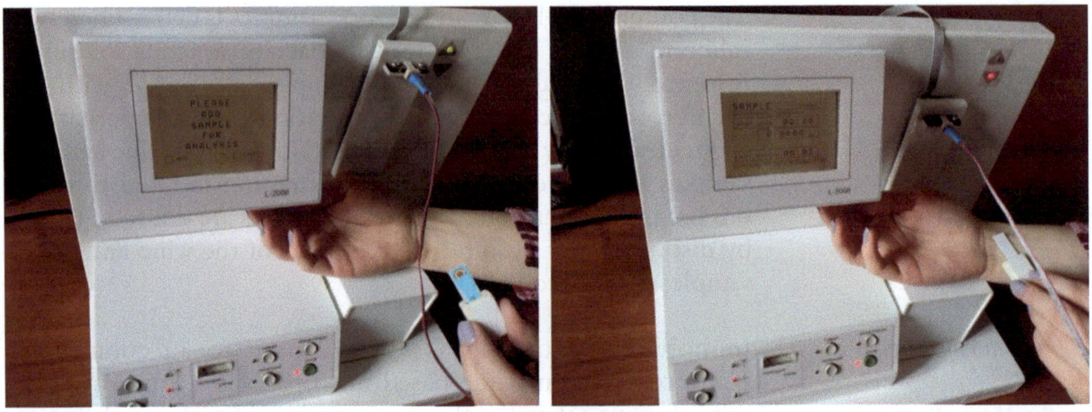

Fig. 6 Noninvasive analysis of L-lactate in sweat using amperometric microcomputer-based analyzer system with integrated FC b_2-based mediatorless L-lactate–selective biosensor

gold electrode DRP-C220AT by reduction of $HAuCl_4$ solution to Au^0. The reaction is performed using 30% H_2O_2 according to Panda and Chattopadhyay [28]. The procedure is as follows: 4 μL 2 mM $HAuCl_4$ is dropped on the top of working electrode; after drying, 4–5 μL 30% H_2O_2 is added. Prior to the modification of the surface, the planar gold electrode is cleaned by 70% ethanol solution. The modification of electrode surface by the nAu is accompanied with changing polished surface structure to scabrous and color from yellow (gold) to high-colored orange (nanogold).

3. The immobilization of FC b_2 on the nAu-modified planar electrode is done as follows: 4 μL FC b_2 suspension (10 U/mL in 200 mM PB, pH 7.8) is dropped on the surface of the gold planar electrode and dried for 2–3 min at room temperature. Then, the electrode is covered with a piece of standard dialysis membrane (with a diameter of about 6 mm) fixed on the electrode surface by an adhesive membrane. The electrode is rinsed with 20 mM PB, pH 7.8, before using.
4. For non-invasive assay of L-lactate in sweat, the electrode connected through 50 cm flexible adaptor is put to a skin of the forearm. The electrode surface is connected with skin and analysis of L-lactate by amperometric microcomputer-based analyzer system is started (Fig. 6). The concentration of L-lactate is calculated automatically according to a sensor output. After repeating the measurement for three times, final L-lactate content in the analyzed skin sweat is calculated as 5.2 ± 0.3 mM.
5. For non-invasive analysis of lactate concentration in saliva, the electrode is put under a tongue for 30 s (5 min before the experiment, a mouth is washed by distilled water). Using the

amperometric microcomputer-based analyzer system L-lactate content in saliva is calculated as 0.36 ± 0.02 mM.

6. The L-lactate content values in the analyzed human liquid samples obtained by the direct noninvasive and reference analysis have a high correlation ($0.7 < r < 1$) [29].

Acknowledgments

This work is financially supported, in part, by the Ministry of Education and Science of Ukraine (projects #0118U000297, #0119U100671 and Ukrainian-Lithuanian R&D project #0120U103398), NAS of Ukraine in the frame of the Scientific-Technical Program "Smart Sensor Devices of a New Generation Based on Modern Materials and Technologies" (project #13), and National Research Foundation of Ukraine (project #48/02.2020).

References

1. Ghrir R, Becam A, Lederer F (1984) Primary structure of flavocytochrome b_2 from Baker's yeast. Purification by reverse-phase high-pressure liquid chromatography and sequencing of fragment a cyanogen-bromide peptides. Eur J Biochem 139:59–65
2. Haumont PY, Thomas MA, Labeyrie F, Lederer F (1987) Amino-acid sequence of the cytochrome-b5-like heme-binding domain from *Hansenula anomala* flavocytochrome b_2. Eur J Biochem 169(3):539–546
3. Silvestrini MC, Teogoni M, Celerier J (1993) Expression in *Escherichia coli* of the flavin and the haem domains of *Hansenula anomala* flavocytochrome b_2 (flavodehydrogenase and b_2 core) and characterization of the recombinant proteins. Biochem J 295(2):501–508
4. Berardi E (1997) Genetics and molecular biology of methylotrophic yeasts. In: Spencer JFT, Spencer DM (eds) Yeasts in natural and artificial habitats. Springer, Berlin, pp 264–294
5. Viola AM, Lodi T, Ferrero I (1999) A Klac null mutant of *Kluyveromices lactis* is complemented by a single copy of the *Saccharomyces cerevisiae AACl* gene. Curr Genet 36(1-2):29–36
6. Alberti A, Goffrini P, Ferrero I, Lodi T (2000) Cloning and characterization of the lactate-specific inducible gene *KlCYB2*, encoding the cytochrome b(2) of *Kluyveromyces lactis.* Yeast 16(7):657–665
7. Jacq C, Lederer F (1974) Cytochrome b_2 from Bakers' Yeast (L-lactate dehydrogenase). A double-headed enzyme. Eur J Biochem 41:311
8. Lederer F (1974) On the first steps of lactate oxidation by Bakers' yeast L-(+)-lactate dehydrogenase (cytochrome b_2). Eur J Biochem 46:393–399
9. Celerier J, Risler Y, Schwencke J (1989) Isolation of the flavodehydrogenase domain of Hansenula anomala flavocytochrome b_2 after mild proteolysis by an *H. anomala* proteinase. Eur J Biochem 182(1):67–75
10. White P, Manson FD, Brunt CE (1993) The importance of the interdomain hinge in intramolecular electron transfer in flavocytochrome b_2. Biochem J 291(1):89–94
11. Smutok O, Gayda G, Gonchar M, Schuhmann W (2005) A novel L-lactate-selective biosensor based on flavocytochrome b_2 from methylotrophic yeast *Hansenula polymorpha.* Biosens Bioelectron 20:1285–1290
12. Appleby C, Morton R (1959) Lactic dehydrogenase and cytochrome b_2 from yeast. Purification and crystallization. J Biochem 71:492–499
13. Smutok OV, Osmak HS, Gayda GZ, Gonchar MV (2006) Screening of yeasts producing stable L-lactate cytochrome c oxidoreductase and study of the regulation of enzyme synthesis. Microbiology (Mosc) 75(1):20–24
14. Gonchar M, Maidan M, Moroz O, Woodward JR, Sibirny AA (1998) Microbial O_2- and H_2O_2-electrode sensors for alcohol assays based on the use of permeabilized mutant yeast cells as the sensitive bioelements. Biosens Bioelectron 13:945–952

15. Gonchar M, Maidan M, Korpan Y, Sibirny V, Kotylak Z, Sibirny A (2002) Metabolically engineered methylotrophic yeast cells and enzymes as sensor biorecognition elements. FEMS Yeast Res 2:307–314
16. Stasyk OV, Stasyk OG, Komduur J, Veenhuis M, Cregg JM, Sibirny AA (2004) A hexose transporter homologue controls glucose repression in the methylotrophic yeast *Hansenula polymorpha*. J Biol Chem 279:8116–8125
17. Gonchar M, Maidan M, Pavlishko H, Sibirny A (2001) A new oxidaseperoxidase kit for ethanol assays in alcoholic beverages. Food Technol Biotechnol 39:37–42
18. Sambrook J, Fritsch EF, Maniatis T (1989) Molecular cloning: a laboratory manual. Cold Spring Harbor Laboratory Press, New York, NY
19. Faber KN, Haima P, Gietl C, Harder W, Ab G, Veenhuis M (1994) The methylotrophic yeast *Hansenula polymorpha* contains an inducible import pathway for peroxisomal matrix proteins with an N-terminal targeting signal (PTS2 proteins). Proc Natl Acad Sci U S A 91 (26):12985–12989
20. Sohn JH, Choi ES, Kang HA, Rhee JS, Agaphonov MO, Ter-Avanesyan MD, Rhee SK (1999) A dominant selection system designed for copy-number-controlled gene integration in *Hansenula polymorpha* DL-1. Appl Microbiol Biotechnol 51(6):800–807
21. Dmitruk KV, Smutok OV, Gonchar MV, Sibirnyĭ AA (2008) Construction of flavocytochrome b_2-overproducing strains of the thermotolerant methylotrophic yeast *Hansenula polymorpha* (*Pichia angusta*). Microbiology (Mosc) 77(2):213–218
22. Smutok O, Karkovska M, Stasyuk N, Gonchar M (2018) Isolation, purification, stabilization and characterisation of flavocytochrome b_2 from overproducing cells of *Ogataea polymorpha* "tr1" (*gcr1 catX CYB2*). Visnyk of L'viv Univ Ser Biol 77:3–15
23. Laemmli UK (1970) Cleavage of structural proteins during the assembly of the head of bacteriophage T4. Nature 227 (5259):680–685
24. Gaida GZ, Stel'mashchuk SY, Smutok OV, Gonchar MV (2003) A new method of visualization of the enzymatic activity of flavocytochrome b_2 in electrophoretograms. Appl Biochem Microbiol (Mosc) 39(2):221–223
25. Gonchar M, Smutok O, Os'mak H (2009) Flavocytochrome b_2-based enzymatic composition, method and kit for L-lactate. U.S Patent WO/2009/009656, 15 Jan 2009
26. Smutok O, Karkovska M, Smutok H, Gonchar M (2013) Flavocytochrome b_2-based enzymatic method of L-lactate assay in food products. Scientific World Journal 2013:461284. https://doi.org/10.1155/2013/461284
27. Sigawi S, Smutok O, Demkiv O, Gayda G, Vus B, Nitzan Y, Gonchar M, Nisnevitch M (2014) Detection of waterborne and airborne formaldehyde: from amperometric chemosensing to a visual biosensor based on alcohol oxidase. Materials (Basel) 7:1055–1068
28. Panda BR, Chattopadhyay A (2007) Synthesis of au nanoparticles at "all" pH by H_2O_2 reduction of $HAuCl_4$. J Nanosci Nanotechnol 7:1911–1915
29. Şmutok O, Karkovska M, Serkiz Y, Vus B, Čenas N, Gonchar M (2017) Development of a new mediatorless biosensor based on flavocytochrome b_2 immobilized onto gold nanolayer for non-invasive *L*-lactate analysis of human liquids. Sens Actuators B 250:469–475

Part V

Significance of Flavoproteome in Humans

Chapter 17

Mammalian Flavoproteome Analysis Using Label-Free Quantitative Mass Spectrometry

Giulia Calloni and R. Martin Vabulas

Abstract

Human flavin cofactor-containing enzymes constitute a small, but highly important flavoproteome. Its stability is required to ensure key metabolic functions, such as oxidative phosphorylation and beta-oxidation of fatty acid. Flavoproteome disfunction due to mutations of individual proteins or because of the lack of FMN and FAD precursor riboflavin (vitamin B2) results in clinically relevant abnormal cellular states and diseases. Current technical possibilities in the field of the quantitative mass spectrometry of proteins allow studying the flavoproteome changes under different stress conditions, including the deficiency of vitamin B2. The biological readouts of flavoenzyme destabilization, such as protein degradation and aggregation, provide important insights into the molecular mechanisms of metabolic adaptation to nutrient deficiency. The proteomic-scale studies of protein stability have significant novelty potential in basic and applied biomedical research.

Key words Flavoproteome, Riboflavin, Mass spectrometry, Protein degradation, Protein aggregation

1 Introduction

Riboflavin, known as vitamin B2 in food, is metabolized in human cells to flavin mononucleotide (FMN) and flavin adenine dinucleotide (FAD), two cofactors that associate with ca. 100 enzymes. The set of flavin cofactor-bound proteins is called the flavoproteome. Despite being only a small part of the whole proteome, the flavoproteome is characterized by its central metabolic importance [1]. For example, mitochondrial NADH dehydrogenase flavoprotein 1 is the key component of the respiratory complex (RC) I and is involved in transferring electrons from NADH to the respiratory chain. A major catalytic subunit of the succinate-ubiquinone oxidoreductase, the RC II of the mitochondrial respiratory chain, is a flavoprotein as well and is responsible for transferring electrons from succinate to ubiquinone. Flavoprotein methylenetetrahydrofolate reductase (MTHFR) is involved in the tetrahydrofolate interconversion as a part of one-carbon metabolism. Mutations of these

Maria Barile (ed.), *Flavins and Flavoproteins: Methods and Protocols*, Methods in Molecular Biology, vol. 2280,
https://doi.org/10.1007/978-1-0716-1286-6_17,

and other flavoproteins lead to severe debilitating diseases, such as dilated cardiomyopathy, homocystinuria, methemoglobinemia, to mention a few. Actually, defects of more than half of human flavoproteins have been associated with clinically significant manifestations [2]. Thus, understanding the flavoproteome stability and turnover is important not only from the metabolic point of view but also in biomedical regard.

The recent technical developments of mass spectrometry (MS) has enabled proteome analyses to an impressive breadth, depth and accuracy [3]. Stable isotope-based quantitative MS has become the standard quantification approach in the high throughput settings [4]. Particularly metabolic labeling methods, such as stable isotope labeling with amino acids in cell culture (SILAC) [5], exemplify the power of the MS-based quantification, which is due to reduced variability in preparation and measurements between samples. From the other side, SILAC and other labeling strategies are impractical or not feasible in some experimental and clinical situations. Therefore, the possibilities of label-free MS quantification has always been of a significant interest and motivated respective analytical developments. As for the label-based quantification, similar factors are important to ensure the reliability of label-free quantitative MS, namely, high mass resolution, high measurement accuracy and high identification rates. At lower mass resolution, counting of assigned MS/MS spectra can be used to estimate protein abundances [6]. At high resolution, the accurate determination of extracted ion currents of peptides is possible and preferable when comparing the abundances of these peptides in different samples [7]. MaxLFQ, an implementation of the latter label-free quantification strategy, uses two inventions, the delayed normalization of the samples and a novel quantification approach, which have ensured the robustness and success of the algorithm [8]. MaxLFQ was used by us to analyze the flavoproteome in murine melanoma cells [9] and the protocol will be described in this chapter.

Lack of flavin cofactor can destabilize flavoproteins and lead to their degradation. For example, a vitamin B2-free diet in rats strongly affects some of the mitochondrial flavoproteins, such as isovaleryl-CoA dehydrogenase and short-chain acyl-CoA dehydrogenase [10]. MS quantification offers a way of simultaneous turnover analyses of most of the flavoproteins upon riboflavin starvation. Obviously, the changes of protein amounts arise not only from their degradation, but from the changed rates of their ribosomal synthesis as well. Metabolic pulse-labeling of nascent polypeptides and their subsequent chase in the label-free medium is a powerful strategy to untangle this synthesis-degradation duality [11]. The comparison of steady-state levels of proteins in cells with active and inhibited protein degradation systems is a simple alternative, which was used to analyze also the stability of the mammalian flavoproteome [9]. In addition to degradation, damaged and misfolded proteins might aggregate, so the isolation of cellular aggregates and their mass spectrometric quantification represent yet another biological readout of proteome stability. If applied in combination, these analyses offer a comprehensive view of the cellular proteome reorganization during different stress situations, such as nutrient deficiency, and thus bear a significant novelty potential in basic and biomedical research.

2 Materials

Common reagents were purchased from Sigma-Aldrich (St. Louis, MO) if not indicated otherwise. Analytical grade (HPLC) water and solvents should be used to prepare buffers. The respective material safety data sheets should be consulted for proper handling of the potentially hazardous materials.

2.1 Soluble Protein Extraction

1. Phosphate-buffered saline (PBS).
2. Lysis buffer: 10% sodium dodecyl sulfate (SDS), 150 mM NaCl, 100 mM 4-(2-hydroxyethyl)-1-piperazineethanesulfonic acid (Hepes) pH 7.6 (*see* **Note 1**).
3. DC Protein Assay (Bio-Rad, Hercules, CA).
4. Bovine serum albumin (BSA) standard in water at concentrations ranging from 0.1 to 2 mg/mL.
5. Microcentrifuge with temperature control.
6. Ultrasonic homogenizer with probe for small volumes.
7. Absorbance microplate reader.

2.2 FASP

1. Lysate dilution buffer: 4% SDS, 150 mM NaCl, 100 mM Hepes pH 7.6 (*see* **Note 1**).
2. Denaturation buffer (DB): 8 M urea, 50 mM Tris/HCl, pH 8.5 (*see* **Note 2**).
3. 1 M dithiothreitol (DTT) stock solution in DB (*see* **Note 3**).
4. 0.05 M iodoacetamide stock solution in DB (*see* **Note 4**).
5. ABC buffer: 50 mM ammonium bicarbonate.
6. Microcon YM-30 filter units (Millipore, Burlington, MA).
7. Thermomixer.

2.3 MS Sample Preparation

1. Trypsin stock: dissolve 20 μg of sequencing grade trypsin (Promega, Madison, WI) in 200 μL of ABC buffer for a 0.1 μg/μL stock solution (*see* **Note 5**).
2. 100 mM $CaCl_2$ stock solution in water.
3. 10% trifluoroacetic acid (TFA) solution.

4. Empore C18-PS solid phase extraction (SPE) disks (3M, Maplewood, MN).
5. Empore PK20 cation SPE disks (3M).
6. C18 stage tips solutions: (a) methanol; (b) buffer A: 0.5% acetic acid; (c) buffer B: 80% acetonitrile (ACN), 0.5% acetic acid.
7. Strong cation exchange (SCX) stage tips solutions: (a) ACN; (b) SCX-1: 50 mM ammonium acetate, 20% ACN, 0.5% acetic acid; (c) SCX-2: 75 mM ammonium acetate, 20% ACN, 0.5% acetic acid; (d) SCX-3: 125 mM ammonium acetate, 20% ACN, 0.5% acetic acid; (e) SCX-4: 200 mM ammonium acetate, 20% ACN, 0.5% acetic acid; (f) SCX-5: 300 mM ammonium acetate, 20% ACN, 0.5% acetic acid; (g) SCX-6: 5% ammonium hydroxide (NH_4OH), 80% ACN; (h) SCX-wash: 0.2% TFA.

2.4 LC-MS/MS

1. Nanoflow liquid chromatography (nLC) system Easy-nLC1000 (Thermo Fisher Scientific, Waltham, MA).
2. Nanospray Flex ion source (Thermo Fisher Scientific).
3. High-resolution mass spectrometer Q Exactive Plus (Thermo Fisher Scientific).
4. C18 capillary columns (*see* **Note 6**).
5. nLC mobile phase A: 4% ACN, 0.1% formic acid (FA).
6. nLC mobile phase B: 80% ACN, 0.1% FA.

3 Methods

Murine melanoma cells B16-F0 (#CRL-6322; ATCC) are analyzed by the described procedure [9]; however, other mammalian cells can be used with minor adaptations. Quick processing of the samples is important. Four to five biological replicates are usually used to ensure robust label-free quantification. All replicates should be processed and measured together in the same session.

3.1 Soluble Protein Extraction

1. Collect 10^6 murine melanoma cells, pellet them, and wash them twice with cold PBS by centrifugation at 200 × *g* for 5 min at 4 °C.
2. Add 200 μL of lysis buffer and mix gently until cells are completely resuspended.
3. Shear DNA by sonicating the lysate on ice. Sonicate for a total of 5 s using the following settings: pulse at 30% amplitude for 0.2 s followed by a time off of 0.8 s, for a total cycle time of 10 s. Repeat if necessary until the lysate is no longer viscous.
4. Clear the cell lysate by centrifugation at 16,000 × *g* for 3 min at 4 °C.

5. Determine protein concentration with the DC Protein Assay in the 96 well plates following the kit instructions. Prepare 1:10 and 1:20 lysate dilutions in water. Perform the measurements using 5 μL sample volume (*see* **Note 7**).

3.2 FASP

The filter aided sample preparation protocol (FASP) is done as described [12]. All the centrifugation steps of this workflow are performed at room temperature.

1. Dilute 100 μg of total protein in lysate dilution buffer to reach a final volume of 45 μL.
2. Reduce proteins by adding 5 μL DTT stock solution and incubating at 95 °C for 5 min.
3. Cool the samples to room temperature.
4. Mix the lysate with 200 μL of denaturation buffer, mix by vortexing, and spin down for 20 s.
5. Load the lysate onto the Microcon spin filter with 30 kDa cut-off.
6. Centrifuge at 9280 × *g* for 15 min (*see* **Note 8**).
7. Add 200 μL of denaturation buffer to the filter unit and centrifuge again at 9280 × *g* for 15 min.
8. Carefully discard the flow-through using a pipet tip.
9. Add 100 μL of IAA solution to the filter unit and incubate for 45 min at room temperature in the dark.
10. Centrifuge at 9280 × *g* for 10 min and carefully discard the flow-through using a pipet tip.
11. Perform two washing steps with 100 μL DB followed by 15 min centrifugation at 15,680 × *g*. Each time carefully discard the flow-through using a pipet tip.
12. Perform two washing steps with 100 μL ABC buffer each followed by 10 min centrifugation at 15,680 × *g*.
13. Transfer the filter unit to a new collection tube.

3.3 MS Sample Preparation

1. Add 100 μL of digestion buffer containing 1 μg sequencing grade trypsin (10 μL of the stock solution) and 10 mM $CaCl_2$ (1 μL of the stock solution) to each sample.
2. Shake gently (600 rpm) for 1 min in a thermomixer and incubate overnight at 37 °C.
3. Recover the peptides by centrifugation at 9280 × *g* for 10 min.
4. Add 40 μL of ABC buffer to increase peptide recovery and centrifuge again at 9280 × *g* for 10 min.
5. Stop digestion by acidifying the peptide solution with 6 μL of 10% TFA (*see* **Note 9**).

6. To desalt the peptide mixture, cut 0.5 mm diameter C18 disks from the Empore C18-PS resin; insert six disks into a 200 μL pipet tip (*see* **Note 10**).
7. Activate the C18 resin in the stage tip by adding 50 μL methanol; drain the liquid by centrifugation at 1600 × *g* for 3–5 min or by applying pressure with a 10 mL syringe (*see* **Note 11**).
8. Wash the stage tip with 50 μL of buffer A and drain the liquid as before.
9. To allow binding of the peptides, add the peptide solution into the stage tip and drain the liquid as before.
10. Wash the stage tip with 50 μL of buffer A and drain the liquid as before (*see* **Note 12**).
11. To fractionate the peptide mixture, cut 0.5 mm diameter strong cation exchange (SCX) disks from the Empore PK20 resin; insert five disks into a 200 μL pipet tip.
12. Activate the SCX resin with 50 μL ACN; drain the liquid as for the C18 stage tips.
13. Wash the stage tip with 100 μL of SCX-wash solution and drain the liquid as before.
14. Load the peptides on the SCX resin by eluting the C18 stage tips with 60 μL buffer B directly into the SCX stage tips and draining the liquid. Collect the C18 transelution into a tube (or a well of a 96-well plate).
15. Wash the SCX stage tip with 100 μL of buffer A and drain the liquid as before.
16. Elute the peptides from the SCX stage tip by applying sequentially 30 μL of the solutions SCX-1 to SCX-6 and collecting the eluates in separate tubes (or wells of a 96-well plate). Combine the SCX fraction 1 (upon elution with the SCX-1) with the C18 trans-eluate.
17. Use a vacuum centrifuge to dry the peptides at 35 °C (*see* **Note 13**). Then dissolve the pellets in 20 μL of 1% ACN, 0.1% FA for subsequent LC-MS/MS measurements.

3.4 LC-MS/MS

1. Inject 5 μL (~20 μg) of the digested protein samples onto the nanoLC column (*see* **Note 14**).
2. A 120 min gradient as detailed in Table 1 is optimized for LC-MS/MS analysis of the obtained fractions. The method uses a gradient from mobile phase A to 30% mobile phase B over 60 min followed by a second step to 60% B over 30 min; flow rate is set at 300 nL/min.

Table 1
Liquid chromatography parameters

Time	LC gradient (%B)
0–60	1–30
61–90	30–60
90–92	60–99
93–98	99
99–101	99–1
102–120	1

Table 2
Mass spectrometry parameters

Full MS	
Microscans	1
Resolution	70,000
Automatic gain control target	3e6
Maximum ion time	120 ms
Number of scans	1
Scan range	300–2000 m/z
Data-dependent MS^2	
Microscans	1
Resolution	17,500
Automatic gain control target	1e5
Maximum ion time	80 ms
Loop count	10
Isolation window	2 m/z
Normalized collision energy	30
dd setting	
Charge exclusion	Unassigned, 1, >8
Exclude isotopes	On
Dynamic exclusion	30 s

3. The parameters for MS analysis vary depending on the type and performance of the MS instrument used. As an example, optimal parameters for a 120 min acquisition method on a Q Exactive Plus mass spectrometer are listed in Table 2.

3.5 Data Analysis

1. Generate an appropriate FASTA database for query. A large number of proteomes can be found for download in FASTA format on the UniProtKB website (https://www.uniprot.org/proteomes/).
2. Search the MS/MS spectra against the FASTA database using MaxQuant suite [13] (https://maxquant.org/) with activated label free quantitation (LFQ) mode [8] (*see* **Note 15**).
3. Use a false discovery rate of 1% for protein identification.
4. Set enzyme specificity to trypsin and allow a maximum of two missed cleavages with a minimal peptide length of seven amino acids.
5. Set carbamidomethylcysteine as a fixed modification and N-terminal acetylation and methionine oxidation as variable modifications.
6. Allow matching of unidentified features between runs (*see* **Note 16**).
7. For label-free quantification set the minimum ratio count to 1.
8. Perform statistical analysis of identified differences using Perseus suite [14] (https://maxquant.org/perseus/) (*see* **Note 17**). For LFQ analysis, a minimum of five biological replicates is recommended.
9. Exclude hits in the following categories from further analysis: *false positives*, *only identified by site*, and *known contaminants*.
10. To evaluate the effects of riboflavin starvation on the cellular proteome, group the samples according to the experimental conditions (e.g., group 1: control cells, group 2: riboflavin-starved cells).
11. Work with the log2 transformed LFQ values.
12. Set the minimum number of measured LFQ values in the data set (e.g., for five biological replicates a minimum of three LFQ values in both experimental conditions can be used).
13. Impute the missing LFQ values on the basis of normal distribution (*see* **Note 18**).
14. Perform a two-sample *t*-test to identified proteins significantly changed upon riboflavin starvation (*see* **Note 19**).
15. Add categorical annotations in Perseus and perform a Fisher's exact test to determine GO term enrichment in the set of significantly changed proteins.

3.6 Additional Bioinformatic Tools for Data Analysis and Interpretation

1. GO term and protein class enrichment analysis can be performed within the online platform Panther [15] (http://www.pantherdb.org/). To this end, the Overrepresentation Test tool and the GO ontology database for genome annotations should be used.

2. Network analysis can be performed online using the STRING platform [16] (https://string-db.org/).

   Use the following settings as starting point for the analysis: network edges represent confidence, exclude textmining from the interaction sources, set confidence to high (score 0.7) and enable "hide disconnected nodes."

3. Large-scale characterization of protein groups can be performed using the cleverSuite platform [17] (http://s.tartaglialab.com/clever_suite). The algorithm predicts secondary structure properties, solubility, chaperone requirements and RNA-binding abilities of polypeptides from different datasets. These and other physicochemical parameters are then used to identify statistically significant differences between the protein groups.

3.7 Stoichiometry Analysis of Oligomeric Complexes

The stoichiometry of protein complex subunits can be analyzed using LFQ data [18]. The stoichiometry analysis is helpful in identifying disturbances in protein stability under given conditions.

1. Estimate the absolute abundance of subunits from a complex of interest using iBAQ calculation (check the iBAQ calculation box in the label free quantitation settings of MaxQuant).
2. Use the average iBAQ of individual subunits across biological replicates to calculate the ratio of the subunit iBAQ over the mean of the iBAQs for all the subunit of the complex.

4 Notes

1. The lysis buffer and lysate dilution buffer can be prepared in advance and stored at room temperature.
2. Dissolve the urea in buffer with gentle heating (<30 °C) and vigorous stirring. Remove from the heat source as soon as the urea dissolves. The solution can be stored at −20 °C for later use.
3. DTT solution is susceptible to oxidation and should be prepared freshly before use.
4. IAA solution is light sensitive. It should be prepared freshly and kept in the dark.
5. Prepare the trypsin solution freshly and store the remaining at −80 °C for subsequent use.
6. To reduce the cost of MS measurements, one could prepare and use self-packed columns. To this end, C18 resin (2.4 μm Reprosil-Pur C18 from Dr. Maisch GmbH, Germany) is filled into empty fused silica capillaries (100 μm inner diameter, 30 cm length) using the high-pressure injection cell from Next Advance (Troy, NY).

7. To decrease errors in protein concentration determination, measure the lysates in triplicate and the BSA standards in duplicate.
8. Check that the filters are not damaged by verifying that only a small volume of buffer has gone through the filter after a short centrifugation (1 min).
9. Make sure that the solution is acidic by spotting 1 μL of the solution on pH paper.
10. A self-made metal cylinder of the respective diameter should be prepared to facilitate cutting of the disks. A pin is used to extrude the cut disk from the channel of the cylinder into the pipette tip.
11. Avoid that the C18 material dries out completely before application of the peptide solution.
12. Peptides bound to stage tips after the washing step can be stored at 4 °C until elution.
13. Not all of the samples dry out equally fast. Check for the completeness and continue until all samples have dried.
14. Usually 15–30 cm-long columns are used, although up to 50 cm-long columns can be employed to improve peptide separation. Commercially available columns seem to be more reliable, yet self-packed columns significantly reduce costs of MS experimentation.
15. MaxQuant offers the possibility of parallel processing of raw files, which strongly reduces computational time. We use a platform of 16 CPU cores, 64 Gb RAM and two hard disks, one being dedicated exclusively for MaxQuant calculations.
16. Adjust the matching window (min) based on the peak shifts observed in the chromatograms during the acquisition of the sample set.
17. The statistical analysis can be performed also with alternative software (e.g., Microsoft Excel or R).
18. Check the distribution of the LFQ values after imputation and if needed adjust the width and downshift of the normal distribution used for the imputation to minimize the appearance of a bimodal distribution.
19. Use different permutation-based false discovery rate (FDR) or *p*-value cutoffs to adjust the stringency of the analysis.

References

1. Lienhart W-D, Gudipati V, Macheroux P (2013) The human flavoproteome. Arch Biochem Biophys 535:150–162. https://doi.org/10.1016/j.abb.2013.02.015
2. Amberger JS, Bocchini CA, Scott AF, Hamosh A (2019) OMIM.org: leveraging knowledge across phenotype-gene relationships. Nucleic Acids Res 47:D1038–D1043. https://doi.org/10.1093/nar/gky1151
3. Aebersold R, Mann M (2016) Mass-spectrometric exploration of proteome structure and function. Nature 537:347–355. https://doi.org/10.1038/nature19949
4. Bantscheff M, Lemeer S, Savitski MM, Kuster B (2012) Quantitative mass spectrometry in proteomics: critical review update from 2007 to the present. Anal Bioanal Chem 404:939–965. https://doi.org/10.1007/s00216-012-6203-4
5. Mann M (2006) Functional and quantitative proteomics using SILAC. Nat Rev Mol Cell Biol 7:952–958. https://doi.org/10.1038/nrm2067
6. Ishihama Y, Oda Y, Tabata T et al (2005) Exponentially modified protein abundance index (emPAI) for estimation of absolute protein amount in proteomics by the number of sequenced peptides per protein. Mol Cell Proteomics 4:1265–1272. https://doi.org/10.1074/mcp.M500061-MCP200
7. Andersen JS, Wilkinson CJ, Mayor T et al (2003) Proteomic characterization of the human centrosome by protein correlation profiling. Nature 426:570–574. https://doi.org/10.1038/nature02166
8. Cox J, Hein MY, Luber CA et al (2014) Accurate proteome-wide label-free quantification by delayed normalization and maximal peptide ratio extraction, termed MaxLFQ. Mol Cell Proteomics 13:2513–2526. https://doi.org/10.1074/mcp.M113.031591
9. Martínez-Limón A, Alriquet M, Lang W-H et al (2016) Recognition of enzymes lacking bound cofactor by protein quality control. Proc Natl Acad Sci U S A 113:12156–12161. https://doi.org/10.1073/pnas.1611994113
10. Nagao M, Tanaka K (1992) FAD-dependent regulation of transcription, translation, post-translational processing, and post-processing stability of various mitochondrial acyl-CoA dehydrogenases and of electron transfer flavoprotein and the site of holoenzyme formation. J Biol Chem 267:17925–17932
11. Doherty MK, Hammond DE, Clague MJ et al (2009) Turnover of the human proteome: determination of protein intracellular stability by dynamic SILAC. J Proteome Res 8:104–112. https://doi.org/10.1021/pr800641v
12. Wiśniewski JR, Zougman A, Nagaraj N, Mann M (2009) Universal sample preparation method for proteome analysis. Nat Methods 6:359–362. https://doi.org/10.1038/nmeth.1322
13. Cox J, Mann M (2011) Quantitative, high-resolution proteomics for data-driven systems biology. Annu Rev Biochem 80:273–299. https://doi.org/10.1146/annurev-biochem-061308-093216
14. Tyanova S, Temu T, Sinitcyn P et al (2016) The Perseus computational platform for comprehensive analysis of (prote)omics data. Nat Methods 13:731–740. https://doi.org/10.1038/nmeth.3901
15. Mi H, Huang X, Muruganujan A et al (2017) PANTHER version 11: expanded annotation data from Gene Ontology and Reactome pathways, and data analysis tool enhancements. Nucleic Acids Res 45:D183–D189. https://doi.org/10.1093/nar/gkw1138
16. Szklarczyk D, Franceschini A, Wyder S et al (2015) STRING v10: protein-protein interaction networks, integrated over the tree of life. Nucleic Acids Res 43:D447–D452. https://doi.org/10.1093/nar/gku1003
17. Klus P, Bolognesi B, Agostini F et al (2014) The cleverSuite approach for protein characterization: predictions of structural properties, solubility, chaperone requirements and RNA-binding abilities. Bioinformatics 30:1601–1608. https://doi.org/10.1093/bioinformatics/btu074
18. Fabre B, Lambour T, Bouyssie D et al (2014) Comparison of label-free quantification methods for the determination of protein complexes subunits stoichiometry. EuPA Open Proteom 4:82–86

Chapter 18

Alteration of Flavin Cofactor Homeostasis in Human Neuromuscular Pathologies

Maria Tolomeo, Alessia Nisco, and Maria Barile

Abstract

The aim of this short review chapter is to provide a brief summary of the relevance of riboflavin (Rf or vitamin B2) and its derived cofactors flavin mononucleotide (FMN) and flavin adenine dinucleotide (FAD) for human neuromuscular bioenergetics.

Therefore, as a completion of this book we would like to summarize what kind of human pathologies could derive from genetic disturbances of Rf transport, flavin cofactor synthesis and delivery to nascent apoflavoproteins, as well as by alteration of vitamin recycling during protein turnover.

Key words *SLC52As*, *FLAD1*, mitochondrial flavoproteome, riboflavin, RTD, MADD, LSMFLAD, RREI

1 Introduction

The scope of this chapter is to stress the relevance of riboflavin (Rf or vitamin B2) and its derived cofactors flavin mononucleotide (FMN) and flavin adenine dinucleotide (FAD) for human neuronal and muscular bioenergetics.

Flavin cofactors are mainly located in mitochondria where ensure the functionality of fundamental biochemical processes, that is, respiratory chain, Krebs cycle, β-oxidation of Acyl-CoA, and so on. Thus, it is not surprising that the first tissues affected by Rf homeostasis alterations are those with a higher bioenergetic demand, like muscles, heart, and nervous tissue [1].

Most human neuromuscular pathologies linked to enzymatic cofactor formation and delivery can be treated with pharmacological doses of vitamin B2, but the molecular reasons of patients' amelioration are still far to be elucidated. In some cases, however, Rf treatment has no effect and this forced us to better investigate about basic concerns in Rf homeostasis in the aim to design alternative therapeutic intervention.

Maria Barile (ed.), *Flavins and Flavoproteins: Methods and Protocols*, Methods in Molecular Biology, vol. 2280, https://doi.org/10.1007/978-1-0716-1286-6_18,

Here we distinguish between inborn errors of metabolism associated to defects in membrane transporters from that associated to cofactor synthesis. Pathologies associated to cofactor delivery necessitate further attention which is, however, out of the scope of this short survey [2, 3].

The reader is, anyway, referred to exhaustive and recent literature for a wider insight into this topic [1, 4–8].

2 Alterations of Flavin Transport

Rf is an essential component of human diet and therefore it must be taken daily from outside [8]. Rf absorption takes place mainly in small intestine and partly in large intestine [9] and it occurs via specific carrier-mediated processes, supported by three members of the solute carrier family SLC52A.

SLC52A1, located on chromosome 17, was the first gene identified which encodes for a plasma membrane Rf translocator, named Rf transporter 1 (RFVT1) a protein consisting of 448 aa [10]. Soon after *SLC52A2* and *SLC52A3* genes were also cloned and characterized, as reported in [11, 12], they encode for RFVT2 and RFVT3, respectively. These two genes are located on chromosomes 8 and 20, respectively. The corresponding sequences of RFVTs are reported in Fig. 1. *SLC52A2* encodes for a 445 aa protein and *SLC52A3* for a protein consisting of 469 aa. All the three RFVTs have been predicted to have 11 transmembrane domains [7, 13, 14]. However, we cannot exclude that human RFVTs might exist in different isoforms produced by the alternative splicing of each gene. The reader is referred, for this aspect to a recent review from our group [1].

In enterocytes, which allow for a transepithelial vectorial Rf movement to portal circulation, the vitamin is taken up by the action of RFVT3 at the apical membrane and it is released in blood by RFVT1 and RFVT2 located at the basolateral membrane [15]. Before crossing the intestinal barrier and entering both the circulating and the peripheral cells, Rf is converted into FMN and FAD via the sequential action of Riboflavin kinase (RFK, EC 2.7.1.26) and FAD synthase (FADS, EC 2.7.7.2) [16, 17]. Excess of vitamin intakes respect to intracellular cofactor requirements is excreted in the urine, as Rf or other metabolites [18]. Due to regulation of both absorption and excretion, the circulating Rf level is subjected to a small circadian variation [19].

Since RFVTs mediate translocation of Rf, rather than FMN or FAD deriving from digested protein food, the activities of extracellular diphospho- (EC 3.6.1._) and monophosphohydrolases (EC 3.1.3.2), are required on the brush-border membrane, before the vitamin is delivered into enterocytes. A comprehensive study on the identity of these "transport-preparing" hydrolases is still

```
hRFVT1      MAAPTLGRLVLTHLLVALFGMGSWAAVNGIWVELPVVVKDLPEGWSLPSYLSVVVALGNLGLLVVTLWRQLAPGKGEQVPIQVV-QVLSV
hRFVT2      MAAPTPARPVLTHLLVALFGMGSWAAVNGIWVELPVVVKELPEGWSLPSYVSVLVALGNLGLLVVTLWRRLAPGKDEQVPIRVV-QVLGM
hRFVT3      -------MAFLMHLLVCVFGMGSWVTINGLWVELPLLVMELPEGWYLPSYLTVVIQLANIGPLLVTLLHHFRPSCLSEVPIIFTLLGVGT
                   .* ****.:******.::**:*****::* :***** ****::*:: *.*:* *:*** ::: *.  .:*** ..   :.

hRFVT1      VGTALLAPLWHHVAPVAGQLHSVAFLTLALVLAMACCTSNVTFLPFLSHLPPPFLRSFFLGQGLSALLPCVLALVQGVGRLECPPAPTNG
hRFVT2      VGTALLASLWHHVAPVAGQLHSVAFLALAFVLALACCASNVTFLPFLSHLPPRFLRSFFLGQGLSALLPCVLALVQGVGRLECPPAPING
hRFVT3      VTCIIFAFLWNMTSWVLDGHHSIAFLVLTFFLALVDCTSSVTFLPFMSRLPTYYLTTFFVGEGLSGLLPALVALAQGSGLTTCVNVTEIS
            *   ::* **: .: * .  **:***.*::.**:. *:*.******:*:**  :* :**:*:***.***.::**.** *   *  .   .

hRFVT1      TS--------------GPPLDF-------------------PERFPASTFFWALTALLVTSAAAFRGLLLLLPSLPSVTTGGSGPELQLG
hRFVT2      TP--------------GPPLDF-------------------LERFPASTFFWALTALLVASAAAFQGLLLLLPPPPSVPTGELGSGLQVG
hRFVT3      DSVPSPVPTRETDIAQGVPRALVSALPGMEAPLSHLESRYLPAHFSPLVFFLLLSIMMACCLVAF---F-VLQRQPRCWEASVEDLLN--
                            * *  :                     :*   .**  *: ::. . .**   : :*   *    .    *:

hRFVT1      SPGAEEEEKEEEEALPLQEPPSQAAGTIPG------PDPEAHQLFSAHGAFLLGLMAFTSAVTNGVLPSVQSFSCLPYGRLAYHLAVVLG
hRFVT2      APGAEEEV---EESSPLQEPPSQAAGTTPG------PDPKAYQLLSARSACLLGLLAATNALTNGVLPAVQSFSCLPYGRLAYHLAVVLG
hRFVT3      ------DQVTLHSIRPREENDLGPAGTVDSSQGQGYLEEKAAPCCPAHLAFIYTLVAFVNALTNGMLPSVQTYSCLSYGPVAYHLAATLS
                  :    ..  * :*     ***  .       : :*     *: * :  *:* ..*:***:**:**::*** ** :*****..*.

hRFVT1      SAANPLACFLAMGVLCRSLAGLVGLSLLGMLFGAYLMALAILSPCPPLVGTTAGVVLVVLSWVLCLCVFSYVKVAASSLLHGGGRPALLA
hRFVT2      SAANPLACFLAMGVLCRSLAGLGGLSLLGVFCGGYLMALAVLSPCPPLVGTSAGVVLVVLSWVLCLGVFSYVKVAASSLLHGGGRPALLA
hRFVT3      IVANPLASLVSMFLPNRSLLFLGVLSVLGTCFGGYNMAMAVMSPCPLLQGHWGGEVLIVASWVLFSGCLSYVKVMLGVVLRDLSRSALLW
             .*****.:::* :   ***  *  **:**   *.* **:*::**** * *  .* **:* ****    :*****  . :*:. .* ***

hRFVT1      AGVAIQVGSLLGAGAMFPPTSIYHVFQSRKDCVDPCGP
hRFVT2      AGVAIQVGSLLGAVAMFPPTSIYHVFHSRKDCADPCDS
hRFVT3      CGAAVQLGSLLGALLMFPLVNVLRLFSSADFCNLHCPA
            .*.*:*:******  *** ..: ::* * . *   *
```

Fig. 1 Alignment of the hRFVT proteins performed by Clustal W. hRFVT1 and hRFVT2 proteins exhibit 87% amino acid identity with each other, and only 44% or 45% identity, respectively, with hRFVT3

missing, even if we think that extracellular hydrolysis of flavin cofactors occurs in the plasma membrane of each specialised cell of peripheral tissues [7, 20].

The three plasma membrane RFVTs have different tissue specific expression profiles, as well as functional and kinetical properties. Investigations on the tissue expression profile of SLC52A members at the transcriptional level revealed that RFVT1 is particularly abundant in small intestine and placenta; RFVT2 is ubiquitously expressed, but it is relevant especially for brain, while RFVT3 is expressed mostly in intestine and testis [12].

As far transport mechanisms and kinetic features of Rf uptake is concerned, all the three transporters were proven to be independent on extracellular Na^+ and Cl^- [12]. The pH sensitivity was also tested, indicating that the sole RFVT3-mediated uptake is pH-dependent [11, 12].

As far as RFVTs' inhibitors, besides the bulk of knowledge obtained with transfected and natural intact cells [10, 12, 21, 22], a piece of work was also made using the recombinant RFVT2 and the native proteins extracted from fibroblasts and reconstituted in proteoliposomes, as described in chapter 4 in this book (Console et al.) and [13]. This is the first example of a study with a recombinant human vitamin translocator.

Since their discovery, alteration of RFVTs have been linked to neuromuscular diseases (see paragraphs below); this is not

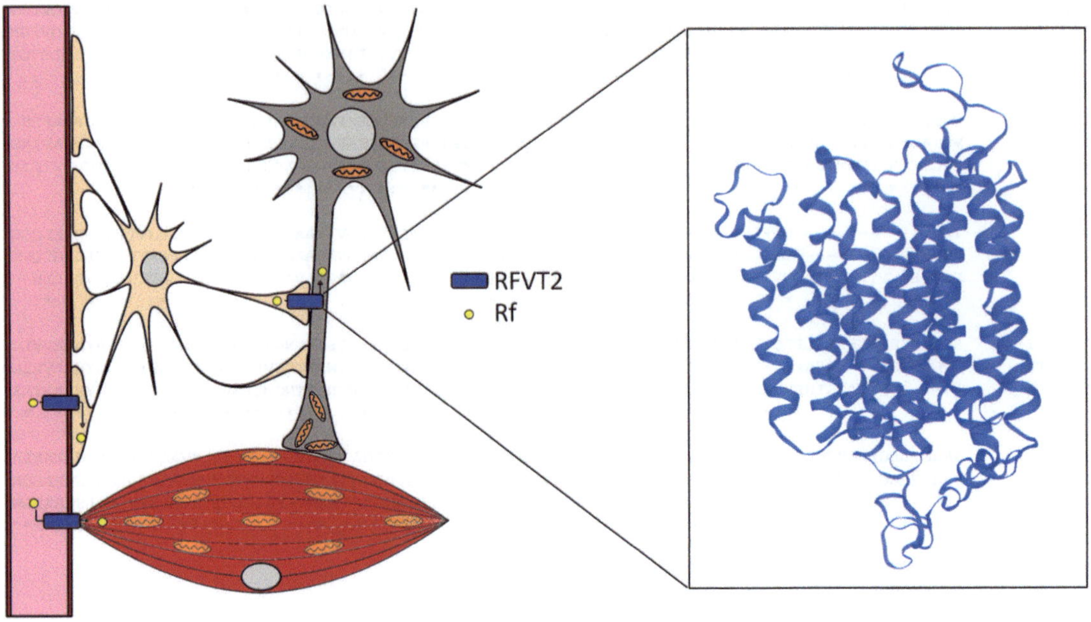

Fig. 2 Rf transport in nervous and in skeletal muscular tissues. At the left of the figure Rf delivery to neuronal and skeletal muscle cells via RFVT2. At the right of the figure a magnification of the transport protein constituted by 11 transmembrane α-helical segments. The homology structural model was built by SWISS-MODEL software using the Equilibrative nucleoside transporter 1 as a template

surprising on the light of the importance of vitamin B2 in flavoenzymes crucial for oxidative metabolism and therefore for nervous system economy. Thus, nervous tissue alterations/degeneration are expected as biochemical consequences of FAD forming deficiencies, as in [23].

Rf enters the brain from blood at the blood–brain barrier (BBB) via a saturable system and this process presumably involves RFVT2 [24, 25] (Fig. 2), but, as far as we know, a direct demonstration of the sole presence of this transporter is lacking. Then, different types of neurons might have different demand for Rf supply and therefore different expression profile of RFVTs, this is a biochemical problem that have never been addressed.

Another underinvestigated point, which necessitates further research, is the expression profile and the subcellular localization of Rf transporters in muscle. RFVT2 seems to be the principal Rf transporter expressed, as reported in "The human protein Atlas database" (https://www.proteinatlas.org/). A systematic study concerning this point in muscle of both control individuals and patients suffering for Rf-responsive myopathies is still missing (see also [1]).

A main question, still completely under investigated, concerns subcellular trafficking of Rf. The existence of an FAD recycling pathway, which implies a carrier-mediated process for Rf across the inner mitochondrial membrane, was demonstrated by our group in yeasts, plants, and mammalian liver and muscle mitochondria [26–29], but the molecular nature of the mitochondrial Rf translocator is still unknown. Conversely, interesting results exist concerning the mitochondrial FAD translocator: in humans, FAD transport across the mitochondrial membrane requires a specific inner membrane carrier encoded by *SLC25A32* gene, which is located on chromosome 8 and encodes for a protein of 315 aa. Initially identified as a mitochondrial folate transporter [30], it is the human ortholog of the yeast mitochondrial FAD transporter FLX1 [31, 32], which allows for cytosolically synthesized FAD to enter mitochondrial membrane, but also (depending on the genetic and metabolic contest) vice versa [33–35].

Among disorders which can be cured with high doses of Rf, correlated to alteration of Rf trafficking, the most recurrent are riboflavin transporter deficiencies (RTDs), characterized by neurological or transient muscular phenotypes, as described below.

Also, mutations in *SLC25A32* are causative of a severe neuromuscular disorder, responsive to Rf treatment [36, 37], named riboflavin responsive exercise intolerance (RREI) whose phenotype is described in further detail below.

The molecular basis of these Rf responsive defects and the molecular reason for their vitamin responsiveness are still not completely elucidated.

Neuromuscular disorders are not the sole pathological status for humans connected to alteration of flavin transport. Recently altered expression patterns of *SLC52As* and *SLC25A32* in different types of human cancer, involving different transcriptional and post-transcriptional mechanisms specific for each transporter, have been also reported [38–44].

2.1 RFVT1 Deficiency

Deficiencies of RFVT1, which is suggested to be responsible for placental transport of Rf [10], have been reported in two women (Table 1) as causing in their newborns biochemical abnormalities resembling that of another disorder, named MADD (multiple acyl-CoA dehydrogenase deficiency), a typical muscular lipid-storage disorder, characterized by β-oxidation impairments [4, 48].

The first case of RFVT1 deficiency has been reported by [45], which described a newborn female, presented on the first day of life with clinical and biochemical findings consistent with MADD. Clinical symptoms were completely corrected after Rf treatment, therefor it was described as a transient form of MADD (OMIM #615026). A possible defect in Rf transport or metabolism in the mother was suspected since she had mild Rf deficiency (erythrocyte glutathione reductase activity coefficient, EGRAC value below control range) and altered plasma acylcarnitine profile without any other symptoms or signs of Rf deficiency. After RFVT1 identification [10], the defects in Rf transport was confirmed by [46],

Table 1
Reported cases with RFVT1 deficiency

Patient	Sex	Ethnicity	Age at onset	Status	Phenotype	Rf treatment	Mutations	References
1	F	Australian	At birth	Alive at 5 y (time of report)	Soon after birth poor suck requiring tube feeding, hypoglycemia and metabolic acidosis.	100 mg/day on the third day of life corrected the biochemicals abnormalities.	Genetic defect of the mother: a heterozygous deletion in *SLC52A1* spanning exons 2 and 3 predicted to result in haploinsufficiency revealed by qPCR analysis. The infant did not carry the deletion.	[45, 46]
2	F	–	NBS diagnosis of MADD	Alive (time of report)	Lethargy, hypotonia, poor peripheral circulation, hypothermia, and metabolic lactic acidosis at 4 days of life.	100 mg two times a day together with other vitamins and carnitine supplementation corrected the clinical and biochemical abnormalities.	Heterozygous intronic variation, c.1134 +11G>A resulted in exon 4 skipping was detected both in the mother and the infant.	[47]

y years, *NBS* newborn screening

who identified a heterozygous genomic deletion in *SLC52A1* in the healthy mother, whereas the infant did not carry the deletion.

The second RFVT1-associated case of transient MADD [47] differs from the case previously reported by Chiong and colleagues since a heterozygous intronic variation in *SLC52A1* has been identified both in the mother and in the child, but only the infant showed biochemical features of Rf deficiency. Rf treatment in association with other vitamins and carnitine supplementation completely corrected the patient's phenotype.

A promising treatment of RFVT1 deficiency could be based on the use of resveratrol, a small polyphenol which has general beneficial effects on human health [49]; this molecule was proven to upregulate *SLC52A1* in human fibroblasts [50].

2.2 RFVT2 and 3 Deficiencies

Since 2010, *SLC52A3* mutations have been correlated to Brown–Vialetto–Van Laere syndrome (BVVLS), as demonstrated by studying a consanguineous family with multiple affected individuals [51]. BVVLS is an early-onset disease characterized by progressive loss of cranial neurons (resembling amyotrophic lateral sclerosis, ALS), degeneration of spinal cord neurons, and respiratory insufficiencies. This disease, now named riboflavin transporter deficiency 3 (RTD3, OMIM #211530), is considered as one with Fazio–Londe disease (OMIM #211500) [52], which differs from BVVLS only for the lack of hearing loss. Mutations in *SLC52A2* have also been associated to BVVLS; this condition is now named RTD2 (OMIM #614707) [53, 54]. Other common clinical features present in both RTD2 and RTD3 patients are dysarthria, weakness, and hypotonia; conversely the most common differences concern facial weakness, typical of RTD3 patients, and, on the other hand, vision loss, typical of RTD2 patients [6, 55].

Since RTD definition, a total of 109 RTD patients have been described, as recently and exhaustively reviewed in [6]; further nine novel cases have been reported in [56–64]. For further information visit the web site http://curertd.org/research/slc52variants/. All the pathogenic mutations in *SLC52A2* and *SLC52A3* are summarized in the scheme in Figs. 3 and 4.

From the biochemical point of view, RTDs are sometimes characterized by abnormal acylcarnitine profiles, resembling that of MADD (see above) [65]. Alteration of EGRAC without acylcarnitine abnormalities has been reported in a single RTD3 patient [66].

A significant reduction in the intracellular levels of FMN and FAD has been observed in RTD2 patient fibroblasts, if grown in low extracellular Rf conditions. Altered co-enzymatic pattern is coupled to impairments of the activities of mitochondrial electron transport chain complex I and complex II [67]. Consistently, in muscle biopsies of both types of RTD patients, mitochondrial respiratory chain deficiencies were revealed [66, 68, 69]. All these

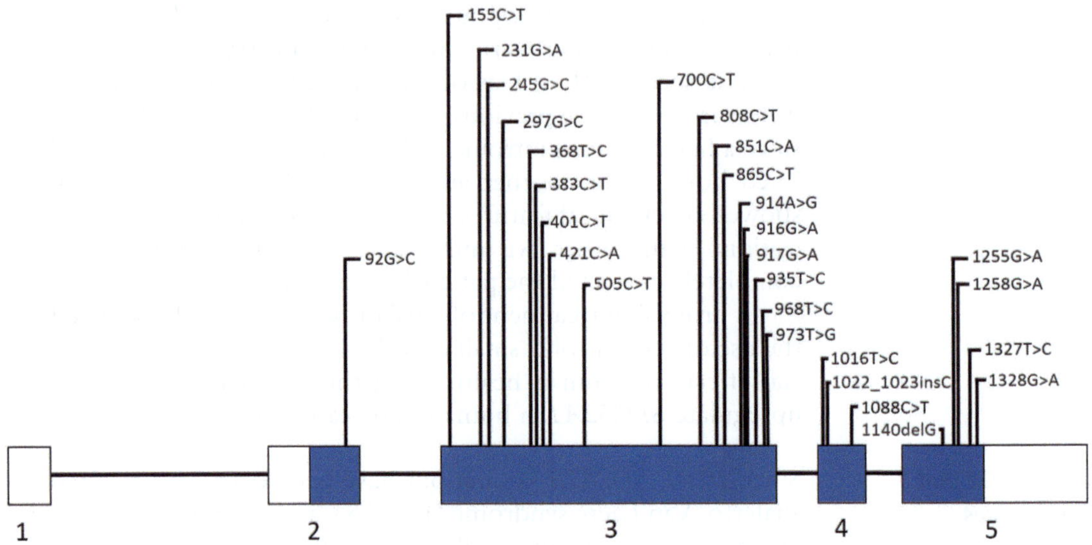

Fig. 3 Spectrum of gene variations in the *SLC52A2* gene in RTD2 patients. Schematic representation of *SLC52A2* with exon coding regions represented as blue boxes, UTRs as white boxes and introns as continuous horizontal lines between the boxes, all reported in scaled proportion. Vertical lines indicate the positions of reported pathogenic mutations

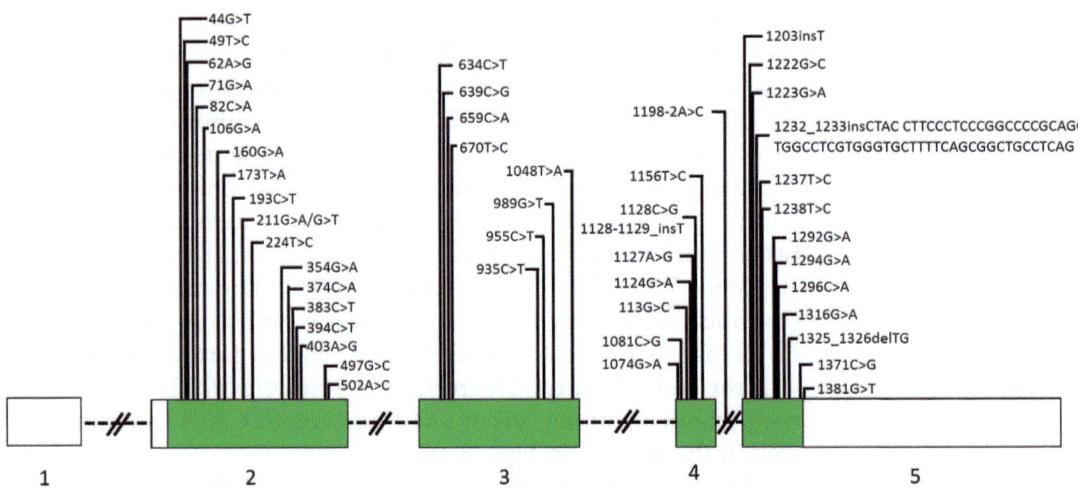

Fig. 4 Spectrum of gene variations in the *SLC52A3* gene in RTD3 patients. Schematic representation of *SLC52A3* with exon coding regions represented as green boxes, UTRs as white boxes and introns as horizontal thin dashed lines (reported with suitable interruptions) between the boxes. Vertical lines indicate the positions of reported pathogenic mutations

data are consistent, not only with neurological damages but also with a mitochondrial myopathy, which was confirmed by muscle histopathology revealing ragged-red fibers.

The question whether RFVT3 can be directly involved in neuronal and muscular supply of Rf or rather a secondary intestinal absorption deficiency causes the myopathy, is still an open question [65].

Several experimental models have been introduced to mimic these human pathologic conditions and to confirm the relationships between flavin supply and mitochondrial functionality. They have been described elsewhere [1].

As discussed before, the only existing cure at present is Rf supplied at high pharmacological doses (10–80 mg/kg/day). Rf therapy has been proven to be successful, especially, but not exclusively [57] in patients treated shortly after disease onset [6].

Very recently a new regimen of therapy was introduced in a patient [59]. Rather than administering Rf three times daily, in 500 mg capsules (the recommended treatment [70]), the dose was divided into 250 mg six times daily; plasma Rf levels remain more constant over time increasing the frequency of administration.

2.3 Riboflavin Responsive Exercise Intolerance or SLC25A32 Deficiency

Riboflavin responsive exercise intolerance (OMIM #616839) is a neuromuscular disorder caused by mutations in *SLC25A32* gene. Only two patients have been reported in the literature to date [36, 37], and their features are summarized in Table 2.

The first patient, described in [36], is a 14-year-old girl harboring two heterozygous variants, who showed abnormalities in the acylcarnitine profile typical of MADD. Faint succinate dehydrogenase staining together with some ragged-red fibers and decreased activities of mitochondrial FAD-dependent succinate dehydrogenase (SDH) and glycerol-3-phosphate dehydrogenase (G3PDH) enzymes have been observed in *SLC25A32*-deficient patient's skeletal-muscle biopsy and fibroblasts, respectively. Clinical and biological abnormalities were dramatically improved following oral supplementation with Rf.

The second patient, described in [37], is a 51-year-old man, also showing biochemical features typical of MADD; this derives from a novel homozygous variant most likely resulting in a complete loss of the mitochondrial protein and consequently in a more severe neuromuscular phenotype compared to patient 1. Since the translocator cannot be synthesized at all in this patient, this suggests that an alternative way of FAD supply in mitochondria, maybe via the mitochondrial isoform of FAD synthase, can operate in this patient.

Biochemical assays of complex I, complex II + III, complex IV and citrate synthetase activities were performed in muscle biopsy of this patient. A reduction of complex II and combined II + III activities compared to control muscle indicated an OXPHOS complex II deficiency. Staining of muscle tissue revealed also for this patient the presence of many ragged-red fibers, as well as multiple cytochrome oxidase-negative muscle fibers. In addition, decreased basal respiration levels and ATP production were observed in patient's fibroblasts. Oral Rf supplementation improved its clinical condition increasing his exercise endurance, presumably because FAD synthesis is increased.

Table 2
Reported cases with *SLC25A32* deficiency

Patient	Sex	Ethnicity	Age at onset	Status	Phenotype	Rf treatment	Mutations	References
1	F	–	Late-onset	Alive at 14 y (time of report)	Recurrent exercise intolerance.	Oral Rf supplementation improved the phenotype.	Compound heterozygous c.425G>A; p.(Trp142*) and c.440G>A p.(Arg147His).	[36]
2	M	Dutch	3 y	Alive at 52 y (time of report)	Muscle weakness first noted at 3 y. Impaired motor development and exercise Intolerance during the childhood. Progressive dysarthria at the age of 20 y, with swallowing difficulties in the last 10 y. Occasionally myoclonic jerks. Now he is in a wheelchair, although not continuously.	Oral Rf supplementation (10 mg three times a day) improved exercise endurance.	Homozygous c.-264_31delinsCTCA CAAATGCTCA variant which deleted the start codon.	[37]

indel insertion/deletion, *y* years

3 Alterations of FAD Synthesis

Following Rf transport, metabolic conversion of the vitamin into enzymatic cofactors is necessary to ensure the biogenesis of hundreds of different flavoenzymes, mainly distributed in mitochondria.

The sequence of intracellular FAD forming reactions and the identity of the enzymes involved are presented in better detail in chapters 6 and 7 of this Book (Leone et al.). Briefly, phosphorylation of Rf to FMN is catalyzed by Riboflavin Kinase (RFK, ATP: riboflavin 5′ phosphotransferase, EC 2.7.1.26), which transfers a phosphoryl group from ATP to Rf to form FMN. The second step of the biosynthetic pathway is catalyzed by FAD synthase (FADS or FMNAT, ATP: FMN adenylyl-transferase, EC 2.7.7.2), which adenylates FMN to FAD.

Human FAD synthase exists in different isoforms with different subcellular localization; they are produced by the alternative splicing of *FLAD1* gene, located on chromosome 1 [7, 8, 17, 71]. Initially our group [17, 72] identified and characterized two protein isoforms as product of *FLAD1* gene (namely, hFADS1 and hFADS2). hFADS1, encoded by the seven exons long transcript variant 1 (GenBank Accession No. NM_025207.5), is a 587 aa protein with a predicted molecular mass of 65.3 kDa; the first 17 residues of this isoform represent a putative mitochondrial targeting peptide [73]; hFADS2 is a 490 aa protein with a predicted molecular mass of 54.2 kDa, which lacks a 97-mer respect to the N-terminal region of hFADS1 and it performs a cytosolic localization. It is the product of transcript variant 2 (GenBank Accession No. NM_201398.3), the most abundant in tissues/cells tested so far, which derives from interruption of exon 1 by an additional intron [73].

Both hFADS1 and 2 contain at their N-terminus a molybdopterin binding (MPTb), recently renamed FADHy [74] domain, performing an FAD hydrolase activity [74, 75], which is fused with a C-terminal 3-phosphoadenosine 5-phosphosulfate reductase domain (PAPS, recently renamed FADSy [74]), which per se catalyzes FMN conversion to FAD [76–78]. As described elsewhere in this Book, hFADS2 has been over-produced and purified in its catalytically active form: the protein contains a couple of redox-sensitive cysteines, which make hFADS2 a putative redox-sensor [79], which takes part in cofactor delivery to the client apoflavoenzyme during holoenzyme biogenesis, operating as an FAD "chaperone" in a flavinylation machinery [80, 81]. Because of its dinucleotide-hydrolyzing properties, hFADS2 [74] creates a link between the flavin and the NAD world [8, 82]. Of course, the two opposite processes, that is, FAD synthesis and hydrolysis, both performed by the same polypeptide (hFADS2), can constitute a "futile cycle," which must be controlled. At the moment, other two RefSeq transcript variants are reported in Entrez Gene database.

The transcript variant 3 (GenBank accession No. NM_00114891.2) differs in 3′ end compared to variant 2 resulting in a 446 aa protein (hFADS3) with a shorter C-terminus. It is very similar to the hFADS2; therefore, it is presumably able to perform both FAD synthesis and FAD hydrolysis. The transcript variant 4 (GenBank accession No. NM_001184892.2) has multiple differences and initiates translation at an alternative start codon, compared to variant 2. The protein product of 294 aa (hFADS4) is shorter, it has distinct N- and C-termini when it is compared to hFADS2 and contains only the FADHy domain.

In 2016 [71] a novel transcript variant, not yet annotated, was revealed by transcriptomic analysis in some patients. The novel transcript encodes for a protein, namely, hFADS6, corresponding to the sole C-terminus domain of the longest isoforms [71]. The protein was produced as a recombinant protein in its FADSy active form, missing as predicted the FAD hydrolase activity [76]. Why a cell necessitates so many isoforms of FAD-forming enzymes is still matter of speculation. The most plausible hypothesis is that they are characteristic of different subcellular compartments.

The alignment of FADS isoforms described is reported in Fig. 5.

```
hFADS1   MGWDLGTRLFQRQEQRSRLSRIWLEKTRVFLEGSTRTPALPHCLFWLLQVPSTQDPLFPGYGPQCPVDLAGPPCLRPLFGGLGGYWRALQRGREGRTMTSRASELSPGRSVTAGIIIVGD
hFADS2   -------------------------------------------------------------------------------------------------MTSRASELSPGRSVTAGIIIVGD
hFADS3   -------------------------------------------------------------------------------------------------MTSRASELSPGRSVTAGIIIVGD
hFADS4   -------------------------------------------------------------------------------------------------MQPSSSTPPLHPYSTDGLIF--P
hAFDS6   ------------------------------------------------------------------------------------------------------------------------

hFADS1   EILKGHTQDTNTFFLCRTLRSLGVQVCRVSVVPDEVATIAAEVTSFSNRFTHVLTAGGIGPTHDDVTFEAVAQAFGDELKPHPKLEAATKALGGEGWEKLSLVPSSARLHYGTDPCTGQP
hFADS2   EILKGHTQDTNTFFLCRTLRSLGVQVCRVSVVPDEVATIAAEVTSFSNRFTHVLTAGGIGPTHDDVTFEAVAQAFGDELKPHPKLEAATKALGGEGWEKLSLVPSSARLHYGTDPCTGQP
hFADS3   EILKGHTQDTNTFFLCRTLRSLGVQVCRVSVVPDEVATIAAEVTSFSNRFTHVLTAGGIGPTHDDVTFEAVAQAFGDELKPHPKLEAATKALGGEGWEKLSLVPSSARLHYGTDPCTGQP
hFADS4   FNPQGHTQDTNTFFLCRTLRSLGVQVCRVSVVPDEVATIAAEVTSFSNRFTHVLTAGGIGPTHDDVTFEAVAQAFGDELKPHPKLEAATKALGGEGWEKLSLVPSSARLHYGTDPCTGQP
hAFDS6   ------------------------------------------------------------------------------------------------------------------------

hFADS1   FRFPLVSVRNVYLFPGIPELLRRVLEGMKGLFQNPAVQFHSKELYVAADEASIAPILAEAQAHFGRRLGLGSYPDWGSNYYQVKLTLDSEEEGPLEECLAYLTARLPQGSLVPYMPNAVE
hFADS2   FRFPLVSVRNVYLFPGIPELLRRVLEGMKGLFQNPAVQFHSKELYVAADEASIAPILAEAQAHFGRRLGLGSYPDWGSNYYQVKLTLDSEEEGPLEECLAYLTARLPQGSLVPYMPNAVE
hFADS3   FRFPLVSVRNVYLFPGIPELLRRVLEGMKGLFQNPAVQFHSKELYVAADEASIAPILAEAQAHFGRRLGLGSYPDWGSNYYQVKLTLDSEEEGPLEECLAYLTARLPQGSLVPYMPNAVE
hFADS4   FRFPLVSVRNVYLFPGIPELLRRVLEGMKGLFQNPAVQFHSKELYVAADEASIAPILAEAQAHFGRRLGLGSYPDWGSNYYQVKLTLDSEEEGPLEECLAYLTARLPQGSLVPYMPNAVE
hAFDS6   ---------------------------MKGLFQNPAVQFHSKELYVAADEASIAPILAEAQAHFGRRLGLGSYPDWGSNYYQVKLTLDSEEEGPLEECLAYLTARLPQGSLVPYMPNAVE
                                    *********************************************************************************************

hFADS1   QASEAVYKLAESGSSLGKKVAGALQTIETSLAQYSLTQLCVGFNGGKDCTALLHLFHAAVQRKLPDVPNPLQILYIRSISPFPELEQFLQDTIKRYNLQMLEAEGSMKQALGELQARHPQ
hFADS2   QASEAVYKLAESGSSLGKKVAGALQTIETSLAQYSLTQLCVGFNGGKDCTALLHLFHAAVQRKLPDVPNPLQILYIRSISPFPELEQFLQDTIKRYNLQMLEAEGSMKQALGELQARHPQ
hFADS3   QASEAVYKLAESGSSLGKKVAGALQTIETSLAQYSLTQLCVGFNGGKDCTALLHLFHAAVQRKLPDVPNPLQILYIRSISPFPELEQFLQDTIKRYNLQMLEAEGSMKQALGELQARHPQ
hFADS4   QASEAVYKLAESGRDLMEEGHYAQSHW----------------------------WHPRSQ-----------------------------------------------------------
hAFDS6   QASEAVYKLAESGSSLGKKVAGALQTIETSLAQYSLTQLCVGFNGGKDCTALLHLFHAAVQRKLPDVPNPLQILYIRSISPFPELEQFLQDTIKRYNLQMLEAEGSMKQALGELQARHPQ
         ************* .* ::   * .                              :*   *

hFADS1   LEAVLMGTRRTDPYSCSLCPFSPTDPGWPAFMRINPLLDWTYRDIWDFLRQLFVPYCILYDRGYTSLGSRENTVRNPALKCLSPGGHPTYRPAYLLENEEEERNSRT
hFADS2   LEAVLMGTRRTDPYSCSLCPFSPTDPGWPAFMRINPLLDWTYRDIWDFLRQLFVPYCILYDRGYTSLGSRENTVRNPALKCLSPGGHPTYRPAYLLENEEEERNSRT
hFADS3   LEAVLMGTRRTDPYSCSLCPFSPTDPGWPAFMRINPLLDWTYRDIWDFLRQLFVPYCILYDRG--------------------------------------------
hFADS4   -----------------------------------------------------------------------------------------------------------
hAFDS6   LEAVLMGTRRTDPYSCSLCPFSPTDPGWPAFMRINPLLDWTYRDIWDFLRQLFVPYCILYDRGYTSLGSRENTVRNPALKCLSPGGHPTYRPAYLLENEEEERNSRT
```

Fig. 5 Alignment of the hFADS isoforms performed by Clustal W. The FADSy (or PAPS, IPR002500) domain present in hFADS1, 2, 3, and 6 at the C-terminal is colored in orange. The FADHy (or MPTb, IPR001453) domain present in hFADS1, 2, 3, and 4 at the N-terminal is colored in blue

3.1 Lipid Storage Myopathy Due to FAD Synthase Deficiency

The relevance of FADS in human neuromuscular disorders was directly demonstrated for the first time in 2016, when *FLAD1* was identified as a novel disease gene in nine patients from seven unrelated families [71]. The affected individuals showed a metabolic myopathy with lipid storage and respiratory-chain deficiency, elevated multiple acylcarnitines and urine organic acids, resembling what is seen in MADD patients. This novel inborn error of metabolism—resulting in Rf responsive and not responsive mitochondrial myopathy—has now named LSMFLAD (lipid storage myopathy due to flavin adenine dinucleotide synthase deficiency, OMIM #255100).

The affected individuals can be distinguished by their symptoms and the identified *FLAD1* variants. Their phenotypic and biochemical characteristics are described in [71]. Subjects carrying missense mutations in the FADSy domain showed a milder clinical course and a remarkable response to Rf, conversely, subjects carrying predicted LOF variants, affecting the FADHy domain, had an early onset and lethal disease. Since *FLAD1* is the sole gene encoding for an enzyme able to perform FAD synthesis, the residual FADS activity in individuals with biallelic frameshift variants in exon 2 was unexpected and led to discover a novel *FLAD1* isoform encoding for a protein named hFADS6 or "emergency protein," which ensures patients' cells not to be completely deprived from FAD and therefore still alive [76].

As a consequence of a reduced FAD synthase enzymatic activity in fibroblasts from these patients, a significant reduction of flavin cofactors was detectable at the mitochondrial level leading to a reduced amount of mitochondrial flavoenzymes ETFQO (Electron Transfer Flavoprotein-ubiquinone Oxidoreductase) and flavoprotein subunit of succinate dehydrogenase (SDHA), whereas only a slight reduction of flavin cofactors at the cellular level was observed probably because of the action of hFADS6, ensuring cofactor synthesis in cytosol [76].

In Table 3, we report Rf responsiveness and phenotypical characteristics of six additional patients discovered thereafter, with heterogeneous pathological *FLAD1* variants.

In fibroblasts of only one among novel patients (here indicated as patient 3) the flavin content and FAD synthase activity were measured; both, unexpectedly, were reduced [84]. The presence of isoform six was confirmed; thus, the therapeutic role of an early treatment with Rf could be explained based on increased amount of FAD available for nascent apoflavoprotein.

In skeletal muscle biopsies of patient 1, besides a lipid storage caused by the derangement of fatty acyl-CoA dehydrogenases, a global decrease of the histochemical staining of SDH was described. Ragged-red fibers have not been reported, so far [83].

The reader is also referred to contemporary comprehensive reviews describing some of these patients [4, 5].

Table 3
Reported cases with FAD synthase deficiency

Patient	Sex	Ethnicity	Age at onset	Status	Phenotype	RR	Mutations	FADS domain affected	References
1	M	Turkish	2 m	Died at 6 m	Progressive weakness and decreased movements, poor suck, tube feeding, persistent vomiting, respiratory insufficiency requiring invasive ventilation from 2.5 to 5 m (extubated after treatment with Rf), died of aspiration pneumonia.	No	Homozygous frameshift deletion c.401_404delTTCT; p.(Phe134Cysfs*8)	FADHy	[83]
2	F	Turkish	2 m	Died at 5 m	Weak, floppy, no head control, distinct hoarse cry, poor suck, swallowing difficulties, tube feeding, vomiting, decreased movements.	No	Homozygous frameshift deletion c.401_404delTTCT; p.(Phe134Cysfs*8)	FADHy	[83]
3	M	Palestinian	NBS diagnosis of MADD	Alive at 8 y	Myopathic facies and nasal speech first noted at 3 y, after cessation of oral Rf. Severe velopharyngeal insufficiency with speech difficulty at 6 y. Evolving myopathy with fatigue on chewing and worsening exercise intolerance at 8 y.	Yes	Homozygous nonsense variant c.745C>T; p.(Arg249*)	FADHy	[84]
4	F	–	6 m	Alive at 7 y	Psychomotor delay associated with loss of gaze fixation and erratic eye movements, hypotonia, low reaction to sound stimuli and reduced ability to object manipulation at 6 m. Scoliosis at 12 m and electromyography confirmed myopathy at 4 y of age.	No	Compound heterozygous c.797_798delAGinsT; p.(Glu266Valfs*3) and c.1555-3C>G; p.(?)	FADHy, FADSy	[85]

5	M	Japanese	ENBS diagnosis of MADD	Alive	Dyspnea with stridor and paralysis of the vocal cords at 3 m. Hypoxic–ischemic encephalopathy due to fulminant respiratory failure associated with aspiration pneumonia and vocal cord paralysis at 4 months of age.	No	Homozygous nonsense variant c.745C>T; p.(Arg249*)	FADHy	[86]
6	F	–	NBS diagnosis of MADD	Alive	Asymptomatic with no clinical decompensation during the first y of life.	NT	Compound heterozygous c.442C>T; p.(Arg148*) and c.1588C>T; p.(Arg530Cys)	FADHy, FADSy	[87]

indel insertion/deletion, *m* months, *y* years, *NSB* newborn screening, *ENBS* expanded newborn screening, *RR* Rf responsivity, *NT* not tested, (?) uncertain protein product

4 Future Perspectives

In conclusion, it is well assessed that a fine coordination among Rf supply, flavin cofactor homeostasis and apoflavoproteome maintenance is necessary for efficient cellular bioenergetics. Disturbances of this coordination due to several inborn errors of metabolism principally affect high-energy requiring tissues, altering muscular and neuronal flavin-dependent bioenergetic pathways.

Further experimental work is still needed to address a relevant number of molecular components and processes, not yet elucidated, which are involved in this coordination.

This is pivotal not only for a better understanding of the molecular rationale of Rf therapy in responsive patients, but, hopefully, also in the aim to potentiate possible therapeutic intervention in Rf-related neuromuscular diseases, in cases, unfortunately, not responding to vitamin therapy.

Acknowledgments

This work was funded by a grant from Cure RTD (Year 2019) http://curertd.org/news/new/ "Alterations of Rf transport and metabolism in Brown-Vialetto-Van-Laere Syndrome (BVVLS)" (to M.B.).

References

1. Tolomeo M, Nisco A, Leone P, Barile M (2020) Development of novel experimental models to study flavoproteome alterations in human neuromuscular diseases: the effect of Rf therapy. Int J Mol Sci 21(15):5310. https://doi.org/10.3390/ijms21155310
2. Rutter J, Winge DR, Schiffman JD (2010) Succinate dehydrogenase - assembly, regulation and role in human disease. Mitochondrion 10(4):393–401. https://doi.org/10.1016/j.mito.2010.03.001
3. Maklashina E, Rajagukguk S, Iverson TM, Cecchini G (2018) The unassembled flavoprotein subunits of human and bacterial complex II have impaired catalytic activity and generate only minor amounts of ROS. J Biol Chem 293(20):7754–7765. https://doi.org/10.1074/jbc.RA118.001977
4. Mosegaard S, Dipace G, Bross P, Carlsen J, Gregersen N, Olsen RKJ (2020) Riboflavin deficiency-implications for general human health and inborn errors of metabolism. Int J Mol Sci 21(11):3847. https://doi.org/10.3390/ijms21113847
5. Balasubramaniam S, Christodoulou J, Rahman S (2019) Disorders of riboflavin metabolism. J Inherit Metab Dis 42:608. https://doi.org/10.1002/jimd.12058
6. O'Callaghan B, Bosch AM, Houlden H (2019) An update on the genetics, clinical presentation, and pathomechanisms of human riboflavin transporter deficiency. J Inherit Metab Dis 42:598. https://doi.org/10.1002/jimd.12053
7. Barile M, Giancaspero TA, Leone P, Galluccio M, Indiveri C (2016) Riboflavin transport and metabolism in humans. J Inherit Metab Dis 39(4):545–557. https://doi.org/10.1007/s10545-016-9950-0
8. Barile M, Giancaspero TA, Brizio C, Panebianco C, Indiveri C, Galluccio M, Vergani L, Eberini I, Gianazza E (2013) Biosynthesis of flavin cofactors in man: implications in health and disease. Curr Pharm Des 19(14):2649–2675
9. Ross NS, Hansen TP (1992) Riboflavin deficiency is associated with selective preservation

of critical flavoenzyme-dependent metabolic pathways. Biofactors 3(3):185–190

10. Yonezawa A, Masuda S, Katsura T, Inui K (2008) Identification and functional characterization of a novel human and rat riboflavin transporter, RFT1. Am J Physiol Cell Physiol 295(3):C632–C641. https://doi.org/10.1152/ajpcell.00019.2008
11. Yamamoto S, Inoue K, Ohta KY, Fukatsu R, Maeda JY, Yoshida Y, Yuasa H (2009) Identification and functional characterization of rat riboflavin transporter 2. J Biochem 145 (4):437–443. https://doi.org/10.1093/jb/mvn181
12. Yao Y, Yonezawa A, Yoshimatsu H, Masuda S, Katsura T, Inui K (2010) Identification and comparative functional characterization of a new human riboflavin transporter hRFT3 expressed in the brain. J Nutr 140 (7):1220–1226. https://doi.org/10.3945/jn.110.122911
13. Console L, Tolomeo M, Colella M, Barile M, Indiveri C (2019) Reconstitution in proteoliposomes of the recombinant human riboflavin transporter 2 (SLC52A2) overexpressed in E. coli. Int J Mol Sci 20(18):4416. https://doi.org/10.3390/ijms20184416
14. Colon-Moran W, Argaw T, Wilson CA (2017) Three cysteine residues of SLC52A1, a receptor for the porcine endogenous retrovirus-A (PERV-A), play a critical role in cell surface expression and infectivity. Virology 507:140–150. https://doi.org/10.1016/j.virol.2017.04.019
15. Subramanian VS, Subramanya SB, Rapp L, Marchant JS, Ma TY, Said HM (2011) Differential expression of human riboflavin transporters -1, -2, and -3 in polarized epithelia: a key role for hRFT-2 in intestinal riboflavin uptake. Biochim Biophys Acta 1808(12):3016–3021. https://doi.org/10.1016/j.bbamem.2011.08.004
16. Karthikeyan S, Zhou Q, Mseeh F, Grishin NV, Osterman AL, Zhang H (2003) Crystal structure of human riboflavin kinase reveals a beta barrel fold and a novel active site arch. Structure 11(3):265–273. https://doi.org/10.1016/s0969-2126(03)00024-8
17. Brizio C, Galluccio M, Wait R, Torchetti EM, Bafunno V, Accardi R, Gianazza E, Indiveri C, Barile M (2006) Over-expression in Escherichia coli and characterization of two recombinant isoforms of human FAD synthetase. Biochem Biophys Res Commun 344 (3):1008–1016. https://doi.org/10.1016/j.bbrc.2006.04.003
18. Chastain JL, McCormick DB (1987) Flavin catabolites: identification and quantitation in human urine. Am J Clin Nutr 46 (5):830–834. https://doi.org/10.1093/ajcn/46.5.830
19. Zempleni J, Galloway JR, McCormick DB (1996) Pharmacokinetics of orally and intravenously administered riboflavin in healthy humans. Am J Clin Nutr 63(1):54–66. https://doi.org/10.1093/ajcn/63.1.54
20. McCormick DB (1989) Two interconnected B vitamins: riboflavin and pyridoxine. Physiol Rev 69(4):1170–1198. https://doi.org/10.1152/physrev.1989.69.4.1170
21. Said HM, Ortiz A, Moyer MP, Yanagawa N (2000) Riboflavin uptake by human-derived colonic epithelial NCM460 cells. Am J Physiol Cell Physiol 278(2):C270–C276. https://doi.org/10.1152/ajpcell.2000.278.2.C270
22. Fujimura M, Yamamoto S, Murata T, Yasujima T, Inoue K, Ohta KY, Yuasa H (2010) Functional characteristics of the human ortholog of riboflavin transporter 2 and riboflavin-responsive expression of its rat ortholog in the small intestine indicate its involvement in riboflavin absorption. J Nutr 140(10):1722–1727. https://doi.org/10.3945/jn.110.128330
23. Udhayabanu T, Manole A, Rajeshwari M, Varalakshmi P, Houlden H, Ashokkumar B (2017) Riboflavin responsive mitochondrial dysfunction in neurodegenerative diseases. J Clin Med 6(5):52. https://doi.org/10.3390/jcm6050052
24. Spector R (1980) Riboflavin homeostasis in the central nervous system. J Neurochem 35 (1):202–209. https://doi.org/10.1111/j.1471-4159.1980.tb12507.x
25. Patel M, Vadlapatla RK, Pal D, Mitra AK (2012) Molecular and functional characterization of riboflavin specific transport system in rat brain capillary endothelial cells. Brain Res 1468:1–10. https://doi.org/10.1016/j.brainres.2012.05.052
26. Barile M, Brizio C, Valenti D, De Virgilio C, Passarella S (2000) The riboflavin/FAD cycle in rat liver mitochondria. Eur J Biochem 267 (15):4888–4900
27. Giancaspero TA, Locato V, de Pinto MC, De Gara L, Barile M (2009) The occurrence of riboflavin kinase and FAD synthetase ensures FAD synthesis in tobacco mitochondria and maintenance of cellular redox status. FEBS J 276(1):219–231. https://doi.org/10.1111/j.1742-4658.2008.06775.x

28. Vergani L, Barile M, Angelini C, Burlina AB, Nijtmans L, Freda MP, Brizio C, Zerbetto E, Dabbeni-Sala F (1999) Riboflavin therapy. Biochemical heterogeneity in two adult lipid storage myopathies. Brain 122(Pt 12):2401–2411
29. Giancaspero TA, Locato V, Barile M (2013) A regulatory role of NAD redox status on flavin cofactor homeostasis in S. cerevisiae mitochondria. Oxid Med Cell Longev 2013:612784. https://doi.org/10.1155/2013/612784
30. Titus SA, Moran RG (2000) Retrovirally mediated complementation of the glyB phenotype. Cloning of a human gene encoding the carrier for entry of folates into mitochondria. J Biol Chem 275(47):36811–36817. https://doi.org/10.1074/jbc.M005163200
31. Tzagoloff A, Jang J, Glerum DM, Wu M (1996) FLX1 codes for a carrier protein involved in maintaining a proper balance of flavin nucleotides in yeast mitochondria. J Biol Chem 271(13):7392–7397. https://doi.org/10.1074/jbc.271.13.7392
32. Spaan AN, Ijlst L, van Roermund CW, Wijburg FA, Wanders RJ, Waterham HR (2005) Identification of the human mitochondrial FAD transporter and its potential role in multiple acyl-CoA dehydrogenase deficiency. Mol Genet Metab 86(4):441–447. https://doi.org/10.1016/j.ymgme.2005.07.014
33. Bafunno V, Giancaspero TA, Brizio C, Bufano D, Passarella S, Boles E, Barile M (2004) Riboflavin uptake and FAD synthesis in Saccharomyces cerevisiae mitochondria: involvement of the Flx1p carrier in FAD export. J Biol Chem 279(1):95–102. https://doi.org/10.1074/jbc.M308230200
34. Giancaspero TA, Wait R, Boles E, Barile M (2008) Succinate dehydrogenase flavoprotein subunit expression in Saccharomyces cerevisiae--involvement of the mitochondrial FAD transporter, Flx1p. FEBS J 275 (6):1103–1117. https://doi.org/10.1111/j.1742-4658.2008.06270.x
35. Giancaspero TA, Dipalo E, Miccolis A, Boles E, Caselle M, Barile M (2014) Alteration of ROS homeostasis and decreased lifespan in S. cerevisiae elicited by deletion of the mitochondrial translocator FLX1. Biomed Res Int 2014:101286. https://doi.org/10.1155/2014/101286
36. Schiff M, Veauville-Merllie A, Su CH, Tzagoloff A, Rak M, Ogier de Baulny H, Boutron A, Smedts-Walters H, Romero NB, Rigal O, Rustin P, Vianey-Saban C, Acquaviva-Bourdain C (2016) SLC25A32 mutations and riboflavin-responsive exercise intolerance. N Engl J Med 374(8):795–797. https://doi.org/10.1056/NEJMc1513610
37. Hellebrekers D, Sallevelt S, Theunissen TEJ, Hendrickx ATM, Gottschalk RW, Hoeijmakers JGJ, Habets DD, Bierau J, Schoonderwoerd KG, Smeets HJM (2017) Novel SLC25A32 mutation in a patient with a severe neuromuscular phenotype. Eur J Hum Genet 25 (7):886–888. https://doi.org/10.1038/ejhg.2017.62
38. Aili A, Hasim A, Kelimu A, Guo X, Mamtimin B, Abudula A, Upur H (2013) Association of the plasma and tissue riboflavin levels with C20orf54 expression in cervical lesions and its relationship to HPV16 infection. PLoS One 8(11):e79937. https://doi.org/10.1371/journal.pone.0079937
39. Tutino V, Defrancesco ML, Tolomeo M, De Nunzio V, Lorusso D, Paleni D, Caruso MG, Notarnicola M, Barile M (2018) The expression of riboflavin transporters in human colorectal cancer. Anticancer Res 38 (5):2659–2667. https://doi.org/10.21873/anticanres.12508
40. Eli M, Li DS, Zhang WW, Kong B, Du CS, Wumar M, Mamtimin B, Sheyhidin I, Hasim A (2012) Decreased blood riboflavin levels are correlated with defective expression of RFT2 gene in gastric cancer. World J Gastroenterol 18(24):3112–3118. https://doi.org/10.3748/wjg.v18.i24.3112
41. Long L, Pang XX, Lei F, Zhang JS, Wang W, Liao LD, Xu XE, He JZ, Wu JY, Wu ZY, Wang LD, Lin DC, Li EM, Xu LY (2018) SLC52A3 expression is activated by NF-kappaB p65/Rel-B and serves as a prognostic biomarker in esophageal cancer. Cell Mol Life Sci 75(14):2643–2661. https://doi.org/10.1007/s00018-018-2757-4
42. Santoro V, Kovalenko I, Vriens K, Christen S, Bernthaler A, Haegebarth A, Fendt SM, Christian S (2020) SLC25A32 sustains cancer cell proliferation by regulating flavin adenine nucleotide (FAD) metabolism. Oncotarget 11 (8):801–812. https://doi.org/10.18632/oncotarget.27486
43. Fu T, Liu Y, Wang Q, Sun Z, Di H, Fan W, Liu M, Wang J (2016) Overexpression of riboflavin transporter 2 contributes toward progression and invasion of glioma. Neuroreport 27(15):1167–1173. https://doi.org/10.1097/WNR.0000000000000674
44. Jiang XR, Yu XY, Fan JH, Guo L, Zhu C, Jiang W, Lu SH (2014) RFT2 is overexpressed in esophageal squamous cell carcinoma and promotes tumorigenesis by sustaining cell proliferation and protecting against cell death. Cancer Lett 353(1):78–86. https://doi.org/10.1016/j.canlet.2014.07.013

45. Chiong MA, Sim KG, Carpenter K, Rhead W, Ho G, Olsen RK, Christodoulou J (2007) Transient multiple acyl-CoA dehydrogenation deficiency in a newborn female caused by maternal riboflavin deficiency. Mol Genet Metab 92(1-2):109–114. https://doi.org/10.1016/j.ymgme.2007.06.017
46. Ho G, Yonezawa A, Masuda S, Inui K, Sim KG, Carpenter K, Olsen RK, Mitchell JJ, Rhead WJ, Peters G, Christodoulou J (2011) Maternal riboflavin deficiency, resulting in transient neonatal-onset glutaric aciduria Type 2, is caused by a microdeletion in the riboflavin transporter gene GPR172B. Hum Mutat 32(1):E1976–E1984. https://doi.org/10.1002/humu.21399
47. Mosegaard S, Bruun GH, Flyvbjerg KF, Bliksrud YT, Gregersen N, Dembic M, Annexstad E, Tangeraas T, Olsen RKJ, Andresen BS (2017) An intronic variation in SLC52A1 causes exon skipping and transient riboflavin-responsive multiple acyl-CoA dehydrogenation deficiency. Mol Genet Metab 122(4):182–188. https://doi.org/10.1016/j.ymgme.2017.10.014
48. Gregersen N, Andresen BS, Pedersen CB, Olsen RK, Corydon TJ, Bross P (2008) Mitochondrial fatty acid oxidation defects--remaining challenges. J Inherit Metab Dis 31(5):643–657. https://doi.org/10.1007/s10545-008-0990-y
49. Baur JA, Sinclair DA (2006) Therapeutic potential of resveratrol: the in vivo evidence. Nat Rev Drug Discov 5(6):493–506. https://doi.org/10.1038/nrd2060
50. Dembic M, Andersen HS, Bastin J, Doktor TK, Corydon TJ, Sass JO, Lopes Costa A, Djouadi F, Andresen BS (2019) Next generation sequencing of RNA reveals novel targets of resveratrol with possible implications for Canavan disease. Mol Genet Metab 126(1):64–76. https://doi.org/10.1016/j.ymgme.2018.10.004
51. Green P, Wiseman M, Crow YJ, Houlden H, Riphagen S, Lin JP, Raymond FL, Childs AM, Sheridan E, Edwards S, Josifova DJ (2010) Brown-Vialetto-Van Laere syndrome, a ponto-bulbar palsy with deafness, is caused by mutations in c20orf54. Am J Hum Genet 86(3):485–489. https://doi.org/10.1016/j.ajhg.2010.02.006
52. Dipti S, Childs AM, Livingston JH, Aggarwal AK, Miller M, Williams C, Crow YJ (2005) Brown-Vialetto-Van Laere syndrome; variability in age at onset and disease progression highlighting the phenotypic overlap with Fazio-Londe disease. Brain Dev 27(6):443–446. https://doi.org/10.1016/j.braindev.2004.10.003
53. Haack TB, Makowski C, Yao Y, Graf E, Hempel M, Wieland T, Tauer U, Ahting U, Mayr JA, Freisinger P, Yoshimatsu H, Inui K, Strom TM, Meitinger T, Yonezawa A, Prokisch H (2012) Impaired riboflavin transport due to missense mutations in SLC52A2 causes Brown-Vialetto-Van Laere syndrome. J Inherit Metab Dis 35(6):943–948. https://doi.org/10.1007/s10545-012-9513-y
54. Johnson JO, Gibbs JR, Megarbane A, Urtizberea JA, Hernandez DG, Foley AR, Arepalli S, Pandraud A, Simon-Sanchez J, Clayton P, Reilly MM, Muntoni F, Abramzon Y, Houlden H, Singleton AB (2012) Exome sequencing reveals riboflavin transporter mutations as a cause of motor neuron disease. Brain 135(Pt 9):2875–2882. https://doi.org/10.1093/brain/aws161
55. Amir F, Atzinger C, Massey K, Greinwald J, Hunter LL, Ulm E, Kettler M (2020) The clinical journey of patients with riboflavin transporter deficiency type 2. J Child Neurol 35(4):283–290. https://doi.org/10.1177/0883073819893159
56. Fan J, Fogel BL (2018) Successful treatment of a genetic childhood ataxia due to riboflavin transporter deficiency. Cereb Atax 5:12. https://doi.org/10.1186/s40673-018-0091-0
57. Bamaga AK, Maamari RN, Culican SM, Shinawi M, Golumbek PT (2018) Child neurology: Brown-Vialetto-Van Laere syndrome: dramatic visual recovery after delayed riboflavin therapy. Neurology 91(20):938–941. https://doi.org/10.1212/WNL.0000000000006498
58. Shi K, Shi Z, Yan H, Wang X, Yang Y, Xiong H, Gu Q, Wu Y, Jiang Y, Wang J (2019) A Chinese pedigree with Brown-Vialetto-Van Laere syndrome due to two novel mutations of SLC52A2 gene: clinical course and response to riboflavin. BMC Med Genet 20(1):76. https://doi.org/10.1186/s12881-019-0811-1
59. Gorcenco S, Vaz FM, Tracewska-Siemiatkowska A, Tranebjaerg L, Cremers FPM, Ygland E, Kicsi J, Rendtorff ND, Moller C, Kjellstrom U, Andreasson S, Puschmann A (2019) Oral therapy for riboflavin transporter deficiency - what is the regimen of choice? Parkinsonism Relat Disord 61:245–247. https://doi.org/10.1016/j.parkreldis.2018.10.017
60. Abbas Q, Jafri SK, Ishaque S, Rahman AJ (2018) Brown-Vialetto-Van Laere syndrome: a novel diagnosis to a common presentation.

BMJ Case Rep 2018:bcr2018224958. https://doi.org/10.1136/bcr-2018-224958

61. Garg M, Kulkarni SD, Hegde AU, Shah KN (2018) Riboflavin treatment in genetically proven Brown-Vialetto-Van Laere syndrome. J Pediatr Neurosci 13(4):471–473. https://doi.org/10.4103/JPN.JPN_131_17
62. Mutlu B, Topcu MT, Ciprut A (2019) A case with Brown-Vialetto-Van Laere syndrome: a sudden onset auditory neuropathy spectrum disorder. Turkish Archiv Otorhinolaryngol 57 (4):201–205. https://doi.org/10.5152/tao.2019.4639
63. Rabbani B, Bakhshandeh MK, Navaeifar MR, Abbaskhanian A, Soveizi M, Geravandpoor S, Mahdieh N (2020) Brown-Vialetto-Van Laere syndrome and Fazio-Londe syndrome: a novel mutation and in silico analyses. J Clin Neurosci 72:342–349. https://doi.org/10.1016/j.jocn.2020.01.040
64. Carreau C, Lenglet T, Mosnier I, Lahlou G, Fargeot G, Weiss N, Demeret S, Salachas F, Veauville-Merllie A, Acquaviva C, Nadjar Y (2020) A juvenile ALS-like phenotype dramatically improved after high-dose riboflavin treatment. Ann Clin Transl Neurol 7(2):250–253. https://doi.org/10.1002/acn3.50977
65. Bosch AM, Abeling NG, Ijlst L, Knoester H, van der Pol WL, Stroomer AE, Wanders RJ, Visser G, Wijburg FA, Duran M, Waterham HR (2011) Brown-Vialetto-Van Laere and Fazio Londe syndrome is associated with a riboflavin transporter defect mimicking mild MADD: a new inborn error of metabolism with potential treatment. J Inherit Metab Dis 34(1):159–164. https://doi.org/10.1007/s10545-010-9242-z
66. Chaya S, Zampoli M, Gray D, Booth J, Riordan G, Ndondo A, Fieggen K, Rusch J, van der Watt G, Pillay K, van der Westhuizen F, Menezes M, Wilmshurst J (2018) The first case of riboflavin transporter deficiency in sub-Saharan Africa. Semin Pediatr Neurol 26:10–14. https://doi.org/10.1016/j.spen.2017.03.002
67. Manole A, Jaunmuktane Z, Hargreaves I, Ludtmann MHR, Salpietro V, Bello OD, Pope S, Pandraud A, Horga A, Scalco RS, Li A, Ashokkumar B, Lourenco CM, Heales S, Horvath R, Chinnery PF, Toro C, Singleton AB, Jacques TS, Abramov AY, Muntoni F, Hanna MG, Reilly MM, Revesz T, Kullmann DM, Jepson JEC, Houlden H (2017) Clinical, pathological and functional characterization of riboflavin-responsive neuropathy. Brain 140(11):2820–2837. https://doi.org/10.1093/brain/awx231
68. Foley AR, Menezes MP, Pandraud A, Gonzalez MA, Al-Odaib A, Abrams AJ, Sugano K, Yonezawa A, Manzur AY, Burns J, Hughes I, McCullagh BG, Jungbluth H, Lim MJ, Lin JP, Megarbane A, Urtizberea JA, Shah AH, Antony J, Webster R, Broomfield A, Ng J, Mathew AA, O'Byrne JJ, Forman E, Scoto M, Prasad M, O'Brien K, Olpin S, Oppenheim M, Hargreaves I, Land JM, Wang MX, Carpenter K, Horvath R, Straub V, Lek M, Gold W, Farrell MO, Brandner S, Phadke R, Matsubara K, McGarvey ML, Scherer SS, Baxter PS, King MD, Clayton P, Rahman S, Reilly MM, Ouvrier RA, Christodoulou J, Zuchner S, Muntoni F, Houlden H (2014) Treatable childhood neuronopathy caused by mutations in riboflavin transporter RFVT2. Brain 137 (Pt 1):44–56. https://doi.org/10.1093/brain/awt315
69. Nimmo GAM, Ejaz R, Cordeiro D, Kannu P, Mercimek-Andrews S (2018) Riboflavin transporter deficiency mimicking mitochondrial myopathy caused by complex II deficiency. Am J Med Genet A 176(2):399–403. https://doi.org/10.1002/ajmg.a.38530
70. Jaeger B, Bosch AM (2016) Clinical presentation and outcome of riboflavin transporter deficiency: mini review after five years of experience. J Inherit Metab Dis 39 (4):559–564. https://doi.org/10.1007/s10545-016-9924-2
71. Olsen RKJ, Konarikova E, Giancaspero TA, Mosegaard S, Boczonadi V, Matakovic L, Veauville-Merllie A, Terrile C, Schwarzmayr T, Haack TB, Auranen M, Leone P, Galluccio M, Imbard A, Gutierrez-Rios P, Palmfeldt J, Graf E, Vianey-Saban C, Oppenheim M, Schiff M, Pichard S, Rigal O, Pyle A, Chinnery PF, Konstantopoulou V, Moslinger D, Feichtinger RG, Talim B, Topaloglu H, Coskun T, Gucer S, Botta A, Pegoraro E, Malena A, Vergani L, Mazza D, Zollino M, Ghezzi D, Acquaviva C, Tyni T, Boneh A, Meitinger T, Strom TM, Gregersen N, Mayr JA, Horvath R, Barile M, Prokisch H (2016) Riboflavin-responsive and -non-responsive mutations in FAD synthase cause multiple acyl-CoA dehydrogenase and combined respiratory-chain deficiency. Am J Hum Genet 98(6):1130–1145. https://doi.org/10.1016/j.ajhg.2016.04.006
72. Galluccio M, Brizio C, Torchetti EM, Ferranti P, Gianazza E, Indiveri C, Barile M (2007) Over-expression in Escherichia coli, purification and characterization of isoform 2 of human FAD synthetase. Protein Expr Purif 52(1):175–181. https://doi.org/10.1016/j.pep.2006.09.002

73. Torchetti EM, Brizio C, Colella M, Galluccio M, Giancaspero TA, Indiveri C, Roberti M, Barile M (2010) Mitochondrial localization of human FAD synthetase isoform 1. Mitochondrion 10(3):263–273. https://doi.org/10.1016/j.mito.2009.12.149
74. Leone P, Galluccio M, Brizio C, Barbiroli A, Iametti S, Indiveri C, Barile M (2019) The hidden side of the human FAD synthase 2. Int J Biol Macromol 138:986–995. https://doi.org/10.1016/j.ijbiomac.2019.07.138
75. Giancaspero TA, Galluccio M, Miccolis A, Leone P, Eberini I, Iametti S, Indiveri C, Barile M (2015) Human FAD synthase is a bi-functional enzyme with a FAD hydrolase activity in the molybdopterin binding domain. Biochem Biophys Res Commun 465(3):443–449. https://doi.org/10.1016/j.bbrc.2015.08.035
76. Leone P, Galluccio M, Barbiroli A, Eberini I, Tolomeo M, Vrenna F, Gianazza E, Iametti S, Bonomi F, Indiveri C, Barile M (2018) Bacterial production, characterization and protein modeling of a novel monofuctional isoform of FAD synthase in humans: an emergency protein? Molecules 23(1):116. https://doi.org/10.3390/molecules23010116
77. Leone P, Galluccio M, Quarta S, Anoz-Carbonell E, Medina M, Indiveri C, Barile M (2019) Mutation of aspartate 238 in FAD synthase isoform 6 increases the specific activity by weakening the FAD binding. Int J Mol Sci 20(24):6203. https://doi.org/10.3390/ijms20246203
78. Miccolis A, Galluccio M, Giancaspero TA, Indiveri C, Barile M (2012) Bacterial over-expression and purification of the 3′phosphoadenosine 5′phosphosulfate (PAPS) reductase domain of human FAD synthase: functional characterization and homology modeling. Int J Mol Sci 13(12):16880–16898. https://doi.org/10.3390/ijms131216880
79. Miccolis A, Galluccio M, Nitride C, Giancaspero TA, Ferranti P, Iametti S, Indiveri C, Bonomi F, Barile M (2014) Significance of redox-active cysteines in human FAD synthase isoform 2. Biochim Biophys Acta 1844(12):2086–2095. https://doi.org/10.1016/j.bbapap.2014.08.005
80. Giancaspero TA, Colella M, Brizio C, Difonzo G, Fiorino GM, Leone P, Brandsch R, Bonomi F, Iametti S, Barile M (2015) Remaining challenges in cellular flavin cofactor homeostasis and flavoprotein biogenesis. Front Chem 3:30. https://doi.org/10.3389/fchem.2015.00030
81. Torchetti EM, Bonomi F, Galluccio M, Gianazza E, Giancaspero TA, Iametti S, Indiveri C, Barile M (2011) Human FAD synthase (isoform 2): a component of the machinery that delivers FAD to apo-flavoproteins. FEBS J 278(22):4434–4449. https://doi.org/10.1111/j.1742-4658.2011.08368.x
82. Lynch JH, Sa N, Saeheng S, Raffaelli N, Roje S (2018) Characterization of a non-nudix pyrophosphatase points to interplay between flavin and NAD(H) homeostasis in Saccharomyces cerevisiae. PLoS One 13(6):e0198787. https://doi.org/10.1371/journal.pone.0198787
83. Yildiz Y, Olsen RKJ, Sivri HS, Akcoren Z, Nygaard HH, Tokatli A (2018) Post-mortem detection of FLAD1 mutations in 2 Turkish siblings with hypotonia in early infancy. Neuromuscul Disord 28(9):787–790. https://doi.org/10.1016/j.nmd.2018.05.009
84. Ryder B, Tolomeo M, Nochi Z, Colella M, Barile M, Olsen RK, Inbar-Feigenberg M (2019) A novel truncating FLAD1 variant, causing multiple acyl-CoA dehydrogenase deficiency (MADD) in an 8-year-old boy. JIMD Rep 45:37–44. https://doi.org/10.1007/8904_2018_139
85. Garcia-Villoria J, De Azua B, Tort F, Mosegaard S, Ugarteburu O, Texido L, Morales-Romero B, Olsen RKJ, Ribes A (2018) FLAD1, encoding FAD synthase, is mutated in a patient with myopathy, scoliosis and cataracts. Clin Genet 94(6):592–593. https://doi.org/10.1111/cge.13452
86. Yamada K, Ito M, Kobayashi H, Hasegawa Y, Fukuda S, Yamaguchi S, Taketani T (2019) Flavin adenine dinucleotide synthase deficiency due to FLAD1 mutation presenting as multiple acyl-CoA dehydrogenation deficiency-like disease: a case report. Brain Dev 41(7):638–642. https://doi.org/10.1016/j.braindev.2019.04.002
87. Muru K, Reinson K, Kunnapas K, Lillevali H, Nochi Z, Mosegaard S, Pajusalu S, Olsen RKJ, Ounap K (2019) FLAD1-associated multiple acyl-CoA dehydrogenase deficiency identified by newborn screening. Mol Genet Genom Med 7(9):e915. https://doi.org/10.1002/mgg3.915

Index

Maria Barile (ed.), *Flavin and Flavoproteins: Methods and Protocols*, Methods in Molecular Biology, vol. 2280, https://doi.org/10.1007/978-1-0716-1286-6, © Springer Science+Business Media, LLC, part of Springer Nature 2021

MIX
Papier aus verantwortungsvollen Quellen
Paper from responsible sources
FSC® C105338

If you have any concerns about our products,
you can contact us on
ProductSafety@springernature.com

In case Publisher is established outside the EU,
the EU authorized representative is:
Springer Nature Customer Service Center GmbH
Europaplatz 3, 69115 Heidelberg, Germany

Printed by Libri Plureos GmbH
in Hamburg, Germany